Food Microbiology

Food Microbiology

Edited by
Garrett Cooper

Larsen & Keller
www.larsen-keller.com

Food Microbiology
Edited by Garrett Cooper
ISBN: 978-1-63549-124-1 (Hardback)

Larsen & Keller
Published by Larsen and Keller Education,
5 Penn Plaza,
19th Floor,
New York, NY 10001, USA

Cataloging-in-Publication Data

Food microbiology / edited by Garrett Cooper.
p. cm.
Includes bibliographical references and index.
ISBN 978-1-63549-124-1
1. Food--Microbiology. 2. Food--Safety measures. 3. Food--Preservation.
4. Microbiology. I. Cooper, Garrett.
QR115 .F66 2017
664.001 579--dc23

Printed and bound in China.

For more information regarding Larsen and Keller Education and its products, please visit the publisher's website www.larsen-keller.com

Table of Contents

Preface

Food microbiology studies microorganisms that cause fundamental changes to food, resulting in either fermentation or spoilage. Topics included in this text concentrate on probiotic microorganism cultures, food safety practices and the benefits of "good" bacteria. The book studies, analyses and upholds the pillars of food microbiology and its utmost significance in modern times. It discusses fundamental processes involved in food microbiology and the food products industry. Selected concepts that redefine the subject have been included in it. Those in search of information to further their knowledge will be greatly assisted by this textbook. It will be of great help to graduates in the fields of environmental microbiology, food science and food preservation and safety.

A short introduction to every chapter is written below to provide an overview of the content of the book:

Chapter 1 - The study of microorganisms that contaminate food resulting in the spoilage of food is termed as food microbiology. Microorganisms are very important for the production of cheese, bread and wine. This chapter will provide an integrated understanding of food microbiology; **Chapter 2** - The microbes found in food are yeast, fungus and bacteria. Yeasts are single-celled microorganisms and their sizes vary greatly and they usually reproduce asexually. The topics discussed in the section are of great importance to broaden the existing knowledge on microbes in food; **Chapter 3** - Probiotics are types of microorganisms whose presence actually benefits the consumer's health. There are a number of benefits of using probiotics, such as the reduction of gastrointestinal discomfort and even in the strengthening of the immune system. The major components of probiotics are discussed in this chapter; **Chapter 4** - Fermentation is the process that is used in converting sugar to gases or to alcohol. It is also used to refer to the growth of microorganisms with the aim of producing a specific chemical product. This section is an overview of the subject matter incorporating all the major aspects of fermentation; **Chapter 5** - Food preservation prevents the growth of microorganisms on food. The traditional techniques mentioned are drying, cooling, freezing, boiling, sugaring and pickling. Other topics discussed in the section are vacuum packing, preservatives, camping food, food safety, ISO 22000, food grading and beef carcass classifications. The aspects elucidates in this chapter are of vital importance, and provides a better understanding of food preservation; **Chapter 6** - The storing of food and then delivering it to the customers is termed as food storage. Food is stored by humans as well as by animals and has several purposes. Some of the examples of food storage are food banks, food storage containers and refrigerators. The following section unfolds its aspects in a critical yet systematic manner.

I extend my sincere thanks to the publisher for considering me worthy of this task. Finally, I thank my family for being a source of support and help.

Editor

1

Introduction to Food Microbiology

The study of microorganisms that contaminate food resulting in the spoilage of food is termed as food microbiology. Microorganisms are very important for the production of cheese, bread and wine. This chapter will provide an integrated understanding of food microbiology.

Food microbiology is the study of the microorganisms that inhabit, create, or contaminatefood, including the study of microorganisms causing food spoilage. "Good" bacteria, however, such as probiotics, are becoming increasingly important in food science. In addition, microorganisms are essential for the production of foods such as cheese, yogurt, bread, beer, wine and, other fermented foods.

Food Safety

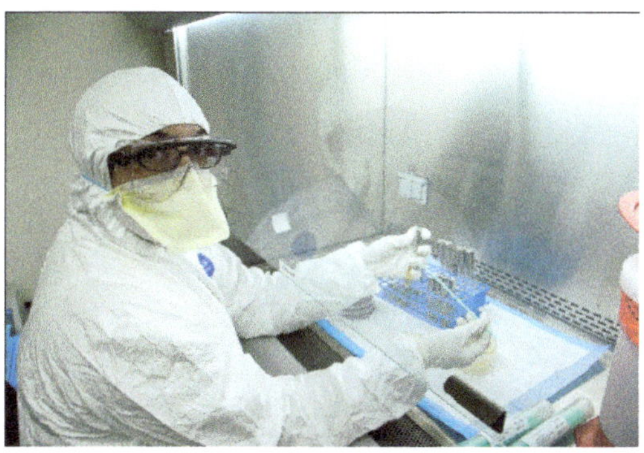

A microbiologist working in a biosafety laboratory tests for high risk pathogens in food.

Food safety is a major focus of food microbiology. Pathogenic bacteria, viruses and toxins produced by microorganisms are all possible contaminants of food. However, microorganisms and their products can also be used to combat these pathogenic microbes. Probiotic bacteria, including those that produce bacteriocins, can kill and inhibit pathogens. Alternatively, purified bacteriocins such as nisin can be added directly to food products. Finally, bacteriophages, viruses that only infect bacteria, can be used to kill bacterial pathogens. Thorough preparation of food, including proper cooking, eliminates most bacteria and viruses. However, *toxins produced* by contaminants may not be liable to change to non-toxic forms by heating or cooking the contaminated food due to other safety conditions.

Fermentation

Fermentation is one of the methods to preserve food and alter its quality. Yeast, especially *Saccharomyces cerevisiae*, is used to leaven bread, brewbeer and make wine. Certain bacteria, including lactic acid bacteria, are used to make yogurt, cheese, hot sauce, pickles, fermented sausages and dishes such as kimchi. A common effect of these fermentations is that the food product is less hospitable to other microorganisms, including pathogens and spoilage-causing microorganisms, thus extending the food's shelf-life. Some cheese varieties also require molds to ripen and develop their characteristic flavors.

Microbial Biopolymers

Several microbially produced biopolymers are used in the food industry.

Alginate

Alginates can be used as thickening agents. Although listed here under the category 'Microbial polysaccharides', commercial alginates are currently only produced by extraction from brown seaweeds such as *Laminaria hyperborea* or *L. japonica*.

Poly-γ-glutamic Acid

Poly-γ-glutamic acid (γ-PGA) produced by various strains of *Bacillus* has potential applications as a thickener in the food industry.

Food Authenticity

It is important to be able to detect microorganisms in food, in particular pathogenic microorganisms or genetically modified microorganisms.

Food Testing

Food microbiology laboratory at the Faculty of Food Technology, Latvia University of Agriculture.

To ensure safety of food products, microbiological tests such as testing for pathogens and spoilage organisms are required. This way the risk of contamination under normal use conditions can be examined and food poisoning outbreaks can be prevented. Testing of food products and ingredients is important along the whole supply chain as possible flaws of products can occur at every stage of production. Apart from detecting spoilage, microbiological tests can also determine germ content, identify yeasts and molds, and salmonella. For salmonella, scientists are also developing rapid and portable technologies capable of identifying unique variants of Salmonella .

Polymerase Chain Reaction (PCR) is a quick and inexpensive method to generate numbers of copies of a DNA fragment at a specific band ("PCR (Polymerase Chain Reaction)," 2008). For that reason, scientists are using PCR to detect different kinds of viruses or bacteria, such as HIV and anthrax based on their unique DNA patterns. Various kits are commercially available to help in food pathogen nucleic acids extraction, PCR detection, and differentiation. The detection of bacterial strands in food products is very important to everyone in the world, for it helps prevent the occurrence of food borne illness. Therefore, PCR is recognized as a DNA detector in order to amplify and trace the presence of pathogenic strands in different processed food.

References

- Rehm BHA, ed. (2009). Microbial Production of Biopolymers and Polymer Precursors: Applications and Perspectives. Caister Academic Press. ISBN 978-1-904455-36-3
- Fratamico PM (2005). Bayles DO, ed. Foodborne Pathogens: Microbiology and Molecular Biology. Caister Academic Press. ISBN 978-1-904455-00-4.
- Tannock GW, ed. (2005). Probiotics and Prebiotics: Scientific Aspects. Caister Academic Press. ISBN 978-1-904455-01-1.
- Ljungh A, Wadstrom T, eds. (2009). Lactobacillus Molecular Biology: From Genomics to Probiotics. Caister Academic Press. ISBN 978-1-904455-41-7.
- Mayo, B (2010). van Sinderen, D, ed. Bifidobacteria: Genomics and Molecular Aspects. Caister Academic Press. ISBN 978-1-904455-68-4.
- Shih & Wu (2009). "Biosynthesis and Application of Poly(gamma-glutamic acid)". Microbial Production of Biopolymers and Polymer Precursors. Caister Academic Press. ISBN 978-1-904455-36-3.

2

Microbes in Food

The microbes found in food are yeast, fungus and bacteria. Yeasts are single-celled microorganisms and their sizes vary greatly and they usually reproduce asexually. The topics discussed in the section are of great importance to broaden the existing knowledge on microbes in food.

Yeast

Yeasts are eukaryotic, single-celled microorganisms classified as members of the fungus kingdom. The yeast lineage originated hundreds of millions of years ago, and 1,500 species are currently identified. They are estimated to constitute 1% of all described fungal species. Yeasts are unicellular organisms which evolved from multicellular ancestors, with some species having the ability to develop multicellular characteristics by forming strings of connected budding cells known as pseudohyphae or false hyphae. Yeast sizes vary greatly, depending on species and environment, typically measuring 3–4 µm in diameter, although some yeasts can grow to 40 µm in size. Most yeasts reproduce asexually by mitosis, and many do so by the asymmetric division process known as budding.

Yeasts, with their single-celled growth habit, can be contrasted with molds, which grow hyphae. Fungal species that can take both forms (depending on temperature or other conditions) are called dimorphic fungi ("dimorphic" means "having two forms").

By fermentation, the yeast species Saccharomyces cerevisiae converts carbohydrates to carbon dioxide and alcohols – for thousands of years the carbon dioxide has been used in baking and the alcohol in alcoholic beverages. It is also a centrally important model organism in modern cell biology research, and is one of the most thoroughly researched eukaryotic microorganisms. Researchers have used it to gather information about the biology of the eukaryotic cell and ultimately human biology. Other species of yeasts, such as Candida albicans, are opportunistic pathogens and can cause infections in humans. Yeasts have recently been used to generate electricity in microbial fuel cells, and produce ethanol for the biofuel industry.

Yeasts do not form a single taxonomic or phylogenetic grouping. The term "yeast" is often taken as a synonym for Saccharomyces cerevisiae, but the phylogenetic diversity of yeasts is shown by their placement in two separate phyla: the Ascomycota and the Basidiomycota. The budding yeasts ("true yeasts") are classified in the order Saccharomycetales, within the phylum Ascomycota.

History

The word "yeast" comes from Old English gist, gyst, and from the Indo-European root yes-, meaning "boil", "foam", or "bubble". Yeast microbes are probably one of the earliest domesticated organisms. Archaeologists digging in Egyptian ruins found early grinding stones and baking chambers for yeast-raised bread, as well as drawings of 4,000-year-old bakeries and breweries. In 1680, Dutch naturalist Anton van Leeuwenhoek first microscopically observed yeast, but at the time did not consider them to be living organisms, but rather globular structures. Researchers were doubtful whether yeasts were algae or fungi, but in 1837 Theodor Schwann recognized them as fungi.

In 1857, French microbiologist Louis Pasteur proved in the paper "Mémoire sur la fermentation alcoolique" that alcoholic fermentation was conducted by living yeasts and not by a chemical catalyst. Pasteur showed that by bubbling oxygen into the yeast broth, cell growth could be increased, but fermentation was inhibited – an observation later called the "Pasteur effect".

By the late 18th century, two yeast strains used in brewing had been identified: Saccharomyces cerevisiae (top-fermenting yeast) and S. carlsbergensis (bottom-fermenting yeast). S. cerevisiae has been sold commercially by the Dutch for bread-making since 1780; while, around 1800, the Germans started producing S. cerevisiae in the form of cream. In 1825, a method was developed to remove the liquid so the yeast could be prepared as solid blocks. The industrial production of yeast blocks was enhanced by the introduction of the filter press in 1867. In 1872, Baron Max de Springer developed a manufacturing process to create granulated yeast, a technique that was used until the first World War. In the United States, naturally occurring airborne yeasts were used almost exclusively until commercial yeast was marketed at the Centennial Exposition in 1876 in Philadelphia, where Charles L. Fleischmann exhibited the product and a process to use it, as well as serving the resultant baked bread.

Nutrition and Growth

Yeasts are chemoorganotrophs, as they use organic compounds as a source of energy and do not require sunlight to grow. Carbon is obtained mostly from hexose sugars, such as glucose and fructose, or disaccharides such as sucrose and maltose. Some species can metabolize pentose sugars such as ribose, alcohols, and organic acids. Yeast species either require oxygen for aerobic cellular respiration (obligate aerobes) or are anaerobic, but also have aerobic methods of energy production (facultative anaerobes). Unlike bacteria, no known yeast species grow only anaerobically (obligate anaerobes). Yeasts grow best in a neutral or slightly acidic pH environment.

Yeasts vary in regard to the temperature range in which they grow best. For example, Leucosporidium frigidum grows at −2 to 20 °C (28 to 68 °F), Saccharomyces telluris at 5 to 35 °C (41 to 95 °F), and Candida slooffi at 28 to 45 °C (82 to 113 °F). The cells can survive freezing under certain conditions, with viability decreasing over time.

In general, yeasts are grown in the laboratory on solid growth media or in liquid broths. Common media used for the cultivation of yeasts include potato dextrose agar or potato dextrose broth, Wallerstein Laboratories nutrient agar, yeast peptone dextrose agar, and yeast mould agar or broth. Home brewers who cultivate yeast frequently use dried malt extract and agar as a solid growth medium. The antibiotic cycloheximide is sometimes added to yeast growth media to inhibit the growth of Saccharomyces yeasts and select for wild/indigenous yeast species. This will change the yeast process.

The appearance of a white, thready yeast, commonly known as kahm yeast, is often a byproduct of the lactofermentation (or pickling) of certain vegetables, usually the result of exposure to air. Although harmless, it can give pickled vegetables a bad flavor and must be removed regularly during fermentation.

Ecology

Yeasts are very common in the environment, and are often isolated from sugar-rich materials. Examples include naturally occurring yeasts on the skins of fruits and berries (such as grapes, apples, or peaches), and exudates from plants (such as plant saps or cacti). Some yeasts are found in association with soil and insects. The ecological function and biodiversity of yeasts are relatively unknown compared to those of other microorganisms. Yeasts, including Candida albicans, Rhodotorula rubra, Torulopsis and Trichosporon cutaneum, have been found living in between people's toes as part of their skin flora. Yeasts are also present in the gut flora of mammals and some insects and even deep-sea environments host an array of yeasts.

An Indian study of seven bee species and 9 plant species found 45 species from 16 genera colonise the nectaries of flowers and honey stomachs of bees. Most were members of the Candida genus; the most common species in honey stomachs was Dekkera intermedia and in flower nectaries, Candida blankii. Yeast colonising nectaries of the stinking hellebore have been found to raise the temperature of the flower, which may aid in attracting pollinators by increasing the evaporation of volatile organic compounds. A black yeast has been recorded as a partner in a complex relationship between ants, their mutualistic fungus, a fungal parasite of the fungus and a bacterium that kills the parasite. The yeast has a negative effect on the bacteria that normally produce antibiotics to kill the parasite, so may affect the ants' health by allowing the parasite to spread.

Certain strains of some species of yeasts produce proteins called yeast killer toxins that allow them to eliminate competing strains. This can cause problems for winemaking but could potentially also be used to advantage by using killer toxin-producing strains to make the wine. Yeast killer toxins may also have medical applications in treating yeast infections .

Reproduction

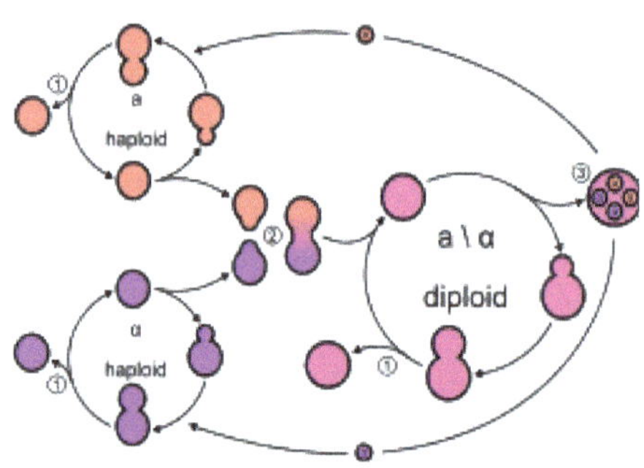

The yeast cell's life cycle:

1. Budding
2. Conjugation
3. Spore

Yeasts, like all fungi, may have asexual and sexual reproductive cycles. The most common mode of vegetative growth in yeast is asexual reproduction by budding. Here, a small bud (also known as a bleb), or daughter cell, is formed on the parent cell. The nucleus of the parent cell splits into a daughter nucleus and migrates into the daughter cell. The bud continues to grow until it separates from the parent cell, forming a new cell. The daughter cell produced during the budding process is generally smaller than the mother cell. Some yeasts, including Schizosaccharomyces pombe, reproduce by fission instead of budding, thereby creating two identically sized daughter cells.

In general, under high-stress conditions such as nutrient starvation, haploid cells will die; under the same conditions, however, diploid cells can undergo sporulation, entering sexual reproduction (meiosis) and producing a variety of haploid spores, which can go on to mate (conjugate), reforming the diploid.

The haploid fission yeast Schizosaccharomyces pombe is a facultative sexual microorganism that can undergo mating when nutrients are limiting. Exposure of S. pombe to hydrogen peroxide, an agent that causes oxidative stress leading to oxidative DNA damage, strongly induces mating and the formation of meiotic spores. The budding yeast Saccharomyces cerevisiae reproduces by mitosis as diploid cells when nutrients are abundant, but when starved, this yeast undergoes meiosis to form haploid spores. Haploid cells may then reproduce asexually by mitosis. Katz Ezov et al. presented evidence that in natural S. cerevisiae populations clonal reproduction and selfing (in the form of intratetrad mating) predominate. In nature, mating of haploid cells to form diploid cells is most often between members of the same clonal population and out-crossing is uncommon. Analysis of the ancestry of natural S. cerevisiae strains led to the conclusion that out-crossing occurs only about once every 50,000 cell divisions. These observations suggest that the possible long-term benefits of outcrossing (e.g. generation of diversity) are likely to be insufficient for generally maintaining sex from one generation to the next. Rather, a short-term benefit, such as recombinational repair during meiosis, may be the key to the maintenance of sex in S. cerevisiae.

Some pucciniomycete yeasts, in particular species of Sporidiobolus and Sporobolomyces, produce aerially dispersed, asexual ballistoconidia.

Uses

The useful physiological properties of yeast have led to their use in the field of biotechnology. Fermentation of sugars by yeast is the oldest and largest application of this technology. Many types of yeasts are used for making many foods: baker's yeast in bread production, brewer's yeast in beer fermentation, and yeast in wine fermentation and for xylitol production. So-called red rice yeast is actually a mold, Monascus purpureus. Yeasts include some of the most widely used model organisms for genetics and cell biology.

Alcoholic Beverages

Alcoholic beverages are defined as beverages that contain ethanol ($C2H5OH$). This ethanol is almost always produced by fermentation – the metabolism of carbohydrates by certain species of yeasts under anaerobic or low-oxygen conditions. Beverages such as mead, wine, beer, or distilled spirits all use yeast at some stage of their production. A distilled beverage is a beverage containing ethanol that has been purified by distillation. Carbohydrate-containing plant material is fermented by yeast, producing a dilute solution of ethanol in the process. Spirits such as whiskey and rum are prepared by distilling these dilute solutions of ethanol. Components other than ethanol are collected in the condensate, including water, esters, and other alcohols, which (in addition to that provided by the oak in which it may be aged) account for the flavour of the beverage.

Beer

Yeast ring used by Swedish homebrewers in the 19th century to preserve yeast between brewing sessions.

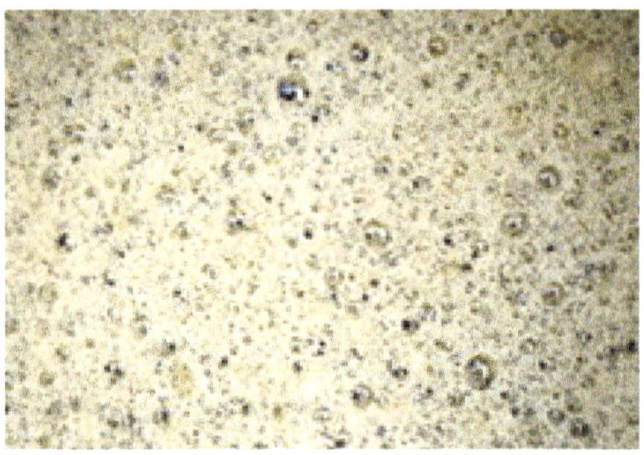

Bubbles of carbon dioxide forming during beer-brewing

Brewing yeasts may be classed as "top-cropping" (or "top-fermenting") and "bottom-cropping" (or "bottom-fermenting"). Top-cropping yeasts are so called because they form a foam at the top of the wort during fermentation. An example of a top-cropping yeast is Saccharomyces cerevisiae, sometimes called an "ale yeast". Bottom-cropping yeasts are typically used to produce lager-type beers, though they can also produce ale-type beers. These yeasts ferment well at low temperatures. An example of bottom-cropping yeast is Saccharomyces pastorianus, formerly known as S. carlsbergensis.

Decades ago, taxonomists reclassified S. carlsbergensis (uvarum) as a member of S. cerevisiae, noting that the only distinct difference between the two is metabolic. Lager strains of S. cerevisiae secrete an enzyme called melibiase, allowing them to hydrolyse melibiose, a disaccharide, into more fermentable monosaccharides. Top- and bottom-cropping and cold- and warm-fermenting distinctions are largely generalizations used by laypersons to communicate to the general public.

The most common top-cropping brewer's yeast, S. cerevisiae, is the same species as the common baking yeast. Brewer's yeast is also very rich in essential minerals and the B vitamins (except B12). However, baking and brewing yeasts typically belong to different strains, cultivated to favour different characteristics: baking yeast strains are more aggressive, to carbonate dough in the shortest amount of time possible; brewing yeast strains act slower but tend to produce fewer off-flavours and tolerate higher alcohol concentrations (with some strains, up to 22%).

Dekkera/Brettanomyces is a genus of yeast known for its important role in the production of 'lambic' and specialty sour ales, along with the secondary conditioning of a particular Belgian Trappist beer. The taxonomy of the genus Brettanomyces has been debated since its early discovery and has seen many re-classifications over the years. Early classification was based on a few species that reproduced asexually (anamorph form) through multipolar budding. Shortly after, the formation of ascospores was observed and the genus Dekkera, which reproduces sexually (teleomorph form), was introduced as part of the taxonomy. The current taxonomy includes five species within the genera of Dekkera/Brettanomyces. Those are the anamorphs Brettanomyces bruxellensis, Brettanomyces anomalus, Brettanomyces custersianus, Brettanomyces naardenensis, and Brettanomyces nanus, with teleomorphs existing for the first two species, Dekkera bruxellensis and Dekkera anomala. The distinction between Dekkera and Brettanomyces is arguable, with Oelofse et al. (2008) citing Loureiro and Malfeito-Ferreira from 2006 when they affirmed that current molecular DNA detection techniques have uncovered no variance between the anamorph and teleomorph states. Over the past decade, Brettanomyces spp. have seen an increasing use in the craft-brewing sector of the industry, with a handful of breweries having produced beers that were primarily fermented with pure cultures of Brettanomyces spp. This has occurred out of experimentation, as very little information exists regarding pure culture fermentative capabilities and the aromatic compounds produced by various strains. Dekkera/Brettanomyces spp. have been the subjects of numerous studies conducted over the past century, although a majority of the recent research has focused on enhancing the knowledge of the wine industry. Recent research on eight Brettanomyces strains available in the brewing industry focused on strain-specific fermentations and identified the major compounds produced during pure culture anaerobic fermentation in wort.

Wine

Yeast is used in winemaking, where it converts the sugars present (glucose and fructose) in grape juice (must) into ethanol. Yeast is normally already present on grape skins. Fermentation can be done with this endogenous "wild yeast", but this procedure gives unpredictable results, which depend upon the exact types of yeast species present. For this reason, a pure yeast culture is usually added to the must; this yeast quickly dominates the fermentation. The wild yeasts are repressed, which ensures a reliable and predictable fermentation.

Most added wine yeasts are strains of S. cerevisiae, though not all strains of the species are suitable. Different S. cerevisiae yeast strains have differing physiological and fermentative properties, therefore the actual strain of yeast selected can have a direct impact on the finished wine. Significant research has been undertaken into the development of novel wine yeast strains that produce atypical flavour profiles or increased complexity in wines.

The growth of some yeasts, such as Zygosaccharomyces and Brettanomyces, in wine can result in wine faults and subsequent spoilage. Brettanomyces produces an array of metabolites when growing in wine, some of which are volatile phenolic compounds. Together, these compounds are often referred to as "Brettanomyces character", and are often described as "antiseptic" or "barnyard" type aromas. Brettanomyces is a significant contributor to wine faults within the wine industry.

Researchers from the University of British Columbia, Canada, have found a new strain of yeast that has reduced amines. The amines in red wine and Chardonnay produce off-flavors and cause headaches and hypertension in some people. About 30% of people are sensitive to biogenic amines, such as histamines.

Baking

Yeast, the most common one being S. cerevisiae, is used in baking as a leavening agent, where it converts the food/fermentable sugars present in dough into the gas carbon dioxide. This causes the dough to expand or rise as gas forms pockets or bubbles. When the dough is baked, the yeast dies and the air pockets "set", giving the baked product a soft and spongy texture. The use of potatoes, water from potato boiling, eggs, or sugar in a bread dough accelerates the growth of yeasts. Most yeasts used in baking are of the same species common in alcoholic fermentation. In addition, Saccharomyces exiguus (also known as S. minor), a wild yeast found on plants, fruits, and grains, is occasionally used for baking. In breadmaking, the yeast initially respires aerobically, producing carbon dioxide and water. When the oxygen is depleted, fermentation begins, producing ethanol as a waste product; however, this evaporates during baking.

A block of compressed fresh yeast

It is not known when yeast was first used to bake bread. The first records that show this use came from Ancient Egypt. Researchers speculate a mixture of flour meal and water was left longer than usual on a warm day and the yeasts that occur in natural contaminants of the flour caused it to ferment before baking. The resulting bread would have been lighter and tastier than the normal flat, hard cake.

Active dried yeast, a granulated form in which yeast is commercially sold

Today, there are several retailers of baker's yeast; one of the earlier developments in North America is Fleischmann's Yeast, in 1868. During World War II, Fleischmann's developed a granulated active dry yeast which did not require refrigeration, had a longer shelf life than fresh yeast, and rose twice as fast. Baker's yeast is also sold as a fresh yeast compressed into a square "cake". This form perishes quickly, so must be used soon after production. A weak solution of water and sugar can be used to determine whether yeast is expired. In the solution, active yeast will foam and bubble as it ferments the sugar into ethanol and carbon dioxide. Some recipes refer to this as proofing the yeast, as it "proves" (tests) the viability of the yeast before the other ingredients are added. When a sourdough starter is used, flour and water are added instead of sugar; this is referred to as proofing the sponge.

When yeast is used for making bread, it is mixed with flour, salt, and warm water or milk. The dough is kneaded until it is smooth, and then left to rise, sometimes until it has doubled in size. The dough is then shaped into loaves. Some bread doughs are knocked back after one rising and left to rise again (this is called dough proofing) and then baked. A longer rising time gives a better flavour, but the yeast can fail to raise the bread in the final stages if it is left for too long initially.

Bioremediation

Some yeasts can find potential application in the field of bioremediation. One such yeast, Yarrowia lipolytica, is known to degrade palm oil mill effluent, TNT (an explosive material), and other hydrocarbons, such as alkanes, fatty acids, fats and oils. It can also tolerate high concentrations of salt and heavy metals, and is being investigated for its potential as a heavy metal biosorbent. Saccharomyces cerevisiae has potential to bioremediate toxic pollutants like arsenic from industrial effluent. Bronze statues are known to be degraded by certain species of yeast. Different yeasts from Brazilian gold mines bioaccumulate free and complexed silver ions.

Industrial Ethanol Production

The ability of yeast to convert sugar into ethanol has been harnessed by the biotechnology industry to produce ethanol fuel. The process starts by milling a feedstock, such as sugar cane, field corn, or other cereal grains, and then adding dilute sulfuric acid, or fungal alpha amylase enzymes, to break down the starches into complex sugars. A glucoamylase is then added to break the complex sugars down into simple sugars. After this, yeasts are added to convert the simple sugars to ethanol, which is then distilled off to obtain ethanol up to 96% in purity.

Saccharomyces yeasts have been genetically engineered to ferment xylose, one of the major fermentable sugars present in cellulosic biomasses, such as agriculture residues, paper wastes, and wood chips. Such a development means ethanol can be efficiently produced from more inexpensive feedstocks, making cellulosic ethanol fuel a more competitively priced alternative to gasoline fuels.

Nonalcoholic Beverages

A kombucha culture fermenting in a jar

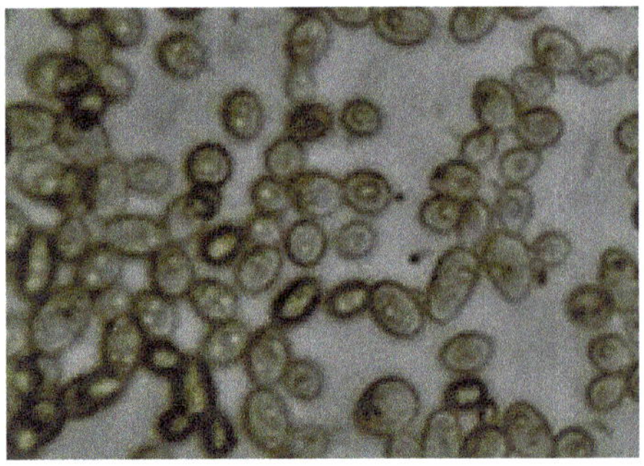

Yeast and bacteria in kombucha at 400X

A number of sweet carbonated beverages can be produced by the same methods as beer, except the fermentation is stopped sooner, producing carbon dioxide, but only trace amounts of alcohol, leaving a significant amount of residual sugar in the drink.

Root beer, originally made by Native Americans, commercialized in the United States by Charles Elmer Hires and especially popular during Prohibition

- Kvass, a fermented drink made from rye, popular in Eastern Europe. It has a recognizable, but low alcoholic content.
- Kombucha, a fermented sweetened tea. Yeast in symbiosis with acetic acid bacteria is used in its preparation. Species of yeasts found in the tea can vary, and may include: Brettanomyces bruxellensis, Candida stellata, Schizosaccharomyces pombe, Torulaspora delbrueckii and Zygosaccharomyces bailii. Also popular in Eastern Europe and some former Soviet republics under the name chajnyj grib (Russian: Чайный гриб), which means "tea mushroom".
- Kefir and kumis are made by fermenting milk with yeast and bacteria.
- Mauby (Spanish: mabí), made by fermenting sugar with the wild yeasts naturally present on the bark of the Colubrina elliptica tree, popular in the Caribbean

Nutritional Supplements

Yeast is used in nutritional supplements,especially those marketed to vegans. It is often referred to as "nutritional yeast" when sold as a dietary supplement. Nutritional yeast is a deactivated yeast, usually S. cerevisiae. It is an excellent source of protein and vitamins, especially the B-complex vitamins, as well as other minerals and cofactors required for growth. It is also naturally low in fat and sodium. Contrary to some claims, it contains little or no vitamin B12. Some brands of nutritional yeast, though not all, are fortified with vitamin B12, which is produced separately by bacteria.

In 1920, the Fleischmann Yeast Company began to promote yeast cakes in a successful "Yeast for Health" campaign. They initially emphasized yeast as a source of vitamins, good for skin and digestion. Their later advertising claimed a much broader range of health benefits, and was censured as misleading by the Federal Trade Commission. The fad for yeast cakes lasted until the late 1930s.

Nutritional yeast has a nutty, cheesy flavor and is often used as an ingredient in cheese substitutes.

Another popular use is as a topping for popcorn. It can also be used in mashed and fried potatoes, as well as in scrambled eggs. It comes in the form of flakes, or as a yellow powder similar in texture to cornmeal. In Australia, it is sometimes sold as "savoury yeast flakes". Though "nutritional yeast" usually refers to commercial products, inadequately fed prisoners have used "home-grown" yeast to prevent vitamin deficiency.

Probiotics

Some probiotic supplements use the yeast S. boulardii to maintain and restore the natural flora in the gastrointestinal tract. S. boulardii has been shown to reduce the symptoms of acute diarrhea, reduce the chance of infection by Clostridium difficile (often identified simply as C. difficile or C. diff), reduce bowel movements in diarrhea-predominant IBS patients, and reduce the incidence of antibiotic-, traveler's-, and HIV/AIDS-associated diarrheas.

Aquarium Hobby

Yeast is often used by aquarium hobbyists to generate carbon dioxide (CO_2) to nourish plants in planted aquaria. CO_2 levels from yeast are more difficult to regulate than those from pressurized CO_2 systems. However, the low cost of yeast makes it a widely used alternative.

Yeast Extract

Marmite and Vegemite, products made from yeast extract

Marmite and Vegemite have a distinctive dark colour

Yeast extract is the common name for various forms of processed yeast products that are used as food additives or flavours. They are often used in the same way that monosodium glutamate (MSG) is used and, like MSG, often contain free glutamic acid. The general method for making yeast extract for food products such as Vegemite and Marmite on a commercial scale is to add salt to a suspension of yeast, making the solution hypertonic, which leads to the cells' shrivelling up.

This triggers autolysis, wherein the yeast's digestive enzymes break their own proteins down into simpler compounds, a process of self-destruction. The dying yeast cells are then heated to complete their breakdown, after which the husks (yeast with thick cell walls that would give poor texture) are separated. Yeast autolysates are used in Vegemite and Promite (Australia); Marmite, Bovril and Oxo (the United Kingdom, Republic of Ireland and South Africa); and Cenovis (Switzerland).

Scientific Research

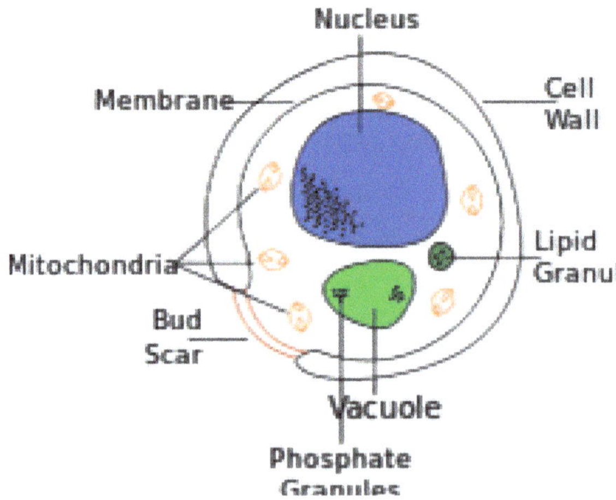

Diagram showing a yeast cell

Several yeasts, in particular S. cerevisiae, have been widely used in genetics and cell biology, largely because S. cerevisiae is a simple eukaryotic cell, serving as a model for all eukaryotes, including humans, for the study of fundamental cellular processes such as the cell cycle, DNA replication, recombination, cell division, and metabolism. Also, yeasts are easily manipulated and cultured in the laboratory, which has allowed for the development of powerful standard techniques, such as yeast two-hybrid, synthetic genetic array analysis, and tetrad analysis. Many proteins important in human biology were first discovered by studying their homologues in yeast; these proteins include cell cycle proteins, signaling proteins, and protein-processing enzymes.

On 24 April 1996, S. cerevisiae was announced to be the first eukaryote to have its genome, consisting of 12 million base pairs, fully sequenced as part of the Genome Project. At the time, it was the most complex organism to have its full genome sequenced, and the work seven years and the involvement of more than 100 laboratories to accomplish. The second yeast species to have its genome sequenced was Schizosaccharomyces pombe, which was completed in 2002. It was the sixth eukaryotic genome sequenced and consists of 13.8 million base pairs. As of 2014, over 50 yeast species have had their genomes sequenced and published.

Genetically Engineered Biofactories

Various yeast species have been genetically engineered to efficiently produce various drugs, a technique called metabolic engineering. S. cerevisiae is easy to genetically engineer; its physiology, metabolism and genetics are well known, and it is amenable for use in harsh industrial conditions. A wide variety of chemical in different classes can be produced by engineered yeast, including phenolics, isoprenoids, alkaloids, and polyketides. About 20% of biopharmaceuticals are produced in S. cerevisiae, including insulin, vaccines for hepatitis, and human serum albumin.

Pathogenic Yeasts

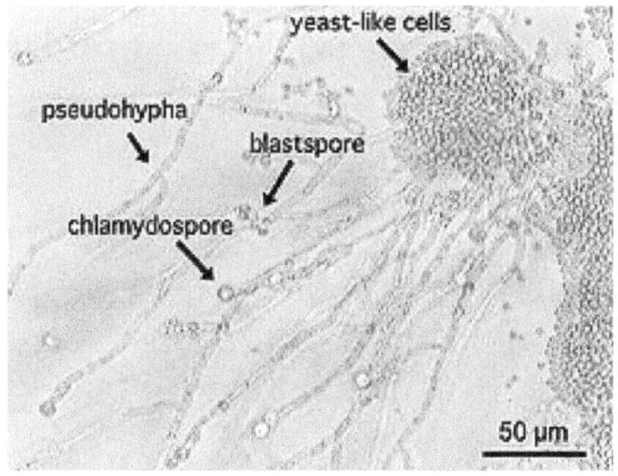

A photomicrograph of Candida albicans showing hyphal outgrowth and other morphological characteristics

Some species of yeast are opportunistic pathogens that can cause infection in people with compromised immune systems. Cryptococcus neoformans and Cryptococcus gattii are significant pathogens of immunocompromised people. They are the species primarily responsible for cryptococcosis, a fungal disease that occurs in about one million HIV/AIDS patients, causing over 600,000 deaths annually. The cells of these yeast are surrounded by a rigid polysaccharide capsule, which helps to prevent them from being recognised and engulfed by white blood cells in the human body.

Yeasts of the Candida genus, another group of opportunistic pathogens, cause oral and vaginal infections in humans, known as candidiasis. Candida is commonly found as a commensal yeast in the mucous membranes of humans and other warm-blooded animals. However, sometimes these same strains can become pathogenic. The yeast cells sprout a hyphal outgrowth, which locally penetrates the mucosal membrane, causing irritation and shedding of the tissues. The pathogenic yeasts of candidiasis in probable descending order of virulence for humans are: C. albicans, C. tropicalis, C. stellatoidea, C. glabrata, C. krusei, C. parapsilosis, C. guilliermondii, C. viswanathii, C. lusitaniae, and Rhodotorula mucilaginosa. Candida glabrata is the second most common Candida pathogen after C. albicans, causing infections of the urogenital tract, and of the bloodstream (candidemia).

Food Spoilage

Yeasts are able to grow in foods with a low pH (5.0 or lower) and in the presence of sugars, organic acids, and other easily metabolized carbon sources. During their growth, yeasts metabolize some food components and produce metabolic end products. This causes the physical, chemical, and sensible properties of a food to change, and the food is spoiled. The growth of yeast within food products is often seen on their surfaces, as in cheeses or meats, or by the fermentation of sugars in beverages, such as juices, and semiliquid products, such as syrups and jams. The yeast of the Zygosaccharomyces genus have had a long history as spoilage yeasts within the food industry. This is mainly because these species can grow in the presence of high sucrose, ethanol, acetic acid, sorbic acid, benzoic acid, and sulphur dioxide concentrations, representing some of the commonly used food preservation methods. Methylene blue is used to test for the presence of live yeast cells. In oenology, the major spoilage yeast is Brettanomyces bruxellensis.

Torulaspora Delbrueckii

*Torulaspora delbrueckii is a u*biquitous yeastspecies with both wild and anthropic habitats. The type strain of *T. delbrueckii* is CBS 1146^T, equivalent to CLIB 230 or ATCC 10662, etc.. The type strain of *T. delbrueckii* CBS 1146 ^{T} was recently sequenced, and is composed of 8 chromosomes in addition to a mitochondrial genome. *T. delbrueckii* was formerly known as *Saccharomyces delbrueckii* or *Saccharomyces rosei* or *Saccharomyces roseus*, and the anamorph is called *Candida colliculosa*.

T. delbrueckii is the most studied species of the *Torulaspora* genus that comprised 8 species to date, of which *T. franciscae, T. pretoriensis, T. microellipsoides, T. globosa, T. indica,T. maleeae*, and *T. quercuum*. The taxonomy of the *Torulaspora* genus is evolving rapidly, and the availability of molecular tools to discriminate *Torulaspora* species will help correcting errors in species assignments.

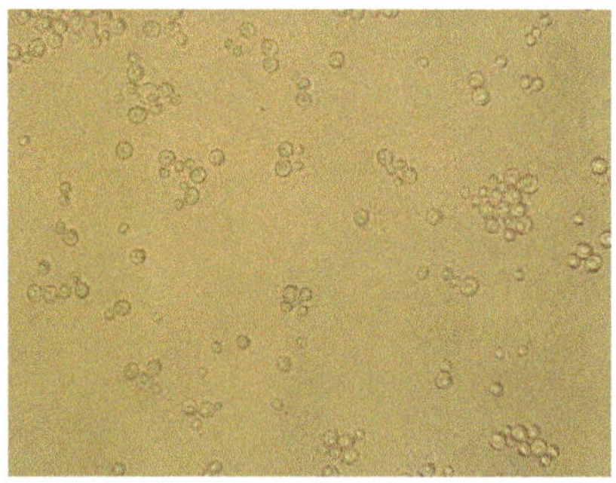

Spores of Torulaspora delbrueckii

Torulaspora Delbrueckii Habitats

T. delbrueckii is isolated from several human bioprocesses, including the bread industry where some *T. delbrueckii* strains are commercialized for frozen dough applications. Other applications include food fermentations of silage, cocoa, olive or cucumber; distilled and traditional fermented beverage production including mescal,colonche,tequila,cider, strawberry tree fruits juice, sugarcane juice or kefir;dairy products' fermentations like traditional cheeses and fermented milk.*T. delbrueckii* can be an opportunistic spoilage yeast for dairy products or soft drinks (fruit juices, etc.).*T.delbrueckii* colonizes several natural environments, ranging from soils, to plants, fruits and insects.*T.delbrueckii* is occasionally found as a clinical isolate, although not considered to be a human pathogen, a state described as opportunistic pathogen.

Torulaspora Delbrueckii and Winemaking

T. delbrueckii has been associated with winemaking for decades and isolated either from grape, must or wine. *T. delbrueckii* is now proposed as starter culture (to be associated with *S. cerevisiae* in mixed cultures) for certain applications, particularly to reduce volatile acidity in high-sugar fermentations like in Sauternes wines. Recent findings show that *T. delbrueckii* species has been domesticated for winemaking and other human uses about 1900 and 4000 years ago respectively.

Torulaspora Delbrueckii Life-cycle

The life-cycle of *T. delbrueckii* remains unclear. Some authors consider *T. delbrueckii* to be a haploid species, while more recent findings suggest *T. delbrueckii* have a mostly diploid homothallic life. To date, the life-cycle of the species is not formally elucidated.

Fungus

A fungus is any member of the group of eukaryotic organisms that includes unicellular microorganisms such as yeasts and molds, as well as multicellular fungi that produce familiar fruiting forms known as mushrooms. These organisms are classified as a kingdom, Fungi, which is separate from the other eukaryotic life kingdoms of plants and animals.

A characteristic that places fungi in a different kingdom from plants, bacteria and some protists, is chitin in their cell walls. Similar to animals, fungi are heterotrophs; they acquire their food by absorbing dissolved molecules, typically by secreting digestive enzymes into their environment. Fungi do not photosynthesise. Growth is their means of mobility, except for spores (a few of which are flagellated), which may travel through the air or water. Fungi are the principal decomposers in ecological systems. These and other differences place fungi in a single group of related organisms, named the Eumycota (true fungi or Eumycetes), which share a common ancestor (form a monophyletic group), an interpretation that is also strongly supported by molecular phylogenetics. This fungal group is distinct from the structurally similar myxomycetes (slime molds) and oomycetes (water molds). The discipline of biology devoted to the study of fungi is known as mycology. In the past, mycology was regarded as a branch of botany, although it is now known fungi are genetically more closely related to animals than to plants.

Abundant worldwide, most fungi are inconspicuous because of the small size of their structures, and their cryptic lifestyles in soil or on dead matter. Fungi include symbionts of plants, animals, or other fungi and also parasites. They may become noticeable when fruiting, either as mushrooms or as molds. Fungi perform an essential role in the decomposition of organic matter and have fundamental roles in nutrient cycling and exchange in the environment. They have long been used as a direct source of human food, in the form of mushrooms and truffles; as a leavening agent for bread; and in the fermentation of various food products, such as wine, beer, and soy sauce. Since the 1940s, fungi have been used for the production of antibiotics, and, more recently, various enzymes produced by fungi are used industrially and in detergents. Fungi are also used as biological pesticides to control weeds, plant diseases and insect pests. Many species produce bioactive compounds called mycotoxins, such as alkaloids and polyketides, that are toxic to animals including humans. The fruiting structures of a few species contain psychotropic compounds and are consumed recreationally or in traditional spiritual ceremonies. Fungi can break down manufactured materials and buildings, and become significant pathogens of humans and other animals. Losses of crops due to fungal diseases (e.g., rice blast disease) or food spoilage can have a large impact on human food supplies and local economies.

The fungus kingdom encompasses an enormous diversity of taxa with varied ecologies, life cycle strategies, and morphologies ranging from unicellular aquatic chytrids to large mushrooms. However, little is known of the true biodiversity of Kingdom Fungi, which has been estimated at 1.5 million to 5 million species, with about 5% of these having been formally classified. Ever since the pioneering 18th and 19th century taxonomical works of Carl Linnaeus, Christian Hendrik Persoon, and Elias Magnus Fries, fungi have been classified according to their morphology (e.g., characteristics such as spore color or microscopic features) or physiology. Advances in molecular genetics have opened the way for DNA analysis to be incorporated into taxonomy, which has sometimes challenged the historical groupings based on morphology and other traits. Phylogenetic studies published in the last decade have helped reshape the classification within Kingdom Fungi, which is divided into one subkingdom, seven phyla, and ten subphyla.

Etymology

The English word fungus is directly adopted from the Latin fungus (mushroom), used in the writings of Horace and Pliny. This in turn is derived from the Greek word sphongos, which refers to the macroscopic structures and morphology of mushrooms and molds; the root is also used in other languages, such as the German Schwamm ("sponge") and Schimmel ("mold"). The use of the word mycology, which is derived from the Greek mykes and logos, to denote the scientific study of fungi is thought to have originated in 1836 with English naturalist Miles Joseph Berkeley's publication The English Flora of Sir James Edward Smith, Vol. 5. A group of all the fungi present in a particular area or geographic region is known as mycobiota (plural noun, no singular), e.g., "the mycobiota of Ireland".

Characteristics

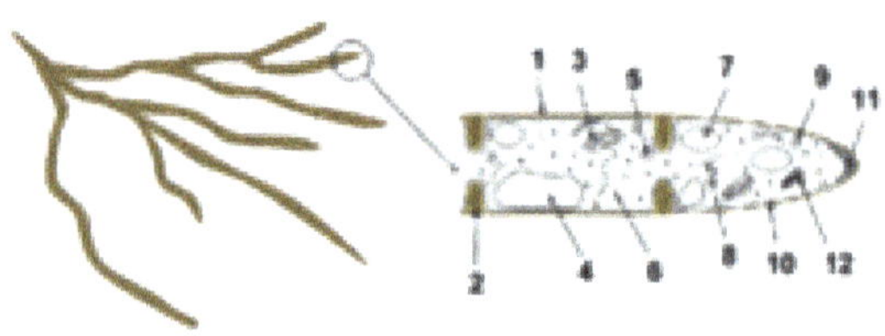

Fungal hyphae cells

1. Hyphal wall
2. Septum
3. Mitochondrion
4. Vacuole
5. Ergosterol crystal
6. Ribosome
7. Nucleus
8. Endoplasmic reticulum
9. Lipid body
10. Plasma membrane
11. Spitzenkörper
12. Golgi apparatus

Before the introduction of molecular methods for phylogenetic analysis, taxonomists considered fungi to be members of the plant kingdom because of similarities in lifestyle: both fungi and plants are mainly immobile, and have similarities in general morphology and growth habitat. Like plants, fungi often grow in soil and, in the case of mushrooms, form conspicuous fruit bodies, which sometimes resemble plants such as mosses. The fungi are now considered a separate kingdom, distinct from both plants and animals, from which they appear to have diverged around one billion years ago. Some morphological, biochemical, and genetic features are shared with other organisms, while others are unique to the fungi, clearly separating them from the other kingdoms:

Shared features:

- With other eukaryotes: Fungal cells contain membrane-bound nuclei with chromosomes that contain DNA with noncoding regions called introns and coding regions called exons. Fungi have membrane-bound cytoplasmic organelles such as mitochondria, sterol-containing membranes, and ribosomes of the 80S type. They have a characteristic range of soluble carbohydrates and storage compounds, including sugar alcohols (e.g., mannitol), disaccharides, (e.g., trehalose), and polysaccharides (e.g., glycogen, which is also found in animals).
- With animals: Fungi lack chloroplasts and are heterotrophic organisms and so require preformed organic compounds as energy sources.
- With plants: Fungi have a cell wall and vacuoles. They reproduce by both sexual and asexual means, and like basal plant groups (such as ferns and mosses) produce spores. Similar to mosses and algae, fungi typically have haploid nuclei.
- With euglenoids and bacteria: Higher fungi, euglenoids, and some bacteria produce the amino acid L-lysine in specific biosynthesis steps, called the α-aminoadipate pathway.
- The cells of most fungi grow as tubular, elongated, and thread-like (filamentous) structures called hyphae, which may contain multiple nuclei and extend by growing at their tips. Each tip contains a set of aggregated vesicles—cellular structures consisting of proteins, lipids, and other organic molecules—called the Spitzenkörper. Both fungi and oomycetes grow as filamentous hyphal cells. In contrast, similar-looking organisms, such as filamentous green algae, grow by repeated cell division within a chain of cells. There are also single-celled fungi (yeasts) that do not form hyphae, and some fungi have both hyphal and yeast forms.
- In common with some plant and animal species, more than 70 fungal species display bioluminescence.

Unique features:

- Some species grow as unicellular yeasts that reproduce by budding or binary fission. Dimorphic fungi can switch between a yeast phase and a hyphal phase in response to environmental conditions.
- The fungal cell wall is composed of glucans and chitin; while glucans are also found in plants and chitin in the exoskeleton of arthropods, fungi are the only organisms that combine these two structural molecules in their cell wall. Unlike those of plants and oomycetes, fungal cell walls do not contain cellulose.

Omphalotus nidiformis, a bioluminescent mushroom

Most fungi lack an efficient system for the long-distance transport of water and nutrients, such as the xylem and phloem in many plants. To overcome this limitation, some fungi, such as Armillaria, form rhizomorphs, which resemble and perform functions similar to the roots of plants. As eukaryotes, fungi possess a biosynthetic pathway for producing terpenes that uses mevalonic acid and pyrophosphate as chemical building blocks. Plants and some other organisms have an additional terpene biosynthesis pathway in their chloroplasts, a structure fungi and animals do not have. Fungi produce several secondary metabolites that are similar or identical in structure to those made by plants. Many of the plant and fungal enzymes that make these compounds differ from each other in sequence and other characteristics, which indicates separate origins and convergent evolution of these enzymes in the fungi and plants.

Diversity

Bracket fungi on a tree stump

Fungi have a worldwide distribution, and grow in a wide range of habitats, including extreme environments such as deserts or areas with high salt concentrations or ionizing radiation, as well as in deep sea sediments. Some can survive the intense UV and cosmic radiation encountered during space travel. Most grow in terrestrial environments, though several species live partly or solely in aquatic habitats, such as the chytrid fungus Batrachochytrium dendrobatidis, a parasite that has been responsible for a worldwide decline in amphibian populations. This organism spends part of its life cycle as a motile zoospore, enabling it to propel itself through water and enter its amphibian host. Other examples of aquatic fungi include those living in hydrothermal areas of the ocean.

Around 100,000 species of fungi have been formally described by taxonomists, but the global biodiversity of the fungus kingdom is not fully understood. On the basis of observations of the ratio of the number of fungal species to the number of plant species in selected environments, the fungal kingdom has been estimated to contain about 1.5 million species. A recent (2011) estimate suggests there may be over 5 million species. In mycology, species have historically been distinguished by a variety of methods and concepts. Classification based on morphological characteristics, such as the size and shape of spores or fruiting structures, has traditionally dominated fungal taxonomy. Species may also be distinguished by their biochemical and physiological characteristics, such as their ability to metabolize certain biochemicals, or their reaction to chemical tests. The biological species concept discriminates species based on their ability to mate. The application of molecular tools, such as DNA sequencing and phylogenetic analysis, to study diversity has greatly enhanced the resolution and added robustness to estimates of genetic diversity within various taxonomic groups.

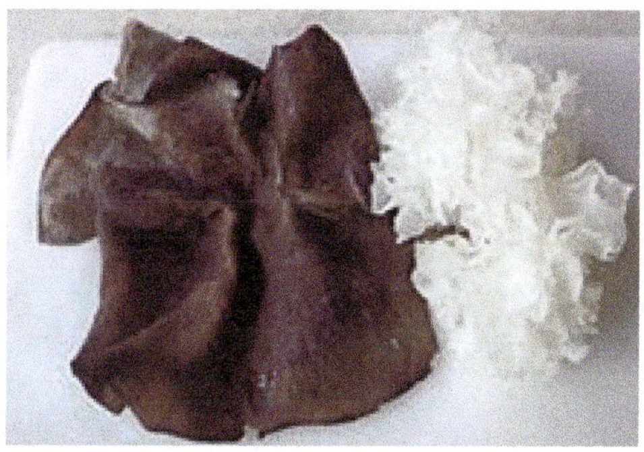

Two types of edible fungi

Mycology

Mycology is the branch of biology concerned with the systematic study of fungi, including their genetic and biochemical properties, their taxonomy, and their use to humans as a source of medicine, food, and psychotropic substances consumed for religious purposes, as well as their dangers, such as poisoning or infection. The field of phytopathology, the study of plant diseases, is closely related because many plant pathogens are fungi.

In 1729, Pier A. Micheli first published descriptions of fungi.

The use of fungi by humans dates back to prehistory; Ötzi the Iceman, a well-preserved mummy of a 5,300-year-old Neolithic man found frozen in the Austrian Alps, carried two species of polypore mushrooms that may have been used as tinder (Fomes fomentarius), or for medicinal purposes

(Piptoporus betulinus). Ancient peoples have used fungi as food sources–often unknowingly–for millennia, in the preparation of leavened bread and fermented juices. Some of the oldest written records contain references to the destruction of crops that were probably caused by pathogenic fungi.

History

Mycology is a relatively new science that became systematic after the development of the microscope in the 17th century. Although fungal spores were first observed by Giambattista della Porta in 1588, the seminal work in the development of mycology is considered to be the publication of Pier Antonio Micheli's 1729 work Nova plantarum genera. Micheli not only observed spores but also showed that, under the proper conditions, they could be induced into growing into the same species of fungi from which they originated. Extending the use of the binomial system of nomenclature introduced by Carl Linnaeus in his Species plantarum (1753), the Dutch Christian Hendrik Persoon (1761–1836) established the first classification of mushrooms with such skill so as to be considered a founder of modern mycology. Later, Elias Magnus Fries (1794–1878) further elaborated the classification of fungi, using spore color and various microscopic characteristics, methods still used by taxonomists today. Other notable early contributors to mycology in the 17th–19th and early 20th centuries include Miles Joseph Berkeley, August Carl Joseph Corda, Anton de Bary, the brothers Louis René and Charles Tulasne, Arthur H. R. Buller, Curtis G. Lloyd, and Pier Andrea Saccardo. The 20th century has seen a modernization of mycology that has come from advances in biochemistry, genetics, molecular biology, and biotechnology. The use of DNA sequencing technologies and phylogenetic analysis has provided new insights into fungal relationships and biodiversity, and has challenged traditional morphology-based groupings in fungal taxonomy.

Morphology

Microscopic Structures

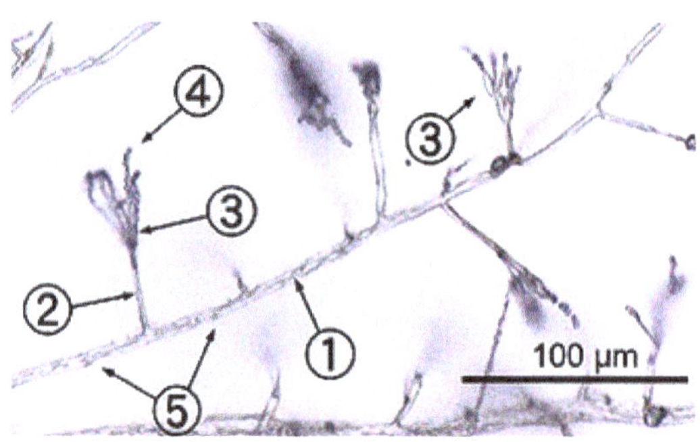

An environmental isolate of Penicillium

1. hypha
2. conidiophore
3. phialide
4. conidia
5. septa

Most fungi grow as hyphae, which are cylindrical, thread-like structures 2–10 μm in diameter and up to several centimeters in length. Hyphae grow at their tips (apices); new hyphae are typically formed by emergence of new tips along existing hyphae by a process called branching, or occasionally growing hyphal tips fork, giving rise to two parallel-growing hyphae. Hyphae also sometimes fuse when they come into contact, a process called hyphal fusion (or anastamosis). These growth processes lead to the development of a mycelium, an interconnected network of hyphae. Hyphae can be either septate or coenocytic. Septate hyphae are divided into compartments separated by cross walls (internal cell walls, called septa, that are formed at right angles to the cell wall giving the hypha its shape), with each compartment containing one or more nuclei; coenocytic hyphae are not compartmentalized. Septa have pores that allow cytoplasm, organelles, and sometimes nuclei to pass through; an example is the dolipore septum in fungi of the phylum Basidiomycota. Coenocytic hyphae are in essence multinucleate supercells.

Many species have developed specialized hyphal structures for nutrient uptake from living hosts; examples include haustoria in plant-parasitic species of most fungal phyla, and arbuscules of several mycorrhizal fungi, which penetrate into the host cells to consume nutrients.

Although fungi are opisthokonts—a grouping of evolutionarily related organisms broadly characterized by a single posterior flagellum—all phyla except for the chytrids have lost their posterior flagella. Fungi are unusual among the eukaryotes in having a cell wall that, in addition to glucans (e.g., β-1,3-glucan) and other typical components, also contains the biopolymer chitin.

Macroscopic Structures

Armillaria solidipes

Fungal mycelia can become visible to the naked eye, for example, on various surfaces and substrates, such as damp walls and spoiled food, where they are commonly called molds. Mycelia grown on solid agar media in laboratory petri dishes are usually referred to as colonies. These colonies can exhibit growth shapes and colors (due to spores or pigmentation) that can be used as diagnostic features in the identification of species or groups. Some individual fungal colonies can reach extraordinary dimensions and ages as in the case of a clonal colony of Armillaria solidipes, which extends over an area of more than 900 ha (3.5 square miles), with an estimated age of nearly 9,000 years.

The apothecium—a specialized structure important in sexual reproduction in the ascomycetes—is a cup-shaped fruit body that is often macroscopic and holds the hymenium, a layer of tissue containing the spore-bearing cells. The fruit bodies of the basidiomycetes (basidiocarps) and some ascomycetes can sometimes grow very large, and many are well known as mushrooms.

Growth and Physiology

Mold growth covering a decaying peach. The frames were taken approximately 12 hours apart over a period of six days.

The growth of fungi as hyphae on or in solid substrates or as single cells in aquatic environments is adapted for the efficient extraction of nutrients, because these growth forms have high surface area to volume ratios. Hyphae are specifically adapted for growth on solid surfaces, and to invade substrates and tissues. They can exert large penetrative mechanical forces; for example, many plant pathogens, including Magnaporthe grisea, form a structure called an appressorium that evolved to puncture plant tissues. The pressure generated by the appressorium, directed against the plant epidermis, can exceed 8 megapascals (1,200 psi). The filamentous fungus Paecilomyces lilacinus uses a similar structure to penetrate the eggs of nematodes.

The mechanical pressure exerted by the appressorium is generated from physiological processes that increase intracellular turgor by producing osmolytes such as glycerol. Adaptations such as these are complemented by hydrolytic enzymes secreted into the environment to digest large organic molecules—such as polysaccharides, proteins, and lipids—into smaller molecules that may then be absorbed as nutrients. The vast majority of filamentous fungi grow in a polar fashion (extending in one direction) by elongation at the tip (apex) of the hypha. Other forms of fungal growth include intercalary extension (longitudinal expansion of hyphal compartments that are below the apex) as in the case of some endophytic fungi, or growth by volume expansion during the development of mushroom stipes and other large organs. Growth of fungi as multicellular structures consisting of somatic and reproductive cells—a feature independently evolved in animals and plants—has several functions, including the development of fruit bodies for dissemination of sexual spores and biofilms for substrate colonization and intercellular communication.

The fungi are traditionally considered heterotrophs, organisms that rely solely on carbon fixed by other organisms for metabolism. Fungi have evolved a high degree of metabolic versatility that allows them to use a diverse range of organic substrates for growth, including simple compounds such as nitrate, ammonia, acetate, or ethanol. In some species the pigment melanin may play a role in extracting energy from ionizing radiation, such as gamma radiation. This form of "radiotrophic" growth has been described for only a few species, the effects on growth rates are small, and the underlying biophysical and biochemical processes are not well known. This process might bear similarity to CO_2 fixation via visible light, but instead uses ionizing radiation as a source of energy.

Reproduction

Polyporus squamosus

Fungal reproduction is complex, reflecting the differences in lifestyles and genetic makeup within this diverse kingdom of organisms. It is estimated that a third of all fungi reproduce using more than one method of propagation; for example, reproduction may occur in two well-differentiated stages within the life cycle of a species, the teleomorph and the anamorph. Environmental conditions trigger genetically determined developmental states that lead to the creation of specialized structures for sexual or asexual reproduction. These structures aid reproduction by efficiently dispersing spores or spore-containing propagules.

Asexual Reproduction

Asexual reproduction occurs via vegetative spores (conidia) or through mycelial fragmentation. Mycelial fragmentation occurs when a fungal mycelium separates into pieces, and each component grows into a separate mycelium. Mycelial fragmentation and vegetative spores maintain clonal populations adapted to a specific niche, and allow more rapid dispersal than sexual reproduction. The "Fungi imperfecti" (fungi lacking the perfect or sexual stage) or Deuteromycota comprise all the species that lack an observable sexual cycle. Deuteromycota is not an accepted taxonomic clade, and is now taken to mean simply fungi that lack a known sexual stage.

Sexual Reproduction

Sexual reproduction with meiosis has been directly observed in all fungal phyla except Glomeromycota (genetic analysis suggests meiosis in Glomeromycota as well). It differs in many aspects from sexual reproduction in animals or plants. Differences also exist between fungal groups and can be used to discriminate species by morphological differences in sexual structures and reproductive strategies. Mating experiments between fungal isolates may identify species on the basis of biological species concepts. The major fungal groupings have initially been delineated based on the morphology of their sexual structures and spores; for example, the spore-containing structures, asci and basidia, can be used in the identification of ascomycetes and basidiomycetes, respectively. Some species may allow mating only between individuals of opposite mating type, whereas others can mate and sexually reproduce with any other individual or itself. Species of the former mating system are called heterothallic, and of the latter homothallic.

Most fungi have both a haploid and a diploid stage in their life cycles. In sexually reproducing fungi, compatible individuals may combine by fusing their hyphae together into an interconnected network; this process, anastomosis, is required for the initiation of the sexual cycle. Many ascomycetes and basidiomycetes go through a dikaryotic stage, in which the nuclei inherited from the two parents do not combine immediately after cell fusion, but remain separate in the hyphal cells .

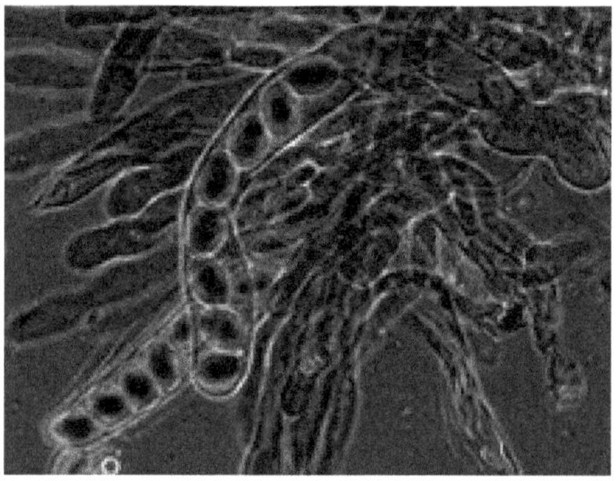

The 8-spore asci of Morchella elata, viewed with phase contrast microscopy

In ascomycetes, dikaryotic hyphae of the hymenium (the spore-bearing tissue layer) form a characteristic hook at the hyphal septum. During cell division, formation of the hook ensures proper distribution of the newly divided nuclei into the apical and basal hyphal compartments. An ascus (plural asci) is then formed, in which karyogamy (nuclear fusion) occurs. Asci are embedded in an ascocarp, or fruiting body. Karyogamy in the asci is followed immediately by meiosis and the production of ascospores. After dispersal, the ascospores may germinate and form a new haploid mycelium.

Sexual reproduction in basidiomycetes is similar to that of the ascomycetes. Compatible haploid hyphae fuse to produce a dikaryotic mycelium. However, the dikaryotic phase is more extensive in the basidiomycetes, often also present in the vegetatively growing mycelium. A specialized anatomical structure, called a clamp connection, is formed at each hyphal septum. As with the structurally similar hook in the ascomycetes, the clamp connection in the basidiomycetes is required for controlled transfer of nuclei during cell division, to maintain the dikaryotic stage with two genetically different nuclei in each hyphal compartment. A basidiocarp is formed in which clublike structures known as basidia generate haploid basidiospores after karyogamy and meiosis. The most commonly known basidiocarps are mushrooms, but they may also take other forms.

In glomeromycetes (formerly zygomycetes), haploid hyphae of two individuals fuse, forming a gametangium, a specialized cell structure that becomes a fertile gamete-producing cell. The gametangium develops into a zygospore, a thick-walled spore formed by the union of gametes. When the zygospore germinates, it undergoes meiosis, generating new haploid hyphae, which may then form asexual sporangiospores. These sporangiospores allow the fungus to rapidly disperse and germinate into new genetically identical haploid fungal mycelia.

Spore Dispersal

Both asexual and sexual spores or sporangiospores are often actively dispersed by forcible ejection from their reproductive structures. This ejection ensures exit of the spores from the reproductive structures as well as traveling through the air over long distances.

The bird's nest fungus Cyathus stercoreus

Specialized mechanical and physiological mechanisms, as well as spore surface structures (such as hydrophobins), enable efficient spore ejection. For example, the structure of the spore-bearing cells in some ascomycete species is such that the buildup of substances affecting cell volume and fluid balance enables the explosive discharge of spores into the air. The forcible discharge of single spores termed ballistospores involves formation of a small drop of water (Buller's drop), which upon contact with the spore leads to its projectile release with an initial acceleration of more than 10,000 g; the net result is that the spore is ejected 0.01–0.02 cm, sufficient distance for it to fall through the gills or pores into the air below. Other fungi, like the puffballs, rely on alternative mechanisms for spore release, such as external mechanical forces. The bird's nest fungi use the force of falling water drops to liberate the spores from cup-shaped fruiting bodies. Another strategy is seen in the stinkhorns, a group of fungi with lively colors and putrid odor that attract insects to disperse their spores.

Other Sexual Processes

Besides regular sexual reproduction with meiosis, certain fungi, such as those in the genera Penicillium and Aspergillus, may exchange genetic material via parasexual processes, initiated by anastomosis between hyphae and plasmogamy of fungal cells. The frequency and relative importance of parasexual events is unclear and may be lower than other sexual processes. It is known to play a role in intraspecific hybridization and is likely required for hybridization between species, which has been associated with major events in fungal evolution.

Evolution

In contrast to plants and animals, the early fossil record of the fungi is meager. Factors that likely contribute to the under-representation of fungal species among fossils include the nature of fungal fruiting bodies, which are soft, fleshy, and easily degradable tissues and the microscopic dimensions of most fungal structures, which therefore are not readily evident. Fungal fossils are difficult to distinguish from those of other microbes, and are most easily identified when they resemble extant fungi. Often recovered from a permineralized plant or animal host, these samples are typically studied by making thin-section preparations that can be examined with light microscopy or transmission electron microscopy. Researchers study compression fossils by dissolving the surrounding matrix with acid and then using light or scanning electron microscopy to examine surface details.

The earliest fossils possessing features typical of fungi date to the Proterozoic eon, some 1,430 million years ago (Ma); these multicellular benthic organisms had filamentous structures with septa, and were capable of anastomosis. More recent studies (2009) estimate the arrival of fungal organisms at about 760–1060 Ma on the basis of comparisons of the rate of evolution in closely related groups. For much of the Paleozoic Era (542–251 Ma), the fungi appear to have been aquatic and consisted of organisms similar to the extant chytrids in having flagellum-bearing spores. The evolutionary adaptation from an aquatic to a terrestrial lifestyle necessitated a diversification of ecological strategies for obtaining nutrients, including parasitism, saprobism, and the development of mutualistic relationships such as mycorrhiza and lichenization. Recent (2009) studies suggest that the ancestral ecological state of the Ascomycota was saprobism, and that independent lichenization events have occurred multiple times.

It is presumed that the fungi colonized the land during the Cambrian (542–488.3 Ma), long before land plants. Fossilized hyphae and spores recovered from the Ordovician of Wisconsin (460 Ma) resemble modern-day Glomerales, and existed at a time when the land flora likely consisted of only non-vascular bryophyte-like plants. Prototaxites, which was probably a fungus or lichen, would have been the tallest organism of the late Silurian. Fungal fossils do not become common and uncontroversial until the early Devonian (416–359.2 Ma), when they occur abundantly in the Rhynie chert, mostly as Zygomycota and Chytridiomycota. At about this same time, approximately 400 Ma, the Ascomycota and Basidiomycota diverged, and all modern classes of fungi were present by the Late Carboniferous (Pennsylvanian, 318.1–299 Ma).

Lichen-like fossils have been found in the Doushantuo Formation in southern China dating back to 635–551 Ma. Lichens formed a component of the early terrestrial ecosystems, and the estimated age of the oldest terrestrial lichen fossil is 400 Ma; this date corresponds to the age of the oldest known sporocarp fossil, a Paleopyrenomycites species found in the Rhynie Chert. The oldest fossil with microscopic features resembling modern-day basidiomycetes is Palaeoancistrus, found permineralized with a fern from the Pennsylvanian. Rare in the fossil record are the Homobasidiomycetes (a taxon roughly equivalent to the mushroom-producing species of the Agaricomycetes). Two amber-preserved specimens provide evidence that the earliest known mushroom-forming fungi (the extinct species Archaeomarasmius leggetti) appeared during the late Cretaceous, 90 Ma.

Some time after the Permian–Triassic extinction event (251.4 Ma), a fungal spike (originally thought to be an extraordinary abundance of fungal spores in sediments) formed, suggesting that fungi were the dominant life form at this time, representing nearly 100% of the available fossil record for this period. However, the relative proportion of fungal spores relative to spores formed by algal species is difficult to assess, the spike did not appear worldwide, and in many places it did not fall on the Permian–Triassic boundary.

Taxonomy

Although commonly included in botany curricula and textbooks, fungi are more closely related to animals than to plants and are placed with the animals in the monophyletic group of opisthokonts. Analyses using molecular phylogenetics support a monophyletic origin of the Fungi. The taxonomy of the Fungi is in a state of constant flux, especially due to recent research based on DNA comparisons. These current phylogenetic analyses often overturn classifications based on older and sometimes less discriminative methods based on morphological features and biological species concepts obtained from experimental matings.

There is no unique generally accepted system at the higher taxonomic levels and there are frequent name changes at every level, from species upwards. Efforts among researchers are now underway to establish and encourage usage of a unified and more consistent nomenclature. Fungal species can also have multiple scientific names depending on their life cycle and mode (sexual or asexual) of reproduction. Web sites such as Index Fungorum and ITIS list current names of fungal species (with cross-references to older synonyms).

The 2007 classification of Kingdom Fungi is the result of a large-scale collaborative research effort involving dozens of mycologists and other scientists working on fungal taxonomy. It recognizes seven phyla, two of which—the Ascomycota and the Basidiomycota—are contained within a branch representing subkingdom Dikarya, the most species rich and familiar group, including all the mushrooms, most food-spoilage molds, most plant pathogenic fungi, and the beer, wine, and bread yeasts. The accompanying cladogram depicts the major fungal taxa and their relationship to opisthokont and unikont organisms, based on the work of Philippe Silar and "The Mycota: A Comprehensive Treatise on Fungi as Experimental Systems for Basic and Applied Research". The lengths of the branches are not proportional to evolutionary distances.

Taxonomic Groups

The major phyla (sometimes called divisions) of fungi have been classified mainly on the basis of characteristics of their sexual reproductive structures. Currently, seven phyla are proposed: Microsporidia, Chytridiomycota, Blastocladiomycota, Neocallimastigomycota, Glomeromycota, Ascomycota, and Basidiomycota.

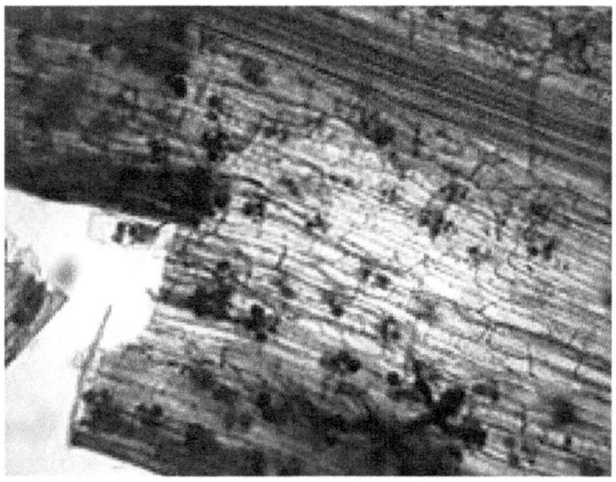

Arbuscular mycorrhiza seen under microscope. Flax root cortical cells containing paired arbuscules.

Phylogenetic analysis has demonstrated that the Microsporidia, unicellular parasites of animals and protists, are fairly recent and highly derived endobiotic fungi (living within the tissue of another species). One 2006 study concludes that the Microsporidia are a sister group to the true fungi; that is, they are each other's closest evolutionary relative. Hibbett and colleagues suggest that this analysis does not clash with their classification of the Fungi, and although the Microsporidia are elevated to phylum status, it is acknowledged that further analysis is required to clarify evolutionary relationships within this group.

The Chytridiomycota are commonly known as chytrids. These fungi are distributed worldwide. Chytrids and their close relatives Neocallimastigomycota and Blastocladiomycota (below) are the only fungi with active motility, producing zoospores that are capable of active movement through aqueous phases with a single flagellum, leading early taxonomists to classify them as protists. Molecular phylogenies, inferred from rRNA sequences in ribosomes, suggest that the Chytrids are a basal group divergent from the other fungal phyla, consisting of four major clades with suggestive evidence for paraphyly or possibly polyphyly.

The Blastocladiomycota were previously considered a taxonomic clade within the Chytridiomycota. Recent molecular data and ultrastructural characteristics, however, place the Blastocladiomycota as a sister clade to the Zygomycota, Glomeromycota, and Dikarya (Ascomycota and Basidiomycota). The blastocladiomycetes are saprotrophs, feeding on decomposing organic matter, and they are parasites of all eukaryotic groups. Unlike their close relatives, the chytrids, most of which exhibit zygotic meiosis, the blastocladiomycetes undergo sporic meiosis.

The Neocallimastigomycota were earlier placed in the phylum Chytridomycota. Members of this small phylum are anaerobic organisms, living in the digestive system of larger herbivorous mammals and in other terrestrial and aquatic environments enriched in cellulose (e.g., domestic waste landfill sites). They lack mitochondria but contain hydrogenosomes of mitochondrial origin. As in the related chrytrids, neocallimastigomycetes form zoospores that are posteriorly uniflagellate or polyflagellate.

Members of the Glomeromycota form arbuscular mycorrhizae, a form of mutualist symbiosis wherein fungal hyphae invade plant root cells and both species benefit from the resulting increased supply of nutrients. All known Glomeromycota species reproduce asexually. The symbiotic association between the Glomeromycota and plants is ancient, with evidence dating to 400 million years ago. Formerly part of the Zygomycota (commonly known as 'sugar' and 'pin' molds), the Glomeromycota were elevated to phylum status in 2001 and now replace the older phylum Zygomycota. Fungi that were placed in the Zygomycota are now being reassigned to the Glomeromycota, or the subphyla incertae sedis Mucoromycotina, Kickxellomycotina, the Zoopagomycotina and the Entomophthoromycotina. Some well-known examples of fungi formerly in the Zygomycota include black bread mold (Rhizopus stolonifer), and Pilobolus species, capable of ejecting spores several meters through the air. Medically relevant genera include Mucor, Rhizomucor, and Rhizopus.

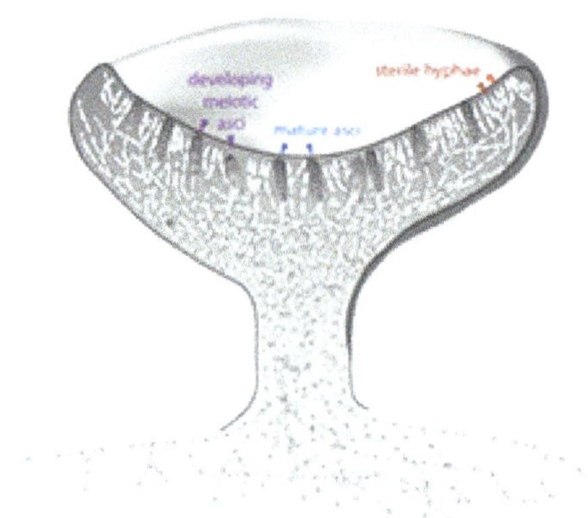

Diagram of an apothecium (the typical cup-like reproductive structure of Ascomycetes) showing sterile tissues as well as developing and mature asci.

The Ascomycota, commonly known as sac fungi or ascomycetes, constitute the largest taxonomic group within the Eumycota. These fungi form meiotic spores called ascospores, which are enclosed in a special sac-like structure called an ascus. This phylum includes morels, a few mushrooms and truffles, unicellular yeasts (e.g., of the genera Saccharomyces, Kluyveromyces, Pichia, and Candida), and many filamentous fungi living as saprotrophs, parasites, and mutualistic symbionts (e.g. lichens). Prominent and important genera of filamentous ascomycetes include Aspergillus, Penicillium, Fusarium, and Claviceps. Many ascomycete species have only been observed undergoing asexual reproduction (called anamorphic species), but analysis of molecular data has often been able to identify their closest teleomorphs in the Ascomycota. Because the products of meiosis are retained within the sac-like ascus, ascomycetes have been used for elucidating principles of genetics and heredity (e.g., Neurospora crassa).

Members of the Basidiomycota, commonly known as the club fungi or basidiomycetes, produce meiospores called basidiospores on club-like stalks called basidia. Most common mushrooms belong to this group, as well as rust and smut fungi, which are major pathogens of grains. Other important basidiomycetes include the maize pathogen Ustilago maydis, human commensal species of the genus Malassezia, and the opportunistic human pathogen, Cryptococcus neoformans.

Fungus-like Organisms

Because of similarities in morphology and lifestyle, the slime molds (mycetozoans, plasmodiophorids, acrasids, Fonticula and labyrinthulids, now in Amoebozoa, Rhizaria, Excavata, Opisthokonta and Stramenopiles, respectively), water molds (oomycetes) and hyphochytrids (both Stramenopiles) were formerly classified in the kingdom Fungi, in groups like Mastigomycotina, Gymnomycota and Phycomycetes. The slime molds were studied also as protozoans, leading to a ambiregnal, duplicated taxonomy.

Unlike true fungi, the cell walls of oomycetes contain cellulose and lack chitin. Hyphochytrids have both chitin and cellulose. Slime molds lack a cell wall during the assimilative phase (except labyrinthulids, which have a wall of scales), and ingest nutrients by ingestion (phagocytosis, except labyrinthulids) rather than absorption (osmotrophy, as fungi, labyrinthulids, oomycetes and hyphochytrids). Neither water molds nor slime molds are closely related to the true fungi, and, therefore, taxonomists no longer group them in the kingdom Fungi. Nonetheless, studies of the oomycetes and myxomycetes are still often included in mycology textbooks and primary research literature.

The Eccrinales and Amoebidiales are opisthokont protists, previously thought to be zygomycete fungi. Other groups now in Opisthokonta (e.g., Corallochytrium, Ichthyosporea) were also at given time classified as fungi. The genus Blastocystis, now in Stramenopiles, was originally classified as a yeast. Ellobiopsis, now in Alveolata, was considered a chytrid. The bacteria were also included in fungi in some classifications, as the group Schizomycetes.

The Rozellida clade, including the "ex-chytrid" Rozella, is a genetically disparate group known mostly from environmental DNA sequences that is a sister group to fungi. Members of the group that have been isolated lack the chitinous cell wall that is characteristic of fungi.

The nucleariids, protists currently grouped in the Choanozoa (Opisthokonta), may be the next sister group to the eumycete clade, and as such could be included in an expanded fungal kingdom.

Many Actinomycetales (Actinobacteria), a group with many filamentous bacteria, were also long believed to be fungi.

Ecology

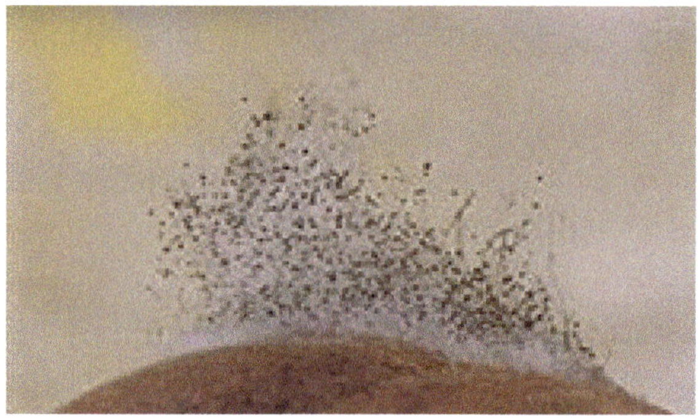

A pin mold decomposing a peach

Although often inconspicuous, fungi occur in every environment on Earth and play very important roles in most ecosystems. Along with bacteria, fungi are the major decomposers in most terrestrial (and some aquatic) ecosystems, and therefore play a critical role in biogeochemical cycles and in many food webs. As decomposers, they play an essential role in nutrient cycling, especially as saprotrophs and symbionts, degrading organic matter to inorganic molecules, which can then re-enter anabolic metabolic pathways in plants or other organisms.

Symbiosis

Many fungi have important symbiotic relationships with organisms from most if not all Kingdoms. These interactions can be mutualistic or antagonistic in nature, or in the case of commensal fungi are of no apparent benefit or detriment to the host.

With Plants

Mycorrhizal symbiosis between plants and fungi is one of the most well-known plant–fungus associations and is of significant importance for plant growth and persistence in many ecosystems; over 90% of all plant species engage in mycorrhizal relationships with fungi and are dependent upon this relationship for survival.

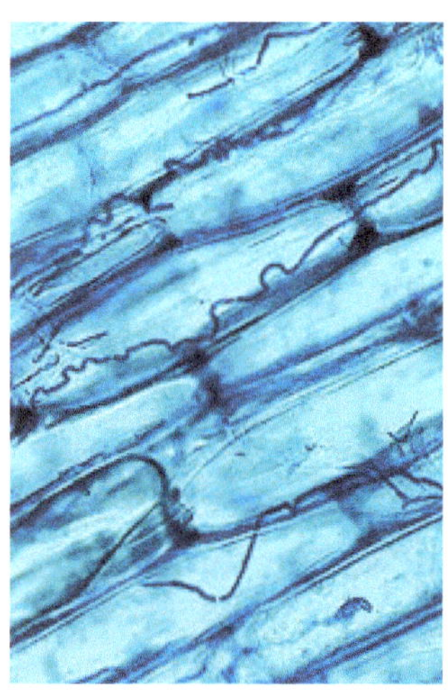

The dark filaments are hyphae of the endophytic fungus Neotyphodium coenophialum in the intercellular spaces of tall fescue leaf sheath tissue

The mycorrhizal symbiosis is ancient, dating to at least 400 million years ago. It often increases the plant's uptake of inorganic compounds, such as nitrate and phosphate from soils having low concentrations of these key plant nutrients. The fungal partners may also mediate plant-to-plant transfer of carbohydrates and other nutrients. Such mycorrhizal communities are called "common mycorrhizal networks". A special case of mycorrhiza is myco-heterotrophy, whereby the plant parasitizes the fungus, obtaining all of its nutrients from its fungal symbiont. Some fungal species inhabit the tissues inside roots, stems, and leaves, in which case they are called endophytes. Similar to mycorrhiza, endophytic colonization by fungi may benefit both symbionts; for example, endophytes of grasses impart to their host increased resistance to herbivores and other environmental stresses and receive food and shelter from the plant in return.

With Algae and Cyanobacteria

The lichen Lobaria pulmonaria, a symbiosis of fungal, algal, and cyanobacterial species

Lichens are a symbiotic relationship between fungi and photosynthetic algae or cyanobacteria. The photosynthetic partner in the relationship is referred to in lichen terminology as a "photobiont". The fungal part of the relationship is composed mostly of various species of ascomycetes and a few basidiomycetes. Lichens occur in every ecosystem on all continents, play a key role in soil formation and the initiation of biological succession, and are prominent in some extreme environments, including polar, alpine, and semiarid desert regions. They are able to grow on inhospitable surfaces, including bare soil, rocks, tree bark, wood, shells, barnacles and leaves. As in mycorrhizas, the photobiont provides sugars and other carbohydrates via photosynthesis to the fungus, while the fungus provides minerals and water to the photobiont. The functions of both symbiotic organisms are so closely intertwined that they function almost as a single organism; in most cases the resulting organism differs greatly from the individual components. Lichenization is a common mode of nutrition for fungi; around 20% of fungi—between 17,500 and 20,000 described species—are lichenized. Characteristics common to most lichens include obtaining organic carbon by photosynthesis, slow growth, small size, long life, long-lasting (seasonal) vegetative reproductive structures, mineral nutrition obtained largely from airborne sources, and greater tolerance of desiccation than most other photosynthetic organisms in the same habitat.

With Insects

Many insects also engage in mutualistic relationships with fungi. Several groups of ants cultivate fungi in the order Agaricales as their primary food source, while ambrosia beetles cultivate various species of fungi in the bark of trees that they infest. Likewise, females of several wood wasp species (genus Sirex) inject their eggs together with spores of the wood-rotting fungus Amylostereum areolatum into the sapwood of pine trees; the growth of the fungus provides ideal nutritional conditions for the development of the wasp larvae. At least one species of stingless bee has a relationship with a fungus in the genus Monascus, where the larvae consume and depend on fungus transferred from old to new nests. Termites on the African savannah are also known to cultivate fungi, and yeasts of the genera Candida and Lachancea inhabit the gut of a wide range of insects, including neuropterans, beetles, and cockroaches; it is not known whether these fungi benefit their hosts. Fungi ingrowing dead wood are essential for xylophagous insects (e.g. woodboring beetles). They deliver nutrients needed by xylophages to nutritionally scarce dead wood. Thanks to this nutritional enrichment the larvae of woodboring insect is able to grow and develop to adulthood. The larvae of many families of fungicolous flies, particularly those within the superfamily Sciaroidea such as the Mycetophilidae and some Keroplatidae feed on fungal fruiting bodies and sterile mycorrhizae.

As Pathogens and Parasites

The plant pathogen Aecidium magellanicum causes calafate rust, seen here on a Berberis shrub in Chile.

Many fungi are parasites on plants, animals (including humans), and other fungi. Serious pathogens of many cultivated plants causing extensive damage and losses to agriculture and forestry include the rice blast fungus Magnaporthe oryzae, tree pathogens such as Ophiostoma ulmi and Ophiostoma novo-ulmi causing Dutch elm disease and Cryphonectria parasitica responsible for chestnut blight, and plant pathogens in the genera Fusarium, Ustilago, Alternaria, and Cochliobolus. Some carnivorous fungi, like Paecilomyces lilacinus, are predators of nematodes, which they capture using an array of specialized structures such as constricting rings or adhesive nets.

Some fungi can cause serious diseases in humans, several of which may be fatal if untreated. These include aspergillosis, candidiasis, coccidioidomycosis, cryptococcosis, histoplasmosis, mycetomas, and paracoccidioidomycosis. Furthermore, persons with immuno-deficiencies are particularly susceptible to disease by genera such as Aspergillus, Candida, Cryptoccocus, Histoplasma, and Pneumocystis. Other fungi can attack eyes, nails, hair, and especially skin, the so-called dermatophytic and keratinophilic fungi, and cause local infections such as ringworm and athlete's foot. Fungal spores are also a cause of allergies, and fungi from different taxonomic groups can evoke allergic reactions.

Mycotoxins

Many fungi produce biologically active compounds, several of which are toxic to animals or plants and are therefore called mycotoxins. Of particular relevance to humans are mycotoxins produced by molds causing food spoilage, and poisonous mushrooms. Particularly infamous are the lethal amatoxins in some Amanita mushrooms, and ergot alkaloids, which have a long history of causing serious epidemics of ergotism (St Anthony's Fire) in people consuming rye or related cereals contaminated with sclerotia of the ergot fungus, Claviceps purpurea. Other notable mycotoxins include the aflatoxins, which are insidious liver toxins and highly carcinogenic metabolites produced by certain Aspergillus species often growing in or on grains and nuts consumed by humans, ochratoxins, patulin, and trichothecenes (e.g., T-2 mycotoxin) and fumonisins, which have significant impact on human food supplies or animal livestock.

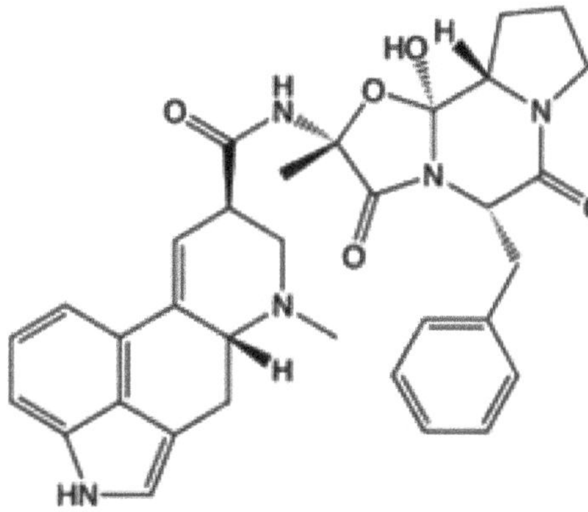

Ergotamine, a major mycotoxin produced by Claviceps species, which if ingested can cause gangrene, convulsions, and hallucinations

Mycotoxins are secondary metabolites (or natural products), and research has established the existence of biochemical pathways solely for the purpose of producing mycotoxins and other natural products in fungi. Mycotoxins may provide fitness benefits in terms of physiological adaptation, competition with other microbes and fungi, and protection from consumption (fungivory). Many fungal secondary metabolites (or derivatives) are used medically, as described under Human Use below.

Pathogenic Mechanisms

Ustilago maydis is a pathogenic plant fungus that causes smut disease in maize and teosinte. Plants have evolved efficient defense systems against pathogenic microbes such as U. maydis. A rapid defense reaction after pathogen attack is the oxidative burst where the plant produces reactive oxygen species at the site of the attempted invasion. U. maydis can respond to the oxidative burst with an oxidative stress response, regulated by the gene YAP1. The response protects U. maydis from the host defense, and is necessary for the pathogen's virulence. Furthermore, U. maydis has a well-established recombinational DNA repair system which acts during mitosis and meiosis. The system may assist the pathogen in surviving DNA damage arising from the host plant's oxidative defensive response to infection.

Cryptococcus neoformans is an encapsulated yeast that can live in both plants and animals. C. neoformans usually infects the lungs, where it is phagocytosed by alveolar macrophages. Some C. neoformans can survive inside macrophages, which appears to be the basis for latency, disseminated disease, and resistance to antifungal agents. One mechanism by which C. neoformans survives the hostile macrophage environment is by up-regulating the expression of genes involved in the oxidative stress response. Another mechanism involves meiosis. The majority of C. neoformans are mating "type a". Filaments of mating "type a" ordinarily have haploid nuclei, but they can become diploid (perhaps by endoduplication or by stimulated nuclear fusion) to form blastospores. The diploid nuclei of blastospores can undergo meiosis, including recombination, to form haploid basidiospores that can be dispersed. This process is referred to as monokaryotic fruiting. this process requires a gene called DMC1, which is a conserved homologue of genes recA in bacteria and RAD51 in eukaryotes, that mediates homologous chromosome pairing during meiosis and repair of DNA double-strand breaks. Thus, C. neoformans can undergo a meiosis, monokaryotic fruiting, that promotes recombinational repair in the oxidative, DNA damaging environment of the host macrophage, and the repair capability may contribute to its virulence.

Human Use

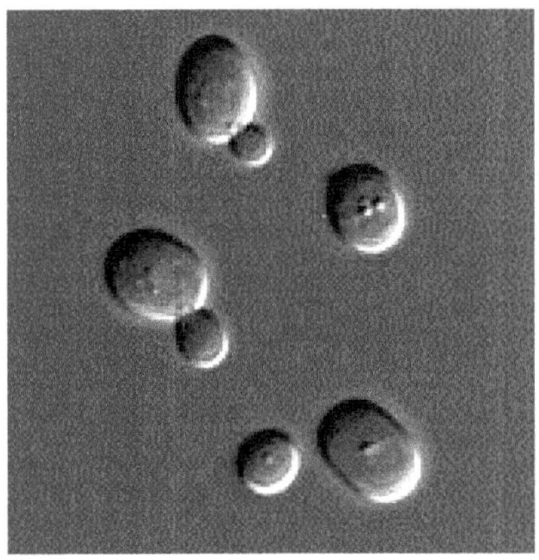

Saccharomyces cerevisiae cells shown with DIC microscopy

The human use of fungi for food preparation or preservation and other purposes is extensive and has a long history. Mushroom farming and mushroom gathering are large industries in many countries. The study of the historical uses and sociological impact of fungi is known as ethnomycology. Because of the capacity of this group to produce an enormous range of natural products with antimicrobial or other biological activities, many species have long been used or are being developed for industrial production of antibiotics, vitamins, and anti-cancer and cholesterol-lowering drugs. More recently, methods have been developed for genetic engineering of fungi, enabling metabolic engineering of fungal species. For example, genetic modification of yeast species—which are easy to grow at fast rates in large fermentation vessels—has opened up ways of pharmaceutical production that are potentially more efficient than production by the original source organisms.

Therapeutic Uses

Modern Chemotherapeutics

Many species produce metabolites that are major sources of pharmacologically active drugs. Particularly important are the antibiotics, including the penicillins, a structurally related group of β-lactam antibiotics that are synthesized from small peptides. Although naturally occurring penicillins such as penicillin G (produced by Penicillium chrysogenum) have a relatively narrow spectrum of biological activity, a wide range of other penicillins can be produced by chemical modification of the natural penicillins. Modern penicillins are semisynthetic compounds, obtained initially from fermentation cultures, but then structurally altered for specific desirable properties. Other antibiotics produced by fungi include: ciclosporin, commonly used as an immunosuppressant during transplant surgery; and fusidic acid, used to help control infection from methicillin-resistant Staphylococcus aureus bacteria. Widespread use of antibiotics for the treatment of bacterial diseases, such as tuberculosis, syphilis, leprosy, and others began in the early 20th century and continues to date. In nature, antibiotics of fungal or bacterial origin appear to play a dual role: at high concentrations they act as chemical defense against competition with other microorganisms in species-rich environments, such as the rhizosphere, and at low concentrations as quorum-sensing molecules for intra- or interspecies signaling. Other drugs produced by fungi include griseofulvin isolated from Penicillium griseofulvum, used to treat fungal infections, and statins (HMG-CoA reductase inhibitors), used to inhibit cholesterol synthesis. Examples of statins found in fungi include mevastatin from Penicillium citrinum and lovastatin from Aspergillus terreus and the oyster mushroom.

Traditional and Folk Medicine

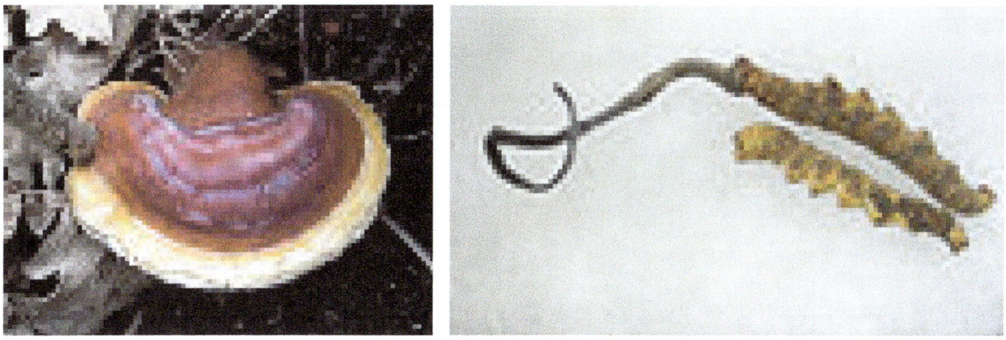

The medicinal fungi Ganoderma lucidum (left) and Ophiocordyceps sinensis (right)

Certain mushrooms enjoy usage as therapeutics in folk medicines, such as Traditional Chinese medicine. Notable medicinal mushrooms with a well-documented history of use include Agaricus subrufescens, Ganoderma lucidum, and Ophiocordyceps sinensis. Research has identified compounds produced by these and other fungi that have inhibitory biological effects against viruses and cancer cells. Specific metabolites, such as polysaccharide-K, ergotamine, and β-lactam antibiotics, are routinely used in clinical medicine. The shiitake mushroom is a source of lentinan, a clinical drug approved for use in cancer treatments in several countries, including Japan. In Europe and Japan, polysaccharide-K (brand name Krestin), a chemical derived from Trametes versicolor, is an approved adjuvant for cancer therapy.

Cultured Foods

Baker's yeast or Saccharomyces cerevisiae, a unicellular fungus, is used to make bread and other wheat-based products, such as pizza dough and dumplings. Yeast species of the genus Saccharomyces are also used to produce alcoholic beverages through fermentation. Shoyu koji mold (Aspergillus oryzae) is an essential ingredient in brewing Shoyu (soy sauce) and sake, and the preparation of miso, while Rhizopus species are used for making tempeh. Several of these fungi are domesticated species that were bred or selected according to their capacity to ferment food without producing harmful mycotoxins, which are produced by very closely related Aspergilli. Quorn, a meat substitute, is made from Fusarium venenatum.

Edible and Poisonous Species

Amanita phalloides accounts for the majority of fatal mushroom poisonings worldwide.

Edible mushrooms are well-known examples of fungi. Many are commercially raised, but others must be harvested from the wild. Agaricus bisporus, sold as button mushrooms when small or Portobello mushrooms when larger, is a commonly eaten species, used in salads, soups, and many other dishes. Many Asian fungi are commercially grown and have increased in popularity in the West. They are often available fresh in grocery stores and markets, including straw mushrooms (Volvariella volvacea), oyster mushrooms (Pleurotus ostreatus), shiitakes (Lentinula edodes), and enokitake (Flammulina spp.).

There are many more mushroom species that are harvested from the wild for personal consumption or commercial sale. Milk mushrooms, morels, chanterelles, truffles, black trumpets, and porcini mushrooms (Boletus edulis) (also known as king boletes) demand a high price on the market. They are often used in gourmet dishes.

Certain types of cheeses require inoculation of milk curds with fungal species that impart a unique flavor and texture to the cheese. Examples include the blue color in cheeses such as Stilton or Roquefort, which are made by inoculation with Penicillium roqueforti. Molds used in cheese production are non-toxic and are thus safe for human consumption; however, mycotoxins (e.g., aflatoxins, roquefortine C, patulin, or others) may accumulate because of growth of other fungi during cheese ripening or storage.

Stilton cheese veined with Penicillium roqueforti

Many mushroom species are poisonous to humans, with toxicities ranging from slight digestive problems or allergic reactions as well as hallucinations to severe organ failures and death.

Genera with mushrooms containing deadly toxins include Conocybe, Galerina, Lepiota, and, the most infamous, Amanita. The latter genus includes the destroying angel (A. virosa) and the death cap (A. phalloides), the most common cause of deadly mushroom poisoning. The false morel (Gyromitra esculenta) is occasionally considered a delicacy when cooked, yet can be highly toxic when eaten raw. Tricholoma equestre was considered edible until it was implicated in serious poisonings causing rhabdomyolysis. Fly agaric mushrooms (Amanita muscaria) also cause occasional non-fatal poisonings, mostly as a result of ingestion for its hallucinogenic properties. Historically, fly agaric was used by different peoples in Europe and Asia and its present usage for religious or shamanic purposes is reported from some ethnic groups such as the Koryak people of north-eastern Siberia.

As it is difficult to accurately identify a safe mushroom without proper training and knowledge, it is often advised to assume that a wild mushroom is poisonous and not to consume it.

Pest Control

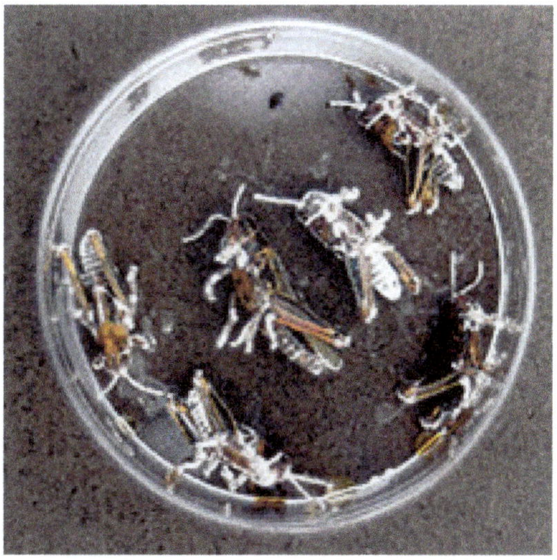

Grasshoppers killed by Beauveria bassiana

In agriculture, fungi may be useful if they actively compete for nutrients and space with pathogenic microorganisms such as bacteria or other fungi via the competitive exclusion principle, or if they are parasites of these pathogens. For example, certain species may be used to eliminate or suppress the growth of harmful plant pathogens, such as insects, mites, weeds, nematodes, and other fungi that cause diseases of important crop plants. This has generated strong interest in practical applications that use these fungi in the biological control of these agricultural pests. Entomopathogenic fungi can be used as biopesticides, as they actively kill insects. Examples that have been used as biological insecticides are Beauveria bassiana, Metarhizium spp, Hirsutella spp, Paecilomyces (Isaria) spp, and Lecanicillium lecanii. Endophytic fungi of grasses of the genus Neotyphodium, such as N. coenophialum, produce alkaloids that are toxic to a range of invertebrate and vertebrate herbivores. These alkaloids protect grass plants from herbivory, but several endophyte alkaloids can poison grazing animals, such as cattle and sheep. Infecting cultivars of pasture or forage grasses with Neotyphodium endophytes is one approach being used in grass breeding programs; the fungal strains are selected for producing only alkaloids that increase resistance to herbivores such as insects, while being non-toxic to livestock.

Bioremediation

Certain fungi, in particular "white rot" fungi, can degrade insecticides, herbicides, pentachlorophenol, creosote, coal tars, and heavy fuels and turn them into carbon dioxide, water, and basic elements. Fungi have been shown to biomineralize uranium oxides, suggesting they may have application in the bioremediation of radioactively polluted sites.

Model Organisms

Several pivotal discoveries in biology were made by researchers using fungi as model organisms, that is, fungi that grow and sexually reproduce rapidly in the laboratory. For example, the one gene-one enzyme hypothesis was formulated by scientists using the bread mold Neurospora crassa to test their biochemical theories. Other important model fungi are Aspergillus nidulans and the yeasts Saccaromyces cerevisiae and Schizosaccharomyces pombe, each of which with a long history of use to investigate issues in eukaryotic cell biology and genetics, such as cell cycle regulation, chromatin structure, and gene regulation. Other fungal models have more recently emerged that address specific biological questions relevant to medicine, plant pathology, and industrial uses; examples include Candida albicans, a dimorphic, opportunistic human pathogen, Magnaporthe grisea, a plant pathogen, and Pichia pastoris, a yeast widely used for eukaryotic protein production.

Others

Fungi are used extensively to produce industrial chemicals like citric, gluconic, lactic, and malic acids, and industrial enzymes, such as lipases used in biological detergents, cellulases used in making cellulosic ethanol and stonewashed jeans, and amylases, invertases, proteases and xylanases. Several species, most notably Psilocybin mushrooms (colloquially known as magic mushrooms), are ingested for their psychedelic properties, both recreationally and religiously.

Aspergillus Niger

*Aspergillus niger or A. niger is a fun*gus and one of the most common species of the genus *Aspergillus.*

It causes a disease called black mould on certain fruits and vegetables such as grapes, apricots, onions, and peanuts, and is a common contaminant of food. It is ubiquitous in soil and is commonly reported from indoor environments, where its black colonies can be confused with those of *Stachybotrys* (species of which have also been called "black mould").

Some strains of *A. niger* have been reported to produce potent mycotoxins called ochratoxins; other sources disagree, claiming this report is based upon misidentification of the fungal species. Recent evidence suggests some true *A. niger* strains do produce ochratoxin A. It also produces the isoflavone orobol.

Taxonomy

A. niger is included in *Aspergillus* subgenus *Circumdati*, section *Nigri*. The section *Nigri* includes 15 related black-spored species that may be confused with *A. niger*, including *A. tubingensis*, *A.*

foetidus, *A. carbonarius*, and *A. awamori*. A number of morphologically similar species were recently described by Samson *et al.*

Recently the strain of ATCC 16404 *Aspergillus niger* has been reclassified at *Aspergillus brasiliensis*. This has required an update to the U.S. Pharmacopoeia and the European Pharmacopoeia which commonly use this strain throughout the pharmaceutical industry.

Pathogenicity

Plant Disease

A. niger growing on onion

A. niger causes black mold of onions and ornamental plants. Infection of onion seedlings by *A. niger* can become systemic, manifesting only when conditions are conducive. *A. niger* causes a common postharvest disease of onions, in which the black conidia can be observed between the scales of the bulb. The fungus also causes disease in peanuts and in grapes.

Human and Animal Disease

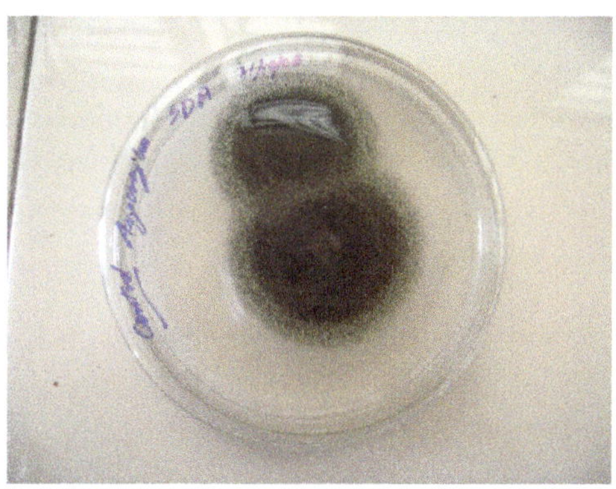

A. niger growing on SDA

A. niger is less likely to cause human disease than some other *Aspergillus* species. In extremely rare instances, humans may become ill, but this is due to a serious lung disease, aspergillosis, that can occur. Aspergillosis is, in particular, frequent among horticultural workers who inhale peat dust, which can be rich in *Aspergillus* spores. It has been found in the mummies of ancient Egyptian tombs and can be inhaled when they are disturbed.

A. niger is one of the most common causes of otomycosis (fungal ear infections), which can cause pain, temporary hearing loss, and, in severe cases, damage to the ear canal and tympanic membrane.

Industrial Uses

A. niger is cultured for the industrial production of many substances. Various strains of *A. niger* are used in the industrial preparation of citric acid (E330) and gluconic acid (E574) and have been assessed as acceptable for daily intake by the World Health Organisation. *A. niger* fermentation is "generally recognized as safe" (GRAS) by the United States Food and Drug Administration under the Federal Food, Drug, and Cosmetic Act.

Many useful enzymes are produced using industrial fermentation of *A. niger*. For example, *A. niger*glucoamylase is used in the production of high fructose corn syrup, and pectinases are used in cider and wine clarification. Alpha-galactosidase, an enzyme that breaks down certain complex sugars, is a component of Beano and other products that decrease flatulence. Another use for *A. niger* within the biotechnology industry is in the production of magnetic isotope-containing variants of biological macromolecules for NMR analysis.

A. niger growing from gold mining solution contained cyano metal complexes, such as gold, silver, copper, iron, and zinc. The fungus also plays a role in the solubilization of heavy metal sulfides. Alkali-treated *A. niger* binds to silver to 10% of dry weight. Silver biosorbtion occurs via stoichiometric exchange with Ca(II) and Mg(II) of the sorbent.

Other Uses

A. niger is the main agent in the fermentation of Pu-erh tea, the Okinawanawamori spirit, and some varieties of shochu.

In 2006, it was reported that a secreted RNase produced by *A. niger* called *actibind* has antiangiogenic and anticarcinogenic characteristics.

A. niger is also cultured for the extraction of the enzymes glucose oxidase (GO) and Alpha-galactosidase (AGS). Glucose oxidase is used in the design of glucose biosensors, due to its high affinity for β-D-glucose. Alpha-galactosidase can be produced by *A. niger* fermentation; it is used to hydrolyze alpha 1-6 bonds found in melibiose, raffinose, and stachyose.

Research published in 2006–2008 investigated *A. niger* prolyl endoprotease (AN-PEP), a microbial-derived prolyl endoprotease which cleaves gluten. This has strong implications in the treatment of coeliac disease or other metabolic gluten sensitivity disease processes. A placebo controlled, double blind study was initiated in December 2008 to determine the efficacy of this enzyme in treating humans with coeliac disease.

In the heyday of the opium trade, *chandu* opium, which was meant to be smoked, was made by long term fermentation of *A. niger* and other moulds on raw opium.

A. niger is often used as a challenge organism for cleaning validation studies performed within sterile manufacturing facilities.

Genetics

Genome

The *A. niger* ATCC 1015 genome was sequenced by the Joint Genome Institute in a collaboration with other institutions.

The genomes of two different *A. niger* strains have been fully sequenced.

Aspergillus Oryzae

Aspergillus oryzae, known in English as koji, is a filamentousfungus (a mold) used in Chinese and other East Asian cuisines to fermentsoybeans for making soy sauce and fermented bean paste, and also to saccharify rice, other grains, and potatoes in the making of alcoholic beverages such as *huangjiu, sake, makgeolli,* and *shōchū*. The domestication of *A. oryzae* occurred at least 2000 years ago.*A. oryzae* is also used for the production of rice vinegars.

Dr. Eiji Ichishima of Tohoku University called the *kōji* fungus a "national fungus" (*kokkin*) in the journal of the Brewing Society of Japan, because of its importance not only for making the *koji* for *sake* brewing, but also for making the *koji* for *miso*, soy sauce, and a range of other traditional Japanese foods. His proposal was approved at the society's annual meeting in 2006.

"Red *kōji-kin*" is a separate species, *Monascus purpureus.*

History of Koji

300 BCE – A. *oryzae* is first mentioned in the *Zhouli* (Rites of the Zhou dynasty) in China. Its development is a milestone in Chinese food technology, for it provides the conceptual framework for three major fermented soy foods: soy sauce, *jiang / miso*, and *douchi*, not to mention grain-based wines (including Japanese *sake*) and *li* (the Chinese forerunner of Japanese *amazake*).

Properties Desirable in Sake Brewing and Testing

These properties of *A. oryzae* strains are important in rice saccharification for sake brewing:

- Growth: rapid mycelial growth on and into the rice kernels
- Enzymes: strong secretion of amylases (α-amylase and glucoamylase); some carboxypeptidase; low tyrosinase
- Aesthetics: pleasant fragrance; accumulation of flavoring compounds
- Color: low production of deferriferrichrome (a siderophore), flavins, and other colored substances

Varieties Used for Shōchū Making

Three varieties of kōji mold are used for making *shōchū*, each with distinct characteristics.

- White was discovered at the beginning of the Taishō period when natural mutation and separation of some black *kōji* to white was observed. This effect was researched and white *kōji* was successfully grown independently. White *kōji* is easy to cultivate and its enzymes promote rapid saccharization; as a result, it is used to produce most *shōchū* today. It gives rise to a drink with a refreshing, gentle, sweet taste.

- Black is mainly used in Okinawa to produce *awamori*. It produces plenty of citric acid which helps to prevent the souring of the *moromi*. Of all three *kōji*, it most effectively extracts the taste and character of the base ingredients, giving its *shōchū* a rich aroma with a slightly sweet, mellow taste. Its spores disperse easily, covering production facilities and workers' clothes in a layer of black. Such issues led to it falling out of favour, but due to the development of *new kuro-kōji* (NK-*kōji*) in the mid-1980s, interest in black *kōji* resurged amongst *honkaku shōchū* makers because of the depth and quality of the taste it produced. Several popular brands now explicitly state they use black *kōji* on their labels.
- Yellow is used to produce sake, and at one time all *honkaku shōchū*. However, yellow *kōji* is extremely sensitive to temperature; its *moromi* can easily sour during fermentation. This makes it difficult to use in warmer regions such as Kyūshū, and gradually black and white *kōji* became more common. Its strength is that it gives rise to a rich, fruity, refreshing taste, so despite the difficulties and great skill required, it is still used by some manufacturers. It is popular amongst young people and women who previously had no interest in typically strong potato *shōchū*, playing a role in its recent revival.

Genome

Initially kept secret, the *A. oryzae*genome was released by a consortium of Japanese biotechnology companies in late 2005. The eight chromosomes together comprise 37 million base pairs and 12 thousand predicted genes. The genome of *A. oryzae* is thus one-third larger than that of two related *Aspergillus* species, the genetics model organism *A. nidulans* and the potentially dangerous *A. fumigatus*. Many of the extra genes present in *A. oryzae* are predicted to be involved in secondary metabolism. The sequenced strain isolated in 1950 is called RIB40 or ATCC 42149; its morphology, growth, and enzyme production are typical of strains used for *sake* brewing.

Use in Biotechnology

Resveratrol can be produced from its glucoside piceid through the process of fermentation by *A. oryzae*.

In Fiction

A. oryzae is a supporting character (of sorts) in the manga series *Moyasimon: Tales of Agriculture* and its anime adaptation.

Candida (Fungus)

Candida is a genus of yeasts and is the most common cause of fungal infections worldwide. Many species are harmless commensals or endosymbionts of hosts including humans; however, when mucosal barriers are disrupted or the immune system is compromised they can invade and cause disease.*Candida albicans* is the most commonly isolated species, and can cause infections (candidiasis or thrush) in humans and other animals. In winemaking, some species of *Candida* can potentially spoil wines.

Many species are found in gut flora, including *C. albicans* in mammalian hosts, whereas others live as endosymbionts in insect hosts.

Systemic infections of the bloodstream and major organs (candidemia or invasive candidiasis), particularly in immunocompromised patients, affect over 90,000 people a year in the U.S.

The DNA of several *Candida* species has been sequenced.

Antibiotics promote yeast infections, including gastrointestinal *Candida* overgrowth, and penetration of the GI mucosa. While women are more susceptible to genital yeast infections, men can also be infected. Certain factors, such as prolonged antibiotic use, increase the risk for both men and women. People with diabetes or impaired immune systems, such as those with HIV, are more susceptible to yeast infections.

Candida antarctica is a source of industrially important lipases.

Biology

When grown in a laboratory, *Candida* appears as large, round, white or cream (*albicans* means "whitish" in Latin) colonies, which emit a yeasty odor on agar plates at room temperature.*C. albicans* ferments glucose and maltose to acid and gas, sucrose to acid, and does not ferment lactose, which help to distinguish it from other *Candida* species.

Many species of *Candida* use a non-standard genetic code in the translation of their nuclear genes into the amino acid sequences of polypeptides. The difference in the genetic code between species possessing the alternative code is that the codon CUG (normally encoding the amino acid leucine) is translated by the yeast as a different amino acid, serine. The alternative translation of the CUG codon in these species is due to a novel nucleic acid sequence in the serine-tRNA (ser-tRNACAG), which has a guanosine located at position 33, 5' to the anticodon. In all other tRNAs, this position is normally is occupied by a pyrimidine (often uridine). This genetic code change is the only such known alteration in cytoplasmic mRNA, in both the prokaryotes, and the eukaryotes, involving the reassignment of a sense codon. This novel genetic code may be a mechanism for more rapid adaptation to the organism's environment, as well as playing an important role in the evolution of the genus *Candida* by creating genetic barriers that encouraged speciation

Pathogen

Candida are almost universal in low numbers on healthy adult skin and *C.albicans* is part of the normal flora of the mucous membranes of the respiratory, gastrointestinal and female genital tracts. The dryness of skin compared to other tissues prevents the growth of the fungus, but damaged skin or skin in intertriginous regions is more amenable to rapid growth.

Overgrowth of several species including *C.albicans* can cause superficial infections such as oropharyngeal candidiasis (thrush) and vulvovaginal candidiasis (vaginal candidiasis). Oral candidiasis is common in elderly denture wearers. In otherwise healthy individuals, these infections can be cured with topical or systemic antifungal medications (commonly over-the-counter antifungal treatments like miconazole or clotrimazole). In debilitated or immunocompromised patients, or if introduced intravenously, candidiasis may become a systemic disease producing abscess, thrombophlebitis, endocarditis, or infections of the eyes or other organs. Typically, relatively severe neutropenia is a prerequisite for *Candida* to pass through the defenses of the skin and cause disease in deeper tissues; in such cases, mechanical disruption of the infected skin sites is typically a factor in the fungal invasion of the deeper tissues.

Applications

C. albicans has been used in combination with carbon nanotubes (CNT) to produce stable electrically conductive bio-nano-composite tissue materials that have been used as temperature sensing elements.

Species

Among *Candida* species, *C. albicans*, which is a normal constituent of the human flora, a commensal of the skin and the gastrointestinal and genitourinary tracts, is responsible for the majority of *Candida* bloodstream infections (candidemia). Yet, there is an increasing incidence of infections caused by *C. glabrata* and *C. rugosa*, which could be because they are frequently less susceptible to the currently used azole antifungals. Other medically important species include *C. parapsilosis*, *C. tropicalis* and *C. dubliniensis*.

Other *Candida* species, such as *C. oleophila* have been used as biological control agents in fruit.

- *C. albicans*
- *C. ascalaphidarum*
- *C. amphixiae*
- *C. antarctica*
- *C. argentea*
- *C. atlantica*
- *C. atmosphaerica*
- *C. auris*
- *C. blattae*
- *C. bromeliacearum*
- *C. carpophila*
- *C. carvajalis*
- *C. cerambycidarum*
- *C. chauliodes*

- *C. corydali*
- *C. dosseyi*
- *C. dubliniensis*
- *C. ergatensis*
- *C. fructus*
- *C. glabrata*
- *C. fermentati*
- *C. guilliermondii*
- *C. haemulonii*
- *C. humilis*
- *C. insectamens*
- *C. insectorum*
- *C. intermedia*
- *C. jeffresii*
- *C. kefyr*
- *C. keroseneae*
- *C. krusei*
- *C. lusitaniae*
- *C. lyxosophila*
- *C. maltosa*
- *C. marina*
- *C. membranifaciens*
- *C. mogii*
- *C. oleophila*
- *C. oregonensis*
- *C. parapsilosis*
- *C. quercitrusa*
- *C. rugosa*
- *C. sake*
- *C. shehatea*

- *C. temnochilae*
- *C. tenuis*
- *C. theae*
- *C. tolerans*
- *C. tropicalis*
- *C. tsuchiyae*
- *C. sinolaborantium*
- *C. sojae*
- *C. subhashii*
- *C. viswanathii*
- *C. utilis*
- *C. ubatubensis*
- *C. zemplinina*

Bacteria

Bacteria constitute a large domain of prokaryotic microorganisms. Typically a few micrometres in length, bacteria have a number of shapes, ranging from spheres to rods and spirals. Bacteria were among the first life forms to appear on Earth, and are present in most of its habitats. Bacteria inhabit soil, water, acidic hot springs, radioactive waste, and the deep portions of Earth's crust. Bacteria also live in symbiotic and parasitic relationships with plants and animals.

There are typically 40 million bacterial cells in a gram of soil and a million bacterial cells in a millilitre of fresh water. There are approximately 5×1030 bacteria on Earth, forming a biomass which exceeds that of all plants and animals. Bacteria are vital in recycling nutrients, with many of the stages in nutrient cycles dependent on these organisms, such as the fixation of nitrogen from the atmosphere and putrefaction. In the biological communities surrounding hydrothermal vents and cold seeps, bacteria provide the nutrients needed to sustain life by converting dissolved compounds, such as hydrogen sulphide and methane, to energy. On 17 March 2013, researchers reported data that suggested bacterial life forms thrive in the Mariana Trench, which with a depth of up to 11 kilometres is the deepest part of the Earth's oceans. Other researchers reported related studies that microbes thrive inside rocks up to 580 metres below the sea floor under 2.6 kilometres of ocean off the coast of the northwestern United States. According to one of the researchers, "You can find microbes everywhere — they're extremely adaptable to conditions, and survive wherever they are."

Most bacteria have not been characterised, and only about half of the bacterial phyla have species that can be grown in the laboratory. The study of bacteria is known as bacteriology, a branch of microbiology.

There are approximately ten times as many bacterial cells in the human flora as there are human cells in the body, with the largest number of the human flora being in the gut flora, and a large number on the skin. The vast majority of the bacteria in the body are rendered harmless by the protective effects of the immune system, and some are beneficial. However, several species of bacteria are pathogenic and cause infectious diseases, including cholera, syphilis, anthrax, leprosy, and bubonic plague. The most common fatal bacterial diseases are respiratory infections, with tuberculosis alone killing about 2 million people per year, mostly in sub-Saharan Africa. In developed countries, antibiotics are used to treat bacterial infections and are also used in farming, making antibiotic resistance a growing problem. In industry, bacteria are important in sewage treatment and the breakdown of oil spills, the production of cheese and yogurt through fermentation, and the recovery of gold, palladium, copper and other metals in the mining sector, as well as in biotechnology, and the manufacture of antibiotics and other chemicals.

Once regarded as plants constituting the class Schizomycetes, bacteria are now classified as prokaryotes. Unlike cells of animals and other eukaryotes, bacterial cells do not contain a nucleus and rarely harbour membrane-bound organelles. Although the term bacteria traditionally included all prokaryotes, the scientific classification changed after the discovery in the 1990s that prokaryotes consist of two very different groups of organisms that evolved from an ancient common ancestor. These evolutionary domains are called Bacteria and Archaea.

Origin and Early Evolution

The ancestors of modern bacteria were unicellular microorganisms that were the first forms of life to appear on Earth, about 4 billion years ago. For about 3 billion years, most organisms were microscopic, and bacteria and archaea were the dominant forms of life. In 2008, fossils of macroorganisms were discovered and named as the Francevillian biota. Although bacterial fossils exist, such as stromatolites, their lack of distinctive morphology prevents them from being used to examine the history of bacterial evolution, or to date the time of origin of a particular bacterial species. However, gene sequences can be used to reconstruct the bacterial phylogeny, and these studies indicate that bacteria diverged first from the archaeal/eukaryotic lineage. Bacteria were also involved in the second great evolutionary divergence, that of the archaea and eukaryotes. Here, eukaryotes resulted from the entering of ancient bacteria into endosymbiotic associations with the ancestors of eukaryotic cells, which were themselves possibly related to the Archaea. This involved the engulfment by proto-eukaryotic cells of alphaproteobacterial symbionts to form either mitochondria or hydrogenosomes, which are still found in all known Eukarya (sometimes in highly reduced form, e.g. in ancient "amitochondrial" protozoa). Later on, some eukaryotes that already contained mitochondria also engulfed cyanobacterial-like organisms. This led to the formation of chloroplasts in algae and plants. There are also some algae that originated from even later endosymbiotic events. Here, eukaryotes engulfed a eukaryotic algae that developed into a "second-generation" plastid. This is known as secondary endosymbiosis.

Morphology

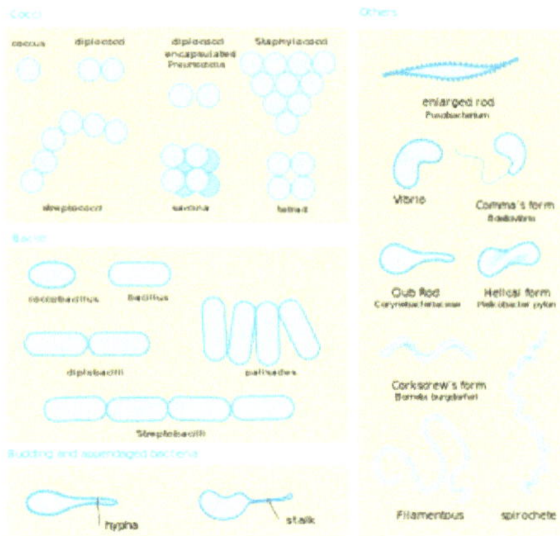

Bacteria display many cell morphologies and arrangements

Bacteria display a wide diversity of shapes and sizes, called morphologies. Bacterial cells are about one-tenth the size of eukaryotic cells and are typically 0.5–5.0 micrometres in length. However, a few species are visible to the unaided eye — for example, Thiomargarita namibiensis is up to half a millimetre long and Epulopiscium fishelsoni reaches 0.7 mm. Among the smallest bacteria are members of the genus Mycoplasma, which measure only 0.3 micrometres, as small as the largest viruses. Some bacteria may be even smaller, but these ultramicrobacteria are not well-studied.

Most bacterial species are either spherical, called cocci, or rod-shaped, called bacilli (sing. bacillus, from Latin baculus, stick). Elongation is associated with swimming. Some bacteria, called vibrio, are shaped like slightly curved rods or comma-shaped; others can be spiral-shaped, called spirilla, or tightly coiled, called spirochaetes. A small number of species even have tetrahedral or cuboidal shapes. More recently, some bacteria were discovered deep under Earth's crust that grow as branching filamentous types with a star-shaped cross-section. The large surface area to volume ratio of this morphology may give these bacteria an advantage in nutrient-poor environments. This wide variety of shapes is determined by the bacterial cell wall and cytoskeleton, and is important because it can influence the ability of bacteria to acquire nutrients, attach to surfaces, swim through liquids and escape predators.

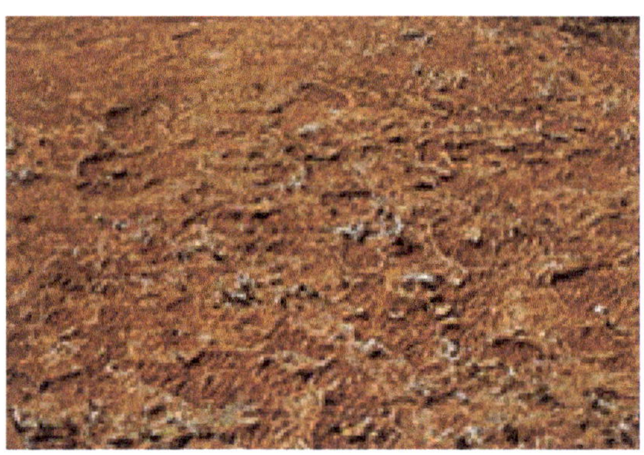

A biofilm of thermophilic bacteria in the outflow of Mickey Hot Springs, Oregon, approximately 20 mm thick.

Many bacterial species exist simply as single cells, others associate in characteristic patterns: Neisseria form diploids (pairs), Streptococcus form chains, and Staphylococcus group together in "bunch of grapes" clusters. Bacteria can also be elongated to form filaments, for example the Actinobacteria. Filamentous bacteria are often surrounded by a sheath that contains many individual cells. Certain types, such as species of the genus Nocardia, even form complex, branched filaments, similar in appearance to fungal mycelia.

Bacteria often attach to surfaces and form dense aggregations called biofilms or bacterial mats. These films can range from a few micrometres in thickness to up to half a metre in depth, and may contain multiple species of bacteria, protists and archaea. Bacteria living in biofilms display a complex arrangement of cells and extracellular components, forming secondary structures, such as microcolonies, through which there are networks of channels to enable better diffusion of nutrients. In natural environments, such as soil or the surfaces of plants, the majority of bacteria are bound to surfaces in biofilms. Biofilms are also important in medicine, as these structures are often present during chronic bacterial infections or in infections of implanted medical devices, and bacteria protected within biofilms are much harder to kill than individual isolated bacteria.

Even more complex morphological changes are sometimes possible. For example, when starved of amino acids, Myxobacteria detect surrounding cells in a process known as quorum sensing, migrate towards each other, and aggregate to form fruiting bodies up to 500 micrometres long and containing approximately 100,000 bacterial cells. In these fruiting bodies, the bacteria perform separate tasks; this type of cooperation is a simple type of multicellular organisation. For example, about one in 10 cells migrate to the top of these fruiting bodies and differentiate into a specialised dormant state called myxospores, which are more resistant to drying and other adverse environmental conditions than are ordinary cells.

Cellular Structure

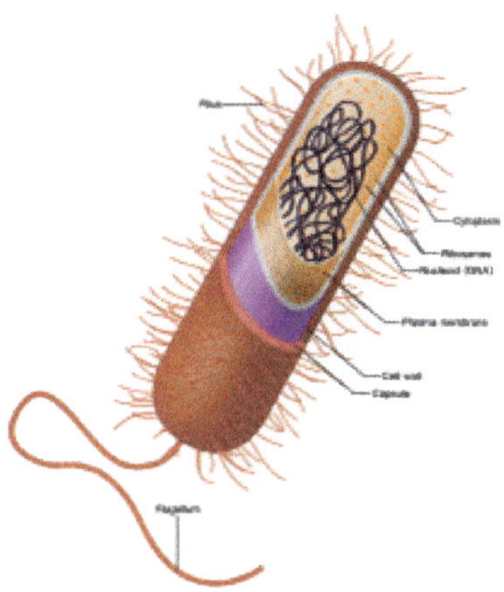

Structure and contents of a typical gram-positive bacterial cell (seen by the fact that only one cell membrane is present).

Intracellular Structures

The bacterial cell is surrounded by a cell membrane (also known as a lipid, cytoplasmic or plasma membrane). This membrane encloses the contents of the cell and acts as a barrier to hold nutrients, proteins and other essential components of the cytoplasm within the cell. As they are prokaryotes, bacteria do not usually have membrane-bound organelles in their cytoplasm, and thus contain few large intracellular structures. They lack a true nucleus, mitochondria, chloroplasts and the other organelles present in eukaryotic cells. Bacteria were once seen as simple bags of cytoplasm, but structures such as the prokaryotic cytoskeleton and the localisation of proteins to specific locations within the cytoplasm that give bacteria some complexity have been discovered. These subcellular levels of organisation have been called "bacterial hyperstructures".

Bacterial microcompartments, such as carboxysomes, provide a further level of organisation; they are compartments within bacteria that are surrounded by polyhedral protein shells, rather than by lipid membranes. These "polyhedral organelles" localise and compartmentalise bacterial metabolism, a function performed by the membrane-bound organelles in eukaryotes.

Many important biochemical reactions, such as energy generation, use concentration gradients across membranes. The general lack of internal membranes in bacteria means reactions such as electron transport occur across the cell membrane between the cytoplasm and the periplasmic space. However, in many photosynthetic bacteria the plasma membrane is highly folded and fills most of the cell with layers of light-gathering membrane. These light-gathering complexes may even form lipid-enclosed structures called chlorosomes in green sulfur bacteria. Other proteins import nutrients across the cell membrane, or expel undesired molecules from the cytoplasm.

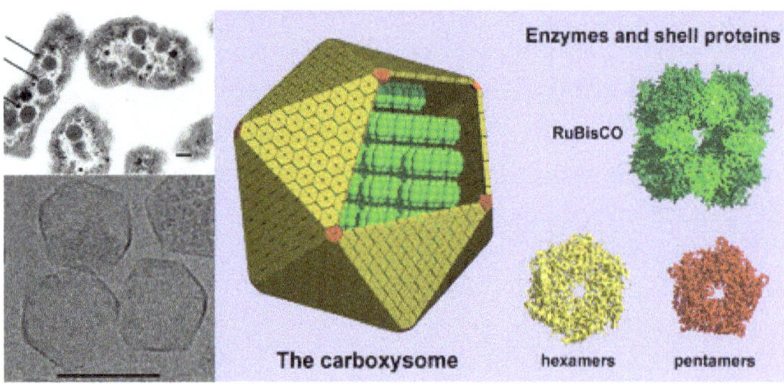

Carboxysomes are protein-enclosed bacterial organelles. Top left is an electron microscope image of carboxysomes in Halothiobacillus neapolitanus, below is an image of purified carboxysomes. On the right is a model of their structure. Scale bars are 100 nm.

Bacteria do not have a membrane-bound nucleus, and their genetic material is typically a single circular DNA chromosome located in the cytoplasm in an irregularly shaped body called the nucleoid. The nucleoid contains the chromosome with its associated proteins and RNA. The phylum Planctomycetes and candidate phylum Poribacteria may be exceptions to the general absence of internal membranes in bacteria, because they appear to have a double membrane around their nucleoids and contain other membrane-bound cellular structures. Like all living organisms, bacteria contain ribosomes, often grouped in chains called polyribosomes, for the production of proteins, but the structure of the bacterial ribosome is different from that of eukaryotes and Archaea. Bacterial ribosomes have a sedimentation rate of 70S (measured in Svedberg units): their subunits have rates of 30S and 50S. Some antibiotics bind specifically to 70S ribosomes and inhibit bacterial protein synthesis. Those antibiotics kill bacteria without affecting the larger 80S ribosomes of eukaryotic cells and without harming the host.

Some bacteria produce intracellular nutrient storage granules for later use, such as glycogen, polyphosphate, sulfur or polyhydroxyalkanoates. Certain bacterial species, such as the photosynthetic Cyanobacteria, produce internal gas vesicles, which they use to regulate their buoyancy – allowing them to move up or down into water layers with different light intensities and nutrient levels.

Intracellular membranes called chromatophores are also found in membranes of phototrophic bacteria. Used primarily for photosynthesis, they contain bacteriochlorophyll pigments and carotenoids. An early idea was that bacteria might contain membrane folds termed mesosomes, but these were later shown to be artefacts produced by the chemicals used to prepare the cells for electron microscopy. Inclusions are considered to be nonliving components of the cell that do not possess metabolic activity and are not bounded by membranes. The most common inclusions are glycogen, lipid droplets, crystals, and pigments. Volutin granules are cytoplasmic inclusions of complexed inorganic polyphosphate. These granules are called metachromatic granules due to their displaying the metachromatic effect; they appear red or blue when stained with the blue dyes methylene blue or toluidine blue. Gas vacuoles, which are freely permeable to gas, are membrane-bound vesicles present in some species of Cyanobacteria. They allow the bacteria to control their buoyancy. Microcompartments are widespread, membrane-bound organelles that are made of a protein shell that surrounds and encloses various enzymes. Carboxysomes are bacterial microcompartments that contain enzymes involved in carbon fixation. Magnetosomes are bacterial microcompartments, present in magnetotactic bacteria, that contain magnetic crystals.

Extracellular Structures

In most bacteria, a cell wall is present on the outside of the cell membrane. The cell membrane and cell wall comprise the cell envelope. A common bacterial cell wall material is peptidoglycan (called "murein" in older sources), which is made from polysaccharide chains cross-linked by peptides containing D-amino acids. Bacterial cell walls are different from the cell walls of plants and fungi, which are made of cellulose and chitin, respectively. The cell wall of bacteria is also distinct from that of Archaea, which do not contain peptidoglycan. The cell wall is essential to the survival of many bacteria, and the antibiotic penicillin is able to kill bacteria by inhibiting a step in the synthesis of peptidoglycan.

There are broadly speaking two different types of cell wall in bacteria, a thick one in the gram-positives and a thinner one in the gram-negatives. The names originate from the reaction of cells to the Gram stain, a long-standing test for the classification of bacterial species.

Gram-positive bacteria possess a thick cell wall containing many layers of peptidoglycan and teichoic acids. In contrast, gram-negative bacteria have a relatively thin cell wall consisting of a few layers of peptidoglycan surrounded by a second lipid membrane containing lipopolysaccharides and lipoproteins. Lipopolysaccharides, also called endotoxins, are composed of polysaccharides and lipid A that is responsible for much of the toxicity of gram-negative bacteria. Most bacteria have the gram-negative cell wall, and only the Firmicutes and Actinobacteria have the alternative gram-positive arrangement. These two groups were previously known as the low G+C and high G+C gram-positive bacteria, respectively. These differences in structure can produce differences in antibiotic susceptibility; for instance, vancomycin can kill only gram-positive bacteria and is ineffective against gram-negative pathogens, such as Haemophilus influenzae or Pseudomonas aeruginosa. If the bacterial cell wall is entirely removed, it is called a protoplast, whereas if it is partially removed, it is called a spheroplast. β-Lactam antibiotics, such as penicillin, inhibit the formation of peptidoglycan cross-links in the bacterial cell wall. The enzyme lysozyme, found in human tears, also digests the cell wall of bacteria and is the body's main defence against eye infections.

Acid-fast bacteria, such as Mycobacteria, are resistant to decolorisation by acids during staining procedures. The high mycolic acid content of Mycobacteria, is responsible for the staining pattern of poor absorption followed by high retention. The most common staining technique used to identify acid-fast bacteria is the Ziehl-Neelsen stain or acid-fast stain, in which the acid-fast bacilli are stained bright-red and stand out clearly against a blue background. L-form bacteria are strains of bacteria that lack cell walls. The main pathogenic bacteria in this class is Mycoplasma.

In many bacteria, an S-layer of rigidly arrayed protein molecules covers the outside of the cell. This layer provides chemical and physical protection for the cell surface and can act as a macromolecular diffusion barrier. S-layers have diverse but mostly poorly understood functions, but are known to act as virulence factors in Campylobacter and contain surface enzymes in Bacillus stearothermophilus.

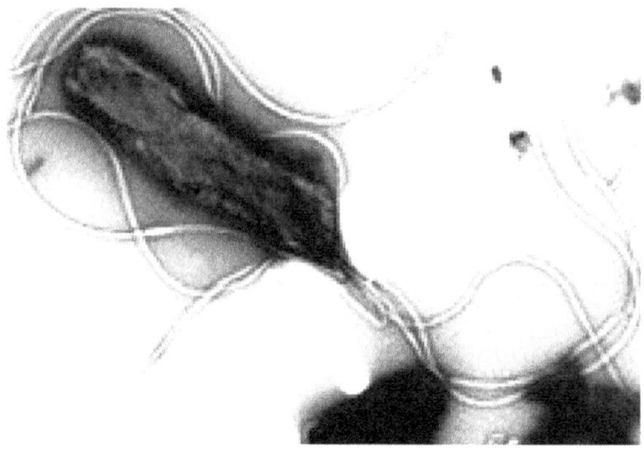

Helicobacter pylori electron micrograph, showing multiple flagella on the cell surface

Flagella are rigid protein structures, about 20 nanometres in diameter and up to 20 micrometres in length, that are used for motility. Flagella are driven by the energy released by the transfer of ions down an electrochemical gradient across the cell membrane.

Fimbriae (sometimes called "attachment pili") are fine filaments of protein, usually 2–10 nanometres in diameter and up to several micrometres in length. They are distributed over the surface of the cell, and resemble fine hairs when seen under the electron microscope. Fimbriae are believed to be involved in attachment to solid surfaces or to other cells, and are essential for the virulence of some bacterial pathogens. Pili (sing. pilus) are cellular appendages, slightly larger than fimbriae, that can transfer genetic material between bacterial cells in a process called conjugation where they are called conjugation pili or "sex pili". They can also generate movement where they are called type IV pili.

Glycocalyx are produced by many bacteria to surround their cells, and vary in structural complexity: ranging from a disorganised slime layer of extra-cellular polymcr to a highly structured capsule. These structures can protect cells from engulfment by eukaryotic cells such as macrophages (part of the human immune system). They can also act as antigens and be involved in cell recognition, as well as aiding attachment to surfaces and the formation of biofilms.

The assembly of these extracellular structures is dependent on bacterial secretion systems. These transfer proteins from the cytoplasm into the periplasm or into the environment around the cell. Many types of secretion systems are known and these structures are often essential for the virulence of pathogens, so are intensively studied.

Endospores

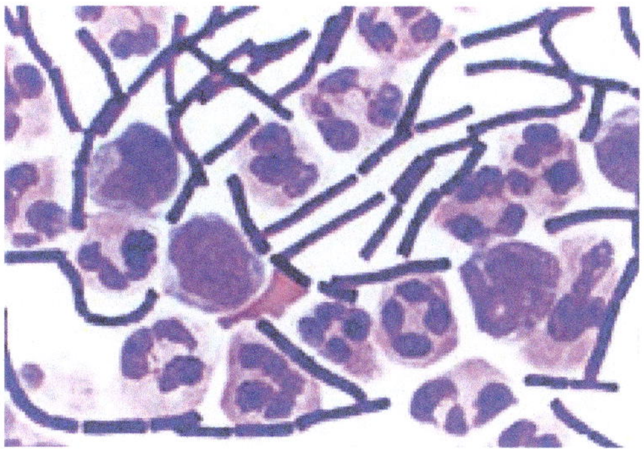

Bacillus anthracis (stained purple) growing in cerebrospinal fluid

Certain genera of gram-positive bacteria, such as Bacillus, Clostridium, Sporohalobacter, Anaerobacter, and Heliobacterium, can form highly resistant, dormant structures called endospores. In almost all cases, one endospore is formed and this is not a reproductive process, although Anaerobacter can make up to seven endospores in a single cell. Endospores have a central core of cytoplasm containing DNA and ribosomes surrounded by a cortex layer and protected by an impermeable and rigid coat. Dipicolinic acid is a chemical compound that composes 5% to 15% of the dry weight of bacterial spores. It is implicated as responsible for the heat resistance of the endospore.

Endospores show no detectable metabolism and can survive extreme physical and chemical stresses, such as high levels of UV light, gamma radiation, detergents, disinfectants, heat, freezing, pressure, and desiccation. In this dormant state, these organisms may remain viable for millions of years, and endospores even allow bacteria to survive exposure to the vacuum and radiation in space. According to scientist Dr. Steinn Sigurdsson, "There are viable bacterial spores that have been found that are 40 million years old on Earth — and we know they're very hardened to radiation." Endospore-forming bacteria can also cause disease: for example, anthrax can be contracted by the inhalation of Bacillus anthracis endospores, and contamination of deep puncture wounds with Clostridium tetani endospores causes tetanus.

Metabolism

Bacteria exhibit an extremely wide variety of metabolic types. The distribution of metabolic traits within a group of bacteria has traditionally been used to define their taxonomy, but these traits often do not correspond with modern genetic classifications. Bacterial metabolism is classified into nutritional groups on the basis of three major criteria: the kind of energy used for growth, the source of carbon, and the electron donors used for growth. An additional criterion of respiratory microorganisms are the electron acceptors used for aerobic or anaerobic respiration.

Nutritional types in bacterial metabolism			
Nutritional type	**Source of energy**	**Source of carbon**	**Examples**
Phototrophs	Sunlight	Organic compounds (photoheterotrophs) or carbon fixation (photoautotrophs)	Cyanobacteria, Green sulfur bacteria, Chloroflexi, or Purple bacteria

Lithotrophs	Inorganic compounds	Organic compounds (litho-heterotrophs) or carbon fixation (lithoautotrophs)	Thermodesulfobacteria, Hydrogenophilaceae, or Nitrospirae
Organotrophs	Organic compounds	Organic compounds (chemoheterotrophs) or carbon fixation (chemoautotrophs)	Bacillus, Clostridium or Enterobacteriaceae

Carbon metabolism in bacteria is either heterotrophic, where organic carbon compounds are used as carbon sources, or autotrophic, meaning that cellular carbon is obtained by fixing carbon dioxide. Heterotrophic bacteria include parasitic types. Typical autotrophic bacteria are phototrophic cyanobacteria, green sulfur-bacteria and some purple bacteria, but also many chemolithotrophic species, such as nitrifying or sulfur-oxidising bacteria. Energy metabolism of bacteria is either based on phototrophy, the use of light through photosynthesis, or based on chemotrophy, the use of chemical substances for energy, which are mostly oxidised at the expense of oxygen or alternative electron acceptors (aerobic/anaerobic respiration).

Filaments of photosynthetic cyanobacteria

Bacteria are further divided into lithotrophs that use inorganic electron donors and organotrophs that use organic compounds as electron donors. Chemotrophic organisms use the respective electron donors for energy conservation (by aerobic/anaerobic respiration or fermentation) and biosynthetic reactions (e.g., carbon dioxide fixation), whereas phototrophic organisms use them only for biosynthetic purposes. Respiratory organisms use chemical compounds as a source of energy by taking electrons from the reduced substrate and transferring them to a terminal electron acceptor in a redox reaction. This reaction releases energy that can be used to synthesise ATP and drive metabolism. In aerobic organisms, oxygen is used as the electron acceptor. In anaerobic organisms other inorganic compounds, such as nitrate, sulfate or carbon dioxide are used as electron acceptors. This leads to the ecologically important processes of denitrification, sulfate reduction, and acetogenesis, respectively.

Another way of life of chemotrophs in the absence of possible electron acceptors is fermentation, wherein the electrons taken from the reduced substrates are transferred to oxidised intermediates to generate reduced fermentation products (e.g., lactate, ethanol, hydrogen, butyric acid). Fermentation is possible, because the energy content of the substrates is higher than that of the products, which allows the organisms to synthesise ATP and drive their metabolism.

These processes are also important in biological responses to pollution; for example, sulfate-reducing bacteria are largely responsible for the production of the highly toxic forms of mercury (methyl- and dimethylmercury) in the environment. Non-respiratory anaerobes use fermentation to generate energy and reducing power, secreting metabolic by-products (such as ethanol in brewing) as waste. Facultative anaerobes can switch between fermentation and different terminal electron acceptors depending on the environmental conditions in which they find themselves.

Lithotrophic bacteria can use inorganic compounds as a source of energy. Common inorganic electron donors are hydrogen, carbon monoxide, ammonia (leading to nitrification), ferrous iron and other reduced metal ions, and several reduced sulfur compounds. In unusual circumstances, the gas methane can be used by methanotrophic bacteria as both a source of electrons and a substrate for carbon anabolism. In both aerobic phototrophy and chemolithotrophy, oxygen is used as a terminal electron acceptor, whereas under anaerobic conditions inorganic compounds are used instead. Most lithotrophic organisms are autotrophic, whereas organotrophic organisms are heterotrophic.

In addition to fixing carbon dioxide in photosynthesis, some bacteria also fix nitrogen gas (nitrogen fixation) using the enzyme nitrogenase. This environmentally important trait can be found in bacteria of nearly all the metabolic types listed above, but is not universal.

Regardless of the type of metabolic process they employ, the majority of bacteria are able to take in raw materials only in the form of relatively small molecules, which enter the cell by diffusion or through molecular channels in cell membranes. The Planctomycetes are the exception (as they are in possessing membranes around their nuclear material). It has recently been shown that Gemmata obscuriglobus is able to take in large molecules via a process that in some ways resembles endocytosis, the process used by eukaryotic cells to engulf external items.

Growth and Reproduction

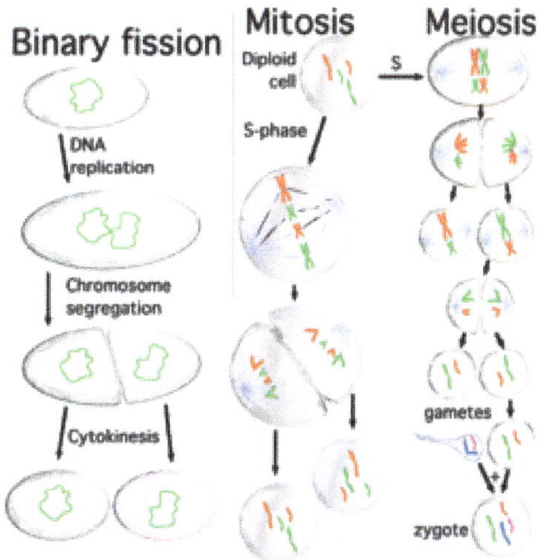

Many bacteria reproduce through binary fission, which is compared to mitosis and meiosis in this image.

Unlike in multicellular organisms, increases in cell size (cell growth) and reproduction by cell division are tightly linked in unicellular organisms. Bacteria grow to a fixed size and then reproduce through binary fission, a form of asexual reproduction. Under optimal conditions, bacteria can grow and divide extremely rapidly, and bacterial populations can double as quickly as every 9.8 minutes. In cell division, two identical clone daughter cells are produced. Some bacteria, while still reproducing asexually, form more complex reproductive structures that help disperse the newly formed daughter cells. Examples include fruiting body formation by Myxobacteria and aerial hyphae formation by Streptomyces, or budding. Budding involves a cell forming a protrusion that breaks away and produces a daughter cell.

A colony of Escherichia coli

In the laboratory, bacteria are usually grown using solid or liquid media. Solid growth media, such as agar plates, are used to isolate pure cultures of a bacterial strain. However, liquid growth media are used when measurement of growth or large volumes of cells are required. Growth in stirred liquid media occurs as an even cell suspension, making the cultures easy to divide and transfer, although isolating single bacteria from liquid media is difficult. The use of selective media (media with specific nutrients added or deficient, or with antibiotics added) can help identify specific organisms.

Most laboratory techniques for growing bacteria use high levels of nutrients to produce large amounts of cells cheaply and quickly. However, in natural environments, nutrients are limited, meaning that bacteria cannot continue to reproduce indefinitely. This nutrient limitation has led the evolution of different growth strategies. Some organisms can grow extremely rapidly when nutrients become available, such as the formation of algal (and cyanobacterial) blooms that often occur in lakes during the summer. Other organisms have adaptations to harsh environments, such as the production of multiple antibiotics by Streptomyces that inhibit the growth of competing microorganisms. In nature, many organisms live in communities (e.g., biofilms) that may allow for increased supply of nutrients and protection from environmental stresses. These relationships can be essential for growth of a particular organism or group of organisms (syntrophy).

Bacterial growth follows four phases. When a population of bacteria first enter a high-nutrient environment that allows growth, the cells need to adapt to their new environment. The first phase of growth is the lag phase, a period of slow growth when the cells are adapting to the high-nutrient environment and preparing for fast growth. The lag phase has high biosynthesis rates, as proteins necessary for rapid growth are produced. The second phase of growth is the log phase, also known as the logarithmic or exponential phase. The log phase is marked by rapid exponential growth. The rate at which cells grow during this phase is known as the growth rate (k), and the time it takes the cells to double is known as the generation time (g). During log phase, nutrients are metabolised at maximum speed until one of the nutrients is depleted and starts limiting growth. The third phase of growth is the stationary phase and is caused by depleted nutrients. The cells reduce their metabolic activity and consume non-essential cellular proteins. The stationary phase is a transition from rapid growth to a stress response state and there is increased expression of genes involved in DNA repair, antioxidant metabolism and nutrient transport. The final phase is the death phase where the bacteria run out of nutrients and die.

Genomes

The genomes of thousands of bacterial species have been sequenced, with at least 9,000 sequences completed and more than 42,000 left as "permanent" drafts (as of Sep 2016).

Most bacteria have a single circular chromosome that can range in size from only 160,000 base pairs in the endosymbiotic bacteria Candidatus Carsonella ruddii, to 12,200,000 base pairs in the soil-dwelling bacteria Sorangium cellulosum. The genes in bacterial genomes are usually a single continuous stretch of DNA and although several different types of introns do exist in bacteria, these are much rarer than in eukaryotes. Some bacteria, including the Spirochaetes of the genus Borrelia are a notable exception to this arrangement. Borrelia burgdorferi, the cause of Lyme disease, contains a single linear chromosome and several linear and circular plasmids.

Plasmids are small extra-chromosomal DNAs that may contain genes for antibiotic resistance or virulence factors. Plasmids replicate independently of chromosomes, so it is possible that plasmids could be lost in bacterial cell division. Against this possibility is the fact that a single bacterium can contain hundreds of copies of a single plasmid.

Genetics

Bacteria, as asexual organisms, inherit identical copies of their parent's genes (i.e., they are clonal). However, all bacteria can evolve by selection on changes to their genetic material DNA caused by genetic recombination or mutations. Mutations come from errors made during the replication of DNA or from exposure to mutagens. Mutation rates vary widely among different species of bacteria and even among different clones of a single species of bacteria. Genetic changes in bacterial genomes come from either random mutation during replication or "stress-directed mutation", where genes involved in a particular growth-limiting process have an increased mutation rate.

DNA Transfer

Some bacteria also transfer genetic material between cells. This can occur in three main ways. First, bacteria can take up exogenous DNA from their environment, in a process called transformation. Genes can also be transferred by the process of transduction, when the integration of a bacteriophage introduces foreign DNA into the chromosome. The third method of gene transfer is conjugation, whereby DNA is transferred through direct cell contact.

Transduction of bacterial genes by bacteriophage appears to be a consequence of infrequent errors during intracellular assembly of virus particles, rather than a bacterial adaptation. Conjugation, in the much-studied E. coli system is determined by plasmid genes, and is an adaptation for transferring copies of the plasmid from one bacterial host to another. It is seldom that a conjugative plasmid integrates into the host bacterial chromosome, and subsequently transfers part of the host bacterial DNA to another bacterium. Plasmid-mediated transfer of host bacterial DNA also appears to be an accidental process rather than a bacterial adaptation.

Transformation, unlike transduction or conjugation, depends on numerous bacterial gene products that specifically interact to perform this complex process, and thus transformation is clearly a bacterial adaptation for DNA transfer. In order for a bacterium to bind, take up and recombine donor DNA into its own chromosome, it must first enter a special physiological state termed competence. In Bacillus subtilis, about 40 genes are required for the development of competence. The length of DNA transferred during B. subtilis transformation can be between a third of a chromosome up to the whole chromosome. Transformation appears to be common among bacterial species, and thus far at least 60 species are known to have the natural ability to become competent for transformation. The development of competence in nature is usually associated with stressful environmental conditions, and seems to be an adaptation for facilitating repair of DNA damage in recipient cells.

In ordinary circumstances, transduction, conjugation, and transformation involve transfer of DNA between individual bacteria of the same species, but occasionally transfer may occur between individuals of different bacterial species and this may have significant consequences, such as the transfer of antibiotic resistance. In such cases, gene acquisition from other bacteria or the environment is called horizontal gene transfer and may be common under natural conditions. Gene transfer is particularly important in antibiotic resistance as it allows the rapid transfer of resistance genes between different pathogens.

Bacteriophages

Bacteriophages are viruses that infect bacteria. Many types of bacteriophage exist, some simply infect and lyse their host bacteria, while others insert into the bacterial chromosome. A bacteriophage can contain genes that contribute to its host's phenotype: for example, in the evolution of Escherichia coli O157:H7 and Clostridium botulinum, the toxin genes in an integrated phage converted a harmless ancestral bacterium into a lethal pathogen. Bacteria resist phage infection through restriction modification systems that degrade foreign DNA, and a system that uses CRISPR sequences to retain fragments of the genomes of phage that the bacteria have come into contact with in the past, which allows them to block virus replication through a form of RNA interference. This CRISPR system provides bacteria with acquired immunity to infection.

Behaviour

Secretion

Bacteria frequently secrete chemicals into their environment in order to modify it favourably. The secretions are often proteins and may act as enzymes that digest some form of food in the environment.

Bioluminescence

A few bacteria have chemical systems that generate light. This bioluminescence often occurs in bacteria that live in association with fish, and the light probably serves to attract fish or other large animals.

Multicellularity

Bacteria often function as multicellular aggregates known as biofilms, exchanging a variety of molecular signals for inter-cell communication, and engaging in coordinated multicellular behaviour.

The communal benefits of multicellular cooperation include a cellular division of labour, accessing resources that cannot effectively be used by single cells, collectively defending against antagonists, and optimising population survival by differentiating into distinct cell types. For example, bacteria in biofilms can have more than 500 times increased resistance to antibacterial agents than individual "planktonic" bacteria of the same species.

One type of inter-cellular communication by a molecular signal is called quorum sensing, which serves the purpose of determining whether there is a local population density that is sufficiently high that it is productive to invest in processes that are only successful if large numbers of similar organisms behave similarly, as in excreting digestive enzymes or emitting light.

Quorum sensing allows bacteria to coordinate gene expression, and enables them to produce, release and detect autoinducers or pheromones which accumulate with the growth in cell population.

Movement

Many bacteria can move using a variety of mechanisms: flagella are used for swimming through fluids; bacterial gliding and twitching motility move bacteria across surfaces; and changes of buoyancy allow vertical motion.

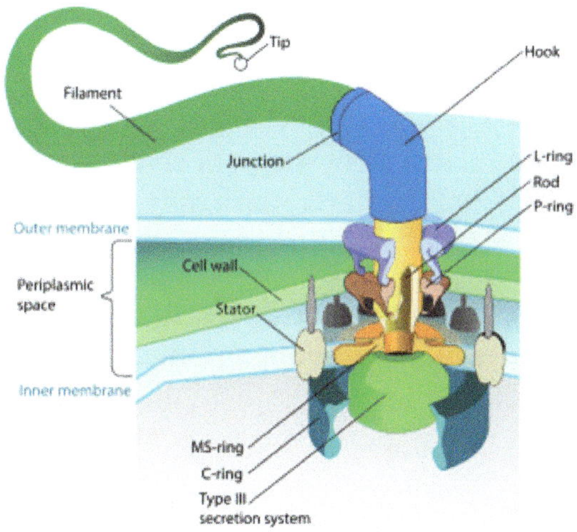

Flagellum of gram-negative bacteria. The base drives the rotation of the hook and filament.

Swimming bacteria frequently move near 10 body lengths per second and a few as fast as 100. This makes them at least as fast as fish, on a relative scale.

In bacterial gliding and twitching motility, bacteria use their type IV pili as a grappling hook, repeatedly extending it, anchoring it and then retracting it with remarkable force (>80 pN).

"Our observations redefine twitching motility as a rapid, highly organized mechanism of bacterial translocation by which Pseudomonas aeruginosa can disperse itself over large areas to colonize new territories. It is also now clear, both morphologically and genetically, that twitching motility and social gliding motility, such as occurs in Myxococcus xanthus, are essentially the same process."

—"A re-examination of twitching motility in Pseudomonas aeruginosa" – Semmler, Whitchurch & Mattick (1999)

Flagella are semi-rigid cylindrical structures that are rotated and function much like the propeller on a ship. Objects as small as bacteria operate a low Reynolds number and cylindrical forms are more efficient than the flat, paddle-like, forms appropriate at human-size scale.

Bacterial species differ in the number and arrangement of flagella on their surface; some have a single flagellum (monotrichous), a flagellum at each end (amphitrichous), clusters of flagella at the poles of the cell (lophotrichous), while others have flagella distributed over the entire surface of the cell (peritrichous). The bacterial flagella is the best-understood motility structure in any organism and is made of about 20 proteins, with approximately another 30 proteins required for its regulation and assembly. The flagellum is a rotating structure driven by a reversible motor at the base that uses the electrochemical gradient across the membrane for power. This motor drives the motion of the filament, which acts as a propeller.

Many bacteria (such as E. coli) have two distinct modes of movement: forward movement (swimming) and tumbling. The tumbling allows them to reorient and makes their movement a three-dimensional random walk. The flagella of a unique group of bacteria, the spirochaetes, are found between two membranes in the periplasmic space. They have a distinctive helical body that twists about as it moves.

Motile bacteria are attracted or repelled by certain stimuli in behaviours called taxes: these include chemotaxis, phototaxis, energy taxis, and magnetotaxis. In one peculiar group, the myxobacteria, individual bacteria move together to form waves of cells that then differentiate to form fruiting bodies containing spores. The myxobacteria move only when on solid surfaces, unlike E. coli, which is motile in liquid or solid media.

Several Listeria and Shigella species move inside host cells by usurping the cytoskeleton, which is normally used to move organelles inside the cell. By promoting actin polymerisation at one pole of their cells, they can form a kind of tail that pushes them through the host cell's cytoplasm.

Classification and Identification

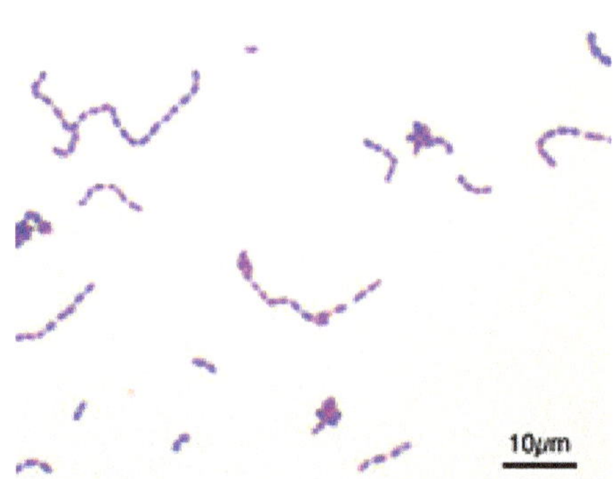

Streptococcus mutans visualised with a Gram stain

Classification seeks to describe the diversity of bacterial species by naming and grouping organisms based on similarities. Bacteria can be classified on the basis of cell structure, cellular metabolism or on differences in cell components, such as DNA, fatty acids, pigments, antigens and quinones. While these schemes allowed the identification and classification of bacterial strains, it was unclear whether these differences represented variation between distinct species or between strains of the same species. This uncertainty was due to the lack of distinctive structures in most bacteria, as well as lateral gene transfer between unrelated species. Due to lateral gene transfer, some closely related bacteria can have very different morphologies and metabolisms. To overcome this uncertainty, modern bacterial classification emphasises molecular systematics, using genetic techniques such as guanine cytosine ratio determination, genome-genome hybridisation, as well as sequencing genes that have not undergone extensive lateral gene transfer, such as the rRNA gene. Classification of bacteria is determined by publication in the International Journal of Systematic Bacteriology, and Bergey's Manual of Systematic Bacteriology. The International Committee on Systematic Bacteriology (ICSB) maintains international rules for the naming of bacteria and taxonomic categories and for the ranking of them in the International Code of Nomenclature of Bacteria.

The term "bacteria" was traditionally applied to all microscopic, single-cell prokaryotes. However, molecular systematics showed prokaryotic life to consist of two separate domains, originally called Eubacteria and Archaebacteria, but now called Bacteria and Archaea that evolved independently from an ancient common ancestor. The archaea and eukaryotes are more closely related to each other than either is to the bacteria. These two domains, along with Eukarya, are the basis of the three-domain system, which is currently the most widely used classification system in microbiolology. However, due to the relatively recent introduction of molecular systematics and a rapid increase in the number of genome sequences that are available, bacterial classification remains a changing and expanding field. For example, a few biologists argue that the Archaea and Eukaryotes evolved from gram-positive bacteria.

The identification of bacteria in the laboratory is particularly relevant in medicine, where the correct treatment is determined by the bacterial species causing an infection. Consequently, the need to identify human pathogens was a major impetus for the development of techniques to identify bacteria.

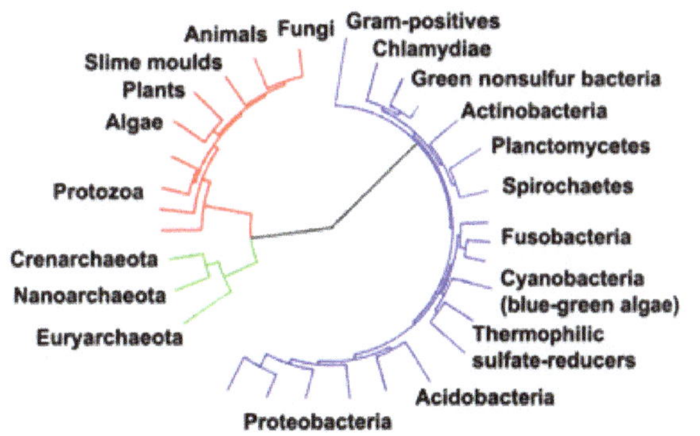

Phylogenetic tree showing the diversity of bacteria, compared to other organisms. Eukaryotes are coloured red, archaea green and bacteria blue.

The Gram stain, developed in 1884 by Hans Christian Gram, characterises bacteria based on the structural characteristics of their cell walls. The thick layers of peptidoglycan in the "gram-positive" cell wall stain purple, while the thin "gram-negative" cell wall appears pink. By combining morphology and Gram-staining, most bacteria can be classified as belonging to one of four groups (gram-positive cocci, gram-positive bacilli, gram-negative cocci and gram-negative bacilli). Some organisms are best identified by stains other than the Gram stain, particularly mycobacteria or Nocardia, which show acid-fastness on Ziehl–Neelsen or similar stains. Other organisms may need to be identified by their growth in special media, or by other techniques, such as serology.

Culture techniques are designed to promote the growth and identify particular bacteria, while restricting the growth of the other bacteria in the sample. Often these techniques are designed for specific specimens; for example, a sputum sample will be treated to identify organisms that cause pneumonia, while stool specimens are cultured on selective media to identify organisms that cause diarrhoea, while preventing growth of non-pathogenic bacteria. Specimens that are normally sterile, such as blood, urine or spinal fluid, are cultured under conditions designed to grow all possible organisms. Once a pathogenic organism has been isolated, it can be further characterised by its morphology, growth patterns (such as aerobic or anaerobic growth), patterns of hemolysis, and staining.

As with bacterial classification, identification of bacteria is increasingly using molecular methods. Diagnostics using DNA-based tools, such as polymerase chain reaction, are increasingly popular due to their specificity and speed, compared to culture-based methods. These methods also allow the detection and identification of "viable but nonculturable" cells that are metabolically active but non-dividing. However, even using these improved methods, the total number of bacterial species is not known and cannot even be estimated with any certainty. Following present classification, there are a little less than 9,300 known species of prokaryotes, which includes bacteria and archaea; but attempts to estimate the true number of bacterial diversity have ranged from 107 to 109 total species – and even these diverse estimates may be off by many orders of magnitude.

Interactions With Other Organisms

Despite their apparent simplicity, bacteria can form complex associations with other organisms. These symbiotic associations can be divided into parasitism, mutualism and commensalism. Due to their small size, commensal bacteria are ubiquitous and grow on animals and plants exactly as they will grow on any other surface. However, their growth can be increased by warmth and sweat, and large populations of these organisms in humans are the cause of body odour.

Predators

Some species of bacteria kill and then consume other microorganisms, these species are called predatory bacteria. These include organisms such as Myxococcus xanthus, which forms swarms of cells that kill and digest any bacteria they encounter. Other bacterial predators either attach to their prey in order to digest them and absorb nutrients, such as Vampirovibrio chlorellavorus, or invade another cell and multiply inside the cytosol, such as Daptobacter. These predatory bacteria are thought to have evolved from saprophages that consumed dead microorganisms, through adaptations that allowed them to entrap and kill other organisms.

Mutualists

Certain bacteria form close spatial associations that are essential for their survival. One such mutualistic association, called interspecies hydrogen transfer, occurs between clusters of anaerobic bacteria that consume organic acids, such as butyric acid or propionic acid, and produce hydrogen, and methanogenic Archaea that consume hydrogen. The bacteria in this association are unable to consume the organic acids as this reaction produces hydrogen that accumulates in their surroundings. Only the intimate association with the hydrogen-consuming Archaea keeps the hydrogen concentration low enough to allow the bacteria to grow.

In soil, microorganisms that reside in the rhizosphere (a zone that includes the root surface and the soil that adheres to the root after gentle shaking) carry out nitrogen fixation, converting nitrogen gas to nitrogenous compounds. This serves to provide an easily absorbable form of nitrogen for many plants, which cannot fix nitrogen themselves. Many other bacteria are found as symbionts in humans and other organisms. For example, the presence of over 1,000 bacterial species in the normal human gut flora of the intestines can contribute to gut immunity, synthesise vitamins, such as folic acid, vitamin K and biotin, convert sugars to lactic acid, as well as fermenting complex undigestible carbohydrates. The presence of this gut flora also inhibits the growth of potentially pathogenic bacteria (usually through competitive exclusion) and these beneficial bacteria are consequently sold as probiotic dietary supplements.

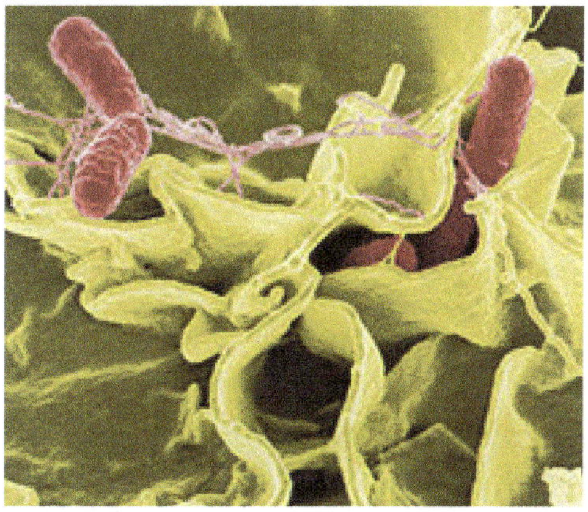

Colour-enhanced scanning electron micrograph showing Salmonella typhimurium (red) invading cultured human cells

Pathogens

If bacteria form a parasitic association with other organisms, they are classed as pathogens. Pathogenic bacteria are a major cause of human death and disease and cause infections such as tetanus, typhoid fever, diphtheria, syphilis, cholera, foodborne illness, leprosy and tuberculosis. A pathogenic cause for a known medical disease may only be discovered many years after, as was the case with Helicobacter pylori and peptic ulcer disease. Bacterial diseases are also important in agriculture, with bacteria causing leaf spot, fire blight and wilts in plants, as well as Johne's disease, mastitis, salmonella and anthrax in farm animals.

Each species of pathogen has a characteristic spectrum of interactions with its human hosts. Some organisms, such as Staphylococcus or Streptococcus, can cause skin infections, pneumonia, meningitis and even overwhelming sepsis, a systemic inflammatory response producing shock, massive vasodilation and death. Yet these organisms are also part of the normal human flora and usually exist on the skin or in the nose without causing any disease at all. Other organisms invariably cause disease in humans, such as the Rickettsia, which are obligate intracellular parasites able to grow and reproduce only within the cells of other organisms. One species of Rickettsia causes typhus, while another causes Rocky Mountain spotted fever. Chlamydia, another phylum of obligate intracellular parasites, contains species that can cause pneumonia, or urinary tract infection and may be involved in coronary heart disease. Finally, some species, such as Pseudomonas aeruginosa, Burkholderia cenocepacia, and Mycobacterium avium, are opportunistic pathogens and cause disease mainly in people suffering from immunosuppression or cystic fibrosis.

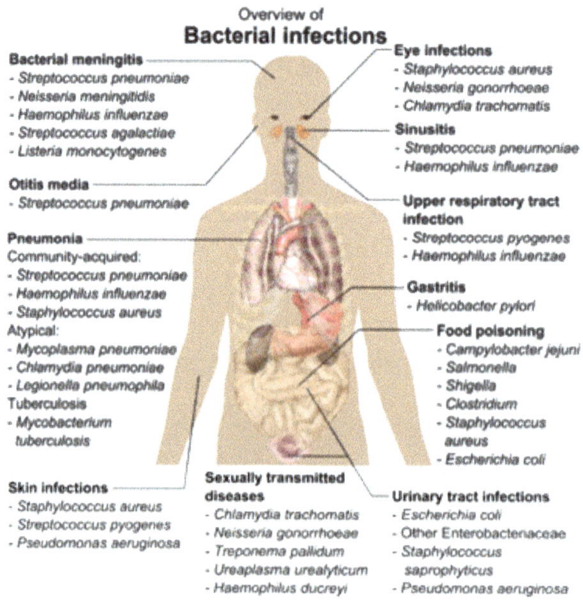

Overview of bacterial infections and main species involved.

Bacterial infections may be treated with antibiotics, which are classified as bacteriocidal if they kill bacteria, or bacteriostatic if they just prevent bacterial growth. There are many types of antibiotics and each class inhibits a process that is different in the pathogen from that found in the host. An example of how antibiotics produce selective toxicity are chloramphenicol and puromycin, which inhibit the bacterial ribosome, but not the structurally different eukaryotic ribosome. Antibiotics are used both in treating human disease and in intensive farming to promote animal growth, where they may be contributing to the rapid development of antibiotic resistance in bacterial populations. Infections can be prevented by antiseptic measures such as sterilising the skin prior to piercing it with the needle of a syringe, and by proper care of indwelling catheters. Surgical and dental instruments are also sterilised to prevent contamination by bacteria. Disinfectants such as bleach are used to kill bacteria or other pathogens on surfaces to prevent contamination and further reduce the risk of infection.

Significance in Technology and Industry

Bacteria, often lactic acid bacteria, such as Lactobacillus and Lactococcus, in combination with yeasts and moulds, have been used for thousands of years in the preparation of fermented foods, such as cheese, pickles, soy sauce, sauerkraut, vinegar, wine and yogurt.

The ability of bacteria to degrade a variety of organic compounds is remarkable and has been used in waste processing and bioremediation. Bacteria capable of digesting the hydrocarbons in petroleum are often used to clean up oil spills. Fertiliser was added to some of the beaches in Prince

William Sound in an attempt to promote the growth of these naturally occurring bacteria after the 1989 Exxon Valdez oil spill. These efforts were effective on beaches that were not too thickly covered in oil. Bacteria are also used for the bioremediation of industrial toxic wastes. In the chemical industry, bacteria are most important in the production of enantiomerically pure chemicals for use as pharmaceuticals or agrichemicals.

Bacteria can also be used in the place of pesticides in the biological pest control. This commonly involves Bacillus thuringiensis (also called BT), a gram-positive, soil dwelling bacterium. Subspecies of this bacteria are used as a Lepidopteran-specific insecticides under trade names such as Dipel and Thuricide. Because of their specificity, these pesticides are regarded as environmentally friendly, with little or no effect on humans, wildlife, pollinators and most other beneficial insects.

Because of their ability to quickly grow and the relative ease with which they can be manipulated, bacteria are the workhorses for the fields of molecular biology, genetics and biochemistry. By making mutations in bacterial DNA and examining the resulting phenotypes, scientists can determine the function of genes, enzymes and metabolic pathways in bacteria, then apply this knowledge to more complex organisms. This aim of understanding the biochemistry of a cell reaches its most complex expression in the synthesis of huge amounts of enzyme kinetic and gene expression data into mathematical models of entire organisms. This is achievable in some well-studied bacteria, with models of Escherichia coli metabolism now being produced and tested. This understanding of bacterial metabolism and genetics allows the use of biotechnology to bioengineer bacteria for the production of therapeutic proteins, such as insulin, growth factors, or antibodies.

Because of their importance for research in general, samples of bacterial strains are isolated and preserved in Biological Resource Centers. This ensures the availability of the strain to scientists worldwide.

History of Bacteriology

Antonie van Leeuwenhoek, the first microbiologist and the first person to observe bacteria using a microscope.

Bacteria were first observed by the Dutch microscopist Antonie van Leeuwenhoek in 1676, using a single-lens microscope of his own design. He then published his observations in a series of letters to the Royal Society of London. Bacteria were Leeuwenhoek's most remarkable microscopic discovery. They were just at the limit of what his simple lenses could make out and, in one of the most striking hiatuses in the history of science, no one else would see them again for over a century. Only then were his by-then-largely-forgotten observations of bacteria — as opposed to his famous "animalcules" (spermatozoa) — taken seriously.

Christian Gottfried Ehrenberg introduced the word "bacterium" in 1828. In fact, his Bacterium was a genus that contained non-spore-forming rod-shaped bacteria, as opposed to Bacillus, a genus of spore-forming rod-shaped bacteria defined by Ehrenberg in 1835.

Louis Pasteur demonstrated in 1859 that the growth of microorganisms causes the fermentation process, and that this growth is not due to spontaneous generation. (Yeasts and moulds, commonly associated with fermentation, are not bacteria, but rather fungi.) Along with his contemporary Robert Koch, Pasteur was an early advocate of the germ theory of disease.

Robert Koch, a pioneer in medical microbiology, worked on cholera, anthrax and tuberculosis. In his research into tuberculosis Koch finally proved the germ theory, for which he received a Nobel Prize in 1905. In Koch's postulates, he set out criteria to test if an organism is the cause of a disease, and these postulates are still used today.

Though it was known in the nineteenth century that bacteria are the cause of many diseases, no effective antibacterial treatments were available. In 1910, Paul Ehrlich developed the first antibiotic, by changing dyes that selectively stained Treponema pallidum — the spirochaete that causes syphilis — into compounds that selectively killed the pathogen. Ehrlich had been awarded a 1908 Nobel Prize for his work on immunology, and pioneered the use of stains to detect and identify bacteria, with his work being the basis of the Gram stain and the Ziehl–Neelsen stain.

A major step forward in the study of bacteria came in 1977 when Carl Woese recognised that archaea have a separate line of evolutionary descent from bacteria. This new phylogenetic taxonomy depended on the sequencing of 16S ribosomal RNA, and divided prokaryotes into two evolutionary domains, as part of the three-domain system.

Bacillus Cereus

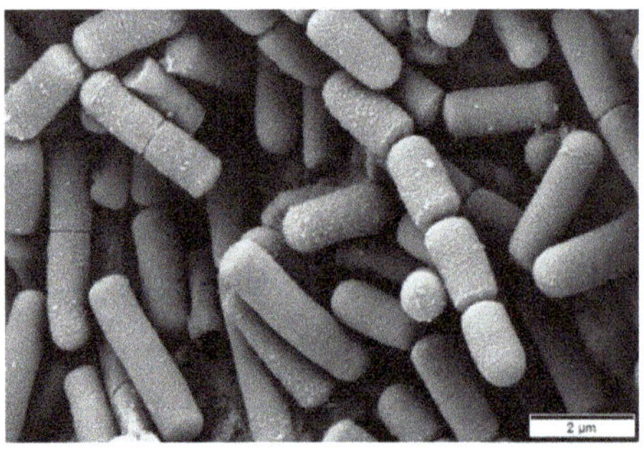

Electron micrograph of *Bacillus cereus*

Bacillus cereus is a Gram-positive, rod-shaped, aerobic, motile, beta hemolyticbacterium commonly found in soil and food. Some strains are harmful to humans and cause foodborne illness, while other strains can be beneficial as probiotics for animals. It is the cause of "fried rice syndrome", as the bacteria are classically contracted from fried rice dishes that have been sitting at room temperature for hours (such as at a buffet).*B. cereus* bacteria are facultative anaerobes, and like other members of the genus *Bacillus*, can produce protective endospores. Its virulence factors include cereolysin and phospholipase C.

It was from this species that two new enzymes, named AlkC and AlkD, which are involved in DNA repair, were discovered in 2006.

History

Colonies of *Bacillus cereus* were originally isolated from an agar plate left exposed to the air in a cow shed. In the 2010s, examination of warning letters issued by the US Food and Drug Administration issued to pharmaceutical manufacturing facilities addressing facility microbial contamination revealed that the most common contaminant was *B. cereus*.

Ecology

B. cereus competes with other microorganisms such as *Salmonella* and *Campylobacter* in the gut, so its presence reduces the numbers of those microorganisms. In food animals such as chickens,rabbits and pigs, some harmless strains of *B. cereus* are used as a probioticfeed additive to reduce *Salmonella* in the intestines and cecum. This improves the animals' growth as well as food safety for humans who eat their meat.

Bacillus cereus and other members of *Bacillus* are not easily killed by alcohol; in fact, they have been known to colonize distilled liquors and alcohol-soaked swabs and pads in numbers sufficient to cause infection.

Some strains of *B. cereus* produce cereins, bacteriocins active against different *B. cereus* strains or other Gram-positive bacteria.

Reproduction

At 30 °C (86 °F), a population of *B. cereus* can double in as little as 20 minutes or as long as 3 hours, depending on the food product.

Food	Minutes to double, 30 °C (86 °F)	Hours to multiply by 1,000,000
Milk	20-36	6.6 - 12
Cooked rice	26-31	8.6 - 10.3
Infant formula	56	18.6

Pathogenesis

B. cereus is responsible for a minority of foodborne illnesses (2–5%), causing severe nausea, vomiting, and diarrhea.*Bacillus* foodborne illnesses occur due to survival of the bacterial endospores when food is improperly cooked. Cooking temperatures less than or equal to 100 °C (212 °F) allow some *B. cereus*spores to survive. This problem is compounded when food is then improperly refrigerated, allowing the endospores to germinate. Cooked foods not meant for either immediate consumption or rapid cooling and refrigeration should be kept at temperatures below 10 °C

(50 °F) or above 50 °C (122 °F). Germination and growth generally occur between 10 °C and 50 °C, though some strains are psychrotrophic. Bacterial growth results in production of enterotoxins, one of which is highly resistant to heat and acids (pH levels between 2 and 11); ingestion leads to two types of illness, diarrheal and emetic (vomiting) syndrome.

- The diarrheal type is associated with a wide range of foods, has an 8.0- to 16-hour incubation time, and is associated with diarrhea and gastrointestinal pain. Also known as the 'long-incubation' form of *B. cereus* food poisoning, it might be difficult to differentiate from poisoning caused by *Clostridium perfringens*. Enterotoxin can be inactivated after heating at 56 °C (133 °F) for 5 minutes however it is unclear whether its presence in food causes the symptom since it degrades in stomach enzymes; its subsequent production by surviving *B. cereus* spores within the small intestine may be the cause of illness.
- The 'emetic' form is commonly caused by rice cooked for a time and temperature insufficient to kill any spores present, then improperly refrigerated. It can produce a toxin, cereulide, which is not inactivated by later reheating. This form leads to nausea and vomiting one to five hours after consumption. It can be difficult to distinguish from other short-term bacterial foodborne intoxications such as by *Staphylococcus aureus*. Emetic toxin can withstand 121 °C (250 °F) for 90 minutes.

The diarrhetic syndromes observed in patients are thought to stem from the three toxins: hemolysin BL (Hbl), nonhemolytic enterotoxin (Nhe) and cytotoxin K (CytK). The *nhe/hbl/cytK* genes are located on the chromosome of the bacteria. Transcription of these genes is controlled by *PlcR*. These genes occur in the taxonomically related *B. thuringiensis* and *B. anthracis*, as well. These enterotoxins are all produced in the small intestine of the host, thus thwarting digestion by host endogenous enzymes. The Hbl and Nhe toxins are pore-forming toxins closely related to ClyA of *E. coli*. The proteins exhibit a conformation known as "beta-barrel" that can insert into cellular membranes due to a hydrophobic exterior, thus creating pores with hydrophilic interiors. The effect is loss of cellular membrane potential and eventually cell death. CytK is a pore-forming protein more related to other hemolysins.

The timing of the toxin production was previously thought to be possibly responsible for the two different courses of disease, but in fact the emetic syndrome is caused by a toxin, cereulide, found only in emetic strains and is not part of the "standard toolbox" of *B. cereus*. Cereulide is a cyclic polypeptide containing three repeats of four amino acids: D-oxy-Leu—D-Ala—L-oxy-Val—L-Val (similar to valinomycin produced by *Streptomyces griseus*) produced by nonribosomal peptide synthesis. Cereulide is believed to bind to 5-hydroxytryptamine 3 (5-HT3) serotonin receptors, activating them and leading to increased afferentvagus nerve stimulation. It was shown independently by two research groups to be encoded on multiple plasmids: pCERE01 or pBCE4810. Plasmid pBCE4810 shares homology with the *Bacillus anthracis* virulence plasmid pXO1, which encodes the anthrax toxin. Periodontal isolates of *B. cereus* also possess distinct pXO1-like plasmids. Like most of cyclic peptides containing nonproteogenic amino acids, cereulid is resistant to heat, proteolysis, and acid conditions.

B. cereus is also known to cause difficult-to-eradicate chronic skin infections, though less aggressive than necrotizing fasciitis. *B. cereus* can also cause keratitis.

Diagnosis

In case of foodborne illness, the diagnosis of *B. cereus* can be confirmed by the isolation of more than 10^5*B. cereus* organisms per gram from epidemiologically implicated food, but such testing is often not done because the illness is relatively harmless and usually self-limiting.

Prognosis

Most emetic patients recover within six to 24 hours, but in some cases, the toxin can be fatal. In 2014, 23 neonates receiving total parenteral nutrition contaminated with *B. cereus* developed septicaemia, with three of the infants later dying as a result of infection.

Bacillus Subtilis

Bacillus subtilis, known also as the hay bacillus or grass bacillus, is a Gram-positive, catalase-positive bacterium, found in soil and the gastrointestinal tract of ruminants and humans. A member of the genus*Bacillus*, *B. subtilis* is rod-shaped, and can form a tough, protective endospore, allowing it to tolerate extreme environmental conditions. *B. subtilis* has historically been classified as an obligate aerobe, though evidence exists that it is a facultative aerobe. *B. subtilis* is considered the best studied Gram-positive bacterium and a model organism to study bacterial chromosome replication and cell differentiation. It is one of the bacterial champions in secreted enzyme production and used on an industrial scale by biotechnology companies.

Description

Bacillus subtilis is a Gram-positive bacterium, rod-shaped and catalase-positive. It was originally named *Vibrio subtilis* by Christian Gottfried Ehrenberg, and renamed *Bacillus subtilis* by Ferdinand Cohn in 1872 (subtilis being the Latin for 'fine'). *B. subtilis* cells are typically rod-shaped, and are about 4-10 micrometers (μm) long and 0.25–1.0 μm in diameter, with a cell volume of about 4.6 fL at stationary phase. As with other members of the genus*Bacillus*, it can form an endospore, to survive extreme environmental conditions of temperature and desiccation.*B. subtilis* is a facultative anaerobe and had been considered as an obligate aerobe until 1998. *B. subtilis* is heavily flagellated, which gives it the ability to move quickly in liquids. *B. subtilis* has proven highly amenable to genetic manipulation, and has become widely adopted as a model organism for laboratory studies, especially of sporulation, which is a simplified example of cellular differentiation. In terms of popularity as a laboratory model organism, *B. subtilis* is often considered as the Gram-positive equivalent of *Escherichia coli*, an extensively studied Gram-negative bacterium.

Habitat

This species is commonly found in the upper layers of the soil, and evidence exists that *B. subtilis* is a normal gut commensal in humans. A 2009 study compared the density of spores found in soil (about 10^6 spores per gram) to that found in human feces (about 10^4 spores per gram). The number of spores found in the human gut was too high to be attributed solely to consumption through food contamination.

Reproduction

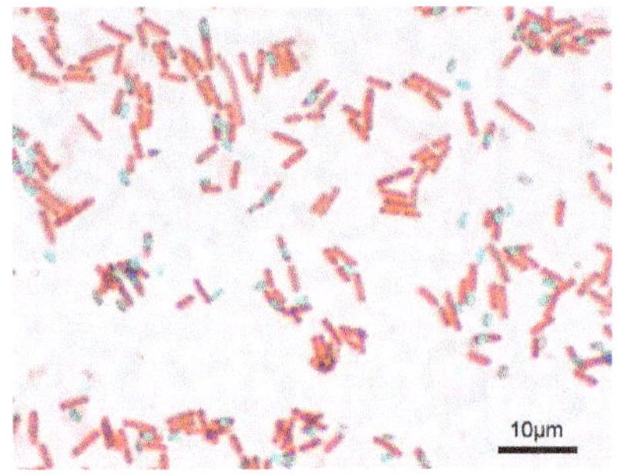

Sporulating *B. subtilis*.

B. subtilis can divide symmetrically to make two daughter cells (binary fission), or asymmetrically, producing a single endospore that can remain viable for decades and is resistant to unfavourable environmental conditions such as drought, salinity, extreme pH, radiation, and solvents. The endospore is formed at times of nutritional stress, allowing the organism to persist in the environment until conditions become favourable. Prior to the process of sporulation the cells might become motile by producing flagella, take up DNA from the environment, or produce antibiotics. These responses are viewed as attempts to seek out nutrients by seeking a more favourable environment, enabling the cell to make use of new beneficial genetic material or simply by killing of competition.

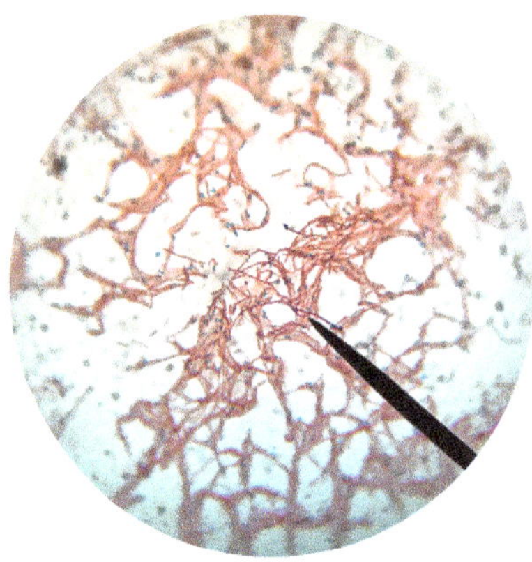

Another endospore stain of *B. subtilis*.

Under stressful conditions, such as nutrient deprivation, *B. subtilis* undergoes the process of sporulation to ensure the survival of the species. This process has been very well studied and has served as a model organism for studying sporulation.

Chromosomal Replication

B. subtilis is a model organism used to study bacterial chromosome replication. Replication of the single circular chromosome initiates at a single locus, the origin (*oriC*). Replication proceeds bidirectionally and two replication forks progress in clockwise and counterclockwise directions along the chromosome. Chromosome replication is completed when the forks reach the terminus region, which is positioned opposite to the origin on the chromosome map. The terminus region contains several short DNA sequences (*Ter* sites) that promote replication arrest. Specific proteins mediate all the steps in DNA replication. Comparison between the proteins involved in chromosomal DNA replication in *B. subtilis* and in *Escherichia coli* reveals similarities and differences. Although the basic components promoting initiation, elongation, and termination of replication are well-conserved, some important differences can be found (such as one bacterium missing proteins essential in the other). These differences underline the diversity in the mechanisms and strategies that various bacterial species have adopted to carry out the duplication of their genomes.

Genome

B. subtilis has about 4,100 genes. Of these, only 192 were shown to be indispensable; another 79 were predicted to be essential, as well. A vast majority of essential genes were categorized in relatively few domains of cell metabolism, with about half involved in information processing, one-fifth involved in the synthesis of cell envelope and the determination of cell shape and division, and one-tenth related to cell energetics.

The complete genome sequence of *B. subtilis* sub-strain QB928 has 4,146,839 DNA base pairs and 4,292 genes. The QB928 strain is widely used in genetic studies due to the presence of various markers [aroI(aroK)906 purE1 dal(alrA)1 trpC2].

Several noncoding RNAs have been characterized in the *B. subtilis* genome in 2009, including Bsr RNAs. Microarray-based comparative genomic analyses have revealed that *B. subtilis* members show considerable genomic diversity.

Transformation

Natural bacterial transformation involves the transfer of DNA from one bacterium to another through the surrounding medium. In *B. subtilis*, length of transferred DNA is greater than 1271 kb (more than 1 million bases). The transferred DNA is likely double-stranded DNA and is often more than a third of the total chromosome length of 4215 kb. It appears that about 7-9% of the recipient cells take up an entire chromosome.

In order for a recipient bacterium to bind, take up exogenous DNA from another bacterium of the same species and recombine it into its chromosome, it must enter a special physiological state called competence. Competence in *B. subtilis* is induced toward the end of logarithmic growth, especially under conditions of amino-acid limitation. Under these stressful conditions of semistarvation, cells typically have just one copy of their chromosome and likely have increased DNA damage. To test whether transformation is an adaptive function for *B. subtilis* to repair its DNA damage, experiments were conducted using UV light as the damaging agent. These experiments led to the conclusion that competence, with uptake of DNA, is specifically induced by DNA-damaging conditions, and that transformation functions as a process for recombinational repair of DNA damage.

Uses

1900s

Cultures of *B. subtilis* were popular worldwide before the introduction of antibiotics as an immunostimulatory agent to aid treatment of gastrointestinal and urinary tract diseases. It was used throughout the 1950s as an alternative medicine, which upon digestion has been found to significantly stimulate broad-spectrum immune activity including activation of secretion of specific antibodiesIgM, IgG and IgA and release of CpG dinucleotides inducing INF A/Y producing activity of leukocytes and cytokines important in the development of cytotoxicity towards tumor cells. It was marketed throughout America and Europe from 1946 as an immunostimulatory aid in the treatment of gut and urinary tract diseases such as Rotavirus and Shigellosis, but declined in popularity after the introduction of antibiotics, despite causing fewer allergic reactions and significantly lower toxicity to normal gut flora. It is still widely used in Western Europe and the Middle East as an alternative medicine.The high stability of *B. subtilis* in harsh environmental conditions makes this microorganism a perfect candidate for probiotics applications either in baked and pasteurized foods/beverages or in other galenic forms like tablets, capsules, and powder.

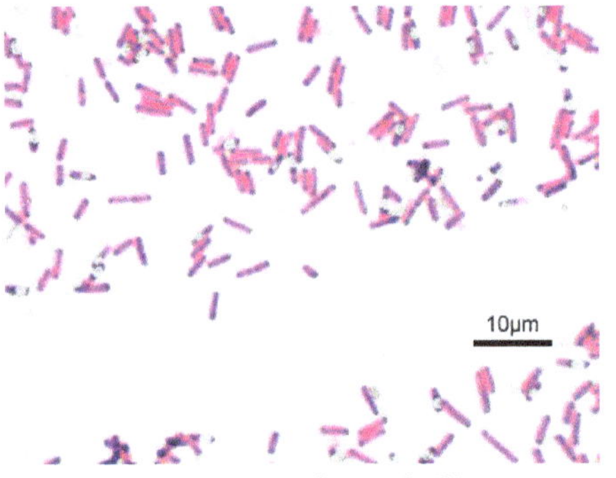

Gram-stained B. subtilis

Since the 1960s *B. subtilis* has had a history as a test species in spaceflight experimentation. Its endospores can survive up to 6 years in space if coated by dust particles protecting it from solar UV rays.* It has been used as an extremophile survival indicator in outer space such as Exobiology Radiation Assembly,EXOSTACK, and EXPOSE orbital missions.

Wild-type natural isolates of *B. subtilis* are difficult to work with compared to laboratory strains that have undergone domestication processes of mutagenesis and selection. These strains often have improved capabilities of transformation (uptake and integration of environmental DNA), growth, and loss of abilities needed "in the wild". And, while dozens of different strains fitting this description exist, the strain designated '168' is the most widely used.

Colonies of *B. subtilis* grown on a culture dish in a molecular biologylaboratory

B. globigii, a closely related but phylogenetically distinct species now known as *Bacillus atrophaeus* was used as a biowarfare simulant during Project SHAD (aka *Project 112*). Subsequent genomic analysis showed that the strains used in those studies were products of deliberate enrichment for strains that exhibited abnormally high rates of sporulation.

A strain of *B. subtilis* formerly known as *Bacillus natto* is used in the commercial production of the Japanese food *natto*, as well as the similar Korean food *cheonggukjang*.

2000s

- As a model organism, *B. subtilis* is commonly used in laboratory studies directed at discovering the fundamental properties and characteristics of Gram-positive spore-forming bacteria. In particular, the basic principles and mechanisms underlying formation of the durable endospore have been deduced from studies of spore formation in *B. subtilis*.
- It can convert some explosives into harmless compounds of nitrogen, carbon dioxide, and water.
- Its surface-binding properties play a role in safe radionuclide waste [e.g. thorium (IV) and plutonium (IV)] disposal.
- Recombinant strains pBE2C1 and pBE2C1AB were used in production of polyhydroxyalkanoates (PHA), and malt waste can be used as their carbon source for lower-cost PHA production.
- Due to its excellent fermentation properties, with high product yields (20 to 25 gram per litre) it is used to produce various enzymes, such as amylase and proteases.
- Other enzymes produced by *B. subtilis* and *B. licheniformis* are widely used as additives in laundry detergents.
- It is used to produce hyaluronic acid, which is used in the joint-care sector in healthcare and cosmetics.
- *B. subtilis* is used as a soil inoculant in horticulture and agriculture.
- It may provide some benefit to saffron growers by speeding corm growth and increasing stigma biomass yield.
- Monsanto has isolated a gene from *B. subtilis* that expresses cold shock protein B and spliced it into their drought-tolerant corn hybrid MON 87460, which was approved for sale in the US in November 2011.
- It is used as an "indicator organism" during gas sterilization procedures, to ensure a sterilization cycle has completed successfully.This is due to the difficulty in sterilizing endospores.
- Novel strains of *B. subtilis* that could use 4-fluorotryptophan (4FTrp) but not canonical tryptophan (Trp) for propagation were isolated. As Trp is only coded by a single codon, there is evidence that Trp can be displaced by 4FTrp in the genetic code. The experiments showed that the canonical genetic code can be mutable.
- Bacillus subtilis strain KPA has the ability to enhance industrially important enzymes, especially amylase under mild stress conditions.

- B. subtilis (natto) enhances the growth and (or) viability of lactobacilli, possibly through production of catalase and subtilisin.
- Bacillus subtilis is currently being used for oral bacteriotherapy and bacterioprophylaxis of gastrointestinal disorders (mostly as a direct result of antibiotic treatment), many of which lead to diarrhea. Ingestion of significant quantities of Bacillus subtilis is thought to restore the normal microbial flora following extensive antibiotic use or illness.
- Bacillus subtilis containing pharmaceutical preparations has been widely used in the treatment and prophylaxis of intestinal disorders associated with antibiotic therapy or diarrhoea of various origin.
- Treatment with Bacillus subtilis probiotics during antibiotic therapy significantly decreased the incidence of Antibiotic associated diarrhoea (AAD) and adverse effects related to the use of antibiotics.
- Bacillus subtilis CU1 supplementation effectively and safely stimulate immune responses in humans by increasing the levels of SIgA in saliva and stools.

Safety

In Animals

B. subtilis was reviewed by the US FDA Center for Veterinary Medicine and found to present no safety concerns when used in direct-fed microbial products, so the Association of American Feed Control Officials has listed it approved for use as an animal feed ingredient under Section 36.14 "Direct-fed Microorganisms". The Canadian Food Inspection Agency Animal Health and Production Feed Section has classified *Bacillus* culture dehydrated approved feed ingredients as a silage additive under Schedule IV-Part 2-Class 8.6 and assigned the International Feed Ingredient number IFN 8-19-119.

In Humans

B. subtilis spores can survive the extreme heat during cooking. Some *B. subtilis* strains are responsible for causing ropiness — a sticky, stringy consistency caused by bacterial production of long-chain polysaccharides — in spoiled bread dough. For a long time, bread ropiness was associated uniquely with *B. subtilis* species by biochemical tests. Molecular assays (randomly amplified polymorphic DNA PCR assay, denaturing gradient gel electrophoresis analysis, and sequencing of the V3 region of 16S ribosomal DNA) revealed greater *Bacillus* species variety in ropy breads, which all seems to have a positive amylase activity and high heat resistance.

B. subtilis and substances derived from it has been evaluated by different authoritative bodies for their safe and beneficial use in food. In the United States, an opinion letter issued in the early 1960s by the Food and Drug Administration (FDA) recognized some substances derived from microorganisms as Generally recognized as safe (GRAS), including carbohydrase and protease enzymes from *B. subtilis*. The opinions were predicated on the use of nonpathogenic and nontoxicogenic strains of the respective organisms and on the use of current good manufacturing practices. The FDA stated the enzymes derived from the *B. subtilis* strain were in common use in food prior to January 1, 1958, and that nontoxigenic and nonpathogenic strains of *B. subtilis* are widely available and have been safely used in a variety of food applications. This includes consumption of Japanese fermented soy bean, in the form of *Natto*, which is commonly consumed in Japan, and contains as many as 10^8 viable cells per gram. The fermented beans are recognized for their contribution to a healthy gut flora and vitamin K_2 intake; during this long history of widespread use, *natto* has not been implicated in adverse events potentially attributable to the presence of *B. subtilis*. The natto product and the *B. subtilis* natto as its principal component are FOSHU (Foods for Specified Health Use) approved by the Japanese Ministry of Health, Labour and Welfare as effective for preservation of health.

B. subtilis has been granted "Qualified Presumption of Safety" status by the European Food Safety Authority.*B. subtilis* is part of the authoritative list of microorganisms with a documented history of safe use in food, established by the International Dairy Federation in collaboration with the European Food and Feed Cultures Association in 2002, and updated in 2012. Bacillus subtilis is Generally Recognized as Safe (GRAS) by the FDA, meaning this bacterium is not harmful to animals or humans. Bacillus subtilis produces 66 anti-microbial substances.

Lactobacillus Crispatus

Lactobacillus crispatus is a common, rod-shaped species of genus*Lactobacillus* and is a hydrogen peroxide (H2O2) producing beneficial micro biota species located in both the vagina, through vaginal discharge, and the vertebrate gastrointestinal. The strain CTV-05 is used as a probiotic that can be used by premenopausal and postmenopausal women that experience recurrent urinary tract infections. It is being evaluated specifically for the prevention and treatment of bacterial vaginosis, which is characterized by the absence of lactobacillus flora, also known as *Lactobacillus acidophilus*; which plays a large role in protecting the host from infection.

Ecology

It is a species in the phylumFirmicutes, in the classBacilli, in the order Lactobacillales, in the family Lactobacillaceae and the genus*Lactobaccillus*. It is one of one hundred twenty two other species identified within the genus. It is made up of about 2,043,161 nucleotides that make up its length of 2,021 protein genes, 76 RNA genes and has a circular chromosomal shape. The strain of *Lactobacillus crispastus* was originally isolated from a pouch in chickens gullet and is considered to be one of the strongest H2O2-producing *Lactobacilli*. Like many other *Lactobacilli* species, it can be severely altered by changes to the immune system, hormone levels and from the use of antimicrobials. *Lactobacillus crispatus* is a normal inhabitant of the lower reproductive tract in healthy women.

How it is Used

CTV-05 gelatin suppository capsules are inserted into the vagina as a probiotic that can help maintain healthy flora. Studies have shown that *L. crispastus* CTV-05 effectively colonized the vagina and helped prevent and treat recurrent bacterial vaginosis and other genital infections. Scientists have stated that evidence from clinical trials proves that these probiotics will safely and effectively treat bacterial vaginosis if used alone or alongside an antibiotic treatment if an infection had already arisen.Condom use has showed increased colonization of *Lactobacillus crispastus* in the vagina because it protects against both bacterial vaginosis (BV) and human immunodeficiency virus (HIV).

Bacterial Vaginosis

Bacterial vaginosis (BV) is a disease of the vagina caused by excessive growth of bacteria. Common symptoms include increased vaginal discharge that often smells like fish. The discharge is usually white or gray in color. Burning with urination may occur. Itching is uncommon. Occasionally there may be no symptoms. Having BV approximately doubles the risk of infection by a number of other sexually transmitted infections including HIV/AIDS. It also increases the risk of early delivery among pregnant women.

BV is caused by an imbalance of the naturally occurring bacteria in the vagina. There is a change in the most common type of bacteria and a hundred to thousandfold increase in total numbers of bacteria present. Typically bacteria other than *Lactobacilli* become more common. Risk factors include douching, new or multiple sex partners, antibiotics, and using an intrauterine device among others. However, it is not considered a sexually transmitted infection. Diagnosis is suspected based on the symptoms and may be verified by testing the vaginal discharge and finding a higher than normal vaginal pH and large numbers of bacteria. BV is often confused with a vaginal yeast infection or infection with Trichomonas.

Usually treatment is with an antibiotic, such as clindamycin or metronidazole. These medications may also be used in the second or third trimesters of pregnancy. However, the condition often recurs following treatment. Probiotics may help prevent re-occurrence. It is unclear if the use of probiotics or antibiotics affects pregnancy outcomes.

BV is the most common vaginal infection in women of reproductive age. The percentage of women affected at any given time varies between 5% and 70%. BV is most common in parts of Africa and least common in Asia and Europe. In the United States about 30% of women between the ages of 14 and 49 are affected. Rates vary considerably between ethnic groups within a country. While BV like symptoms have been described for much of recorded history, the first clearly documented case occurred in 1894.

Signs and Symptoms

Common symptoms include increased vaginal discharge that usually smells like fish. The discharge is often white or gray in color. There may be burning with urination. Occasionally there may be no symptoms.

The discharge coats the walls of the vagina, and is usually without significant irritation, pain, or erythema (redness), although mild itching can sometimes occur. By contrast, the normal vaginal discharge will vary in consistency and amount throughout the menstrual cycle and is at its clearest at ovulation—about 2 weeks before the period starts. Some practitioners claim that BV can be asymptomatic in almost half of affected women, though others argue that this is often a misdiagnosis.

Causes

Healthy vaginal microbiota consists of species which do not cause symptoms, infections, or negatively affect pregnancy. It is dominated mainly by Lactobacillus species. BV is defined by the disequilibrium in the vaginal microbiota with decline in the number of lactobacilli. While the infection involves a number of bacteria, it is believed that most infections start with Gardnerella vaginalis creating a biofilm which allows other opportunistic bacteria to thrive.

One of the main risks for developing BV is douching, which alters the vaginal flora and predisposes women to developing BV. Douching is strongly discouraged by the U.S. Department of Health and Human Services and various medical authorities, for this and other reasons.

BV is a risk factor for pelvic inflammatory disease, HIV, sexually transmitted infections (STIs), and reproductive and obstetric disorders or negative outcomes. It is possible for sexually inactive persons to develop bacterial vaginosis.

Bacterial vaginosis may sometimes affect women after menopause. Subclinical iron deficiency may correlate with bacterial vaginosis in early pregnancy. A longitudinal study published in February 2006 in the American Journal of Obstetrics and Gynecology showed a link between psychosocial stress and bacterial vaginosis persisted even when other risk factors were taken into account. Exposure to the spermicide nonoxynol-9 does not affect the risk of developing bacterial vaginosis.

Having a female partner increases the risk of BV by 60%. The bacteria associated with BV have been isolated from male genitalia. BV microbiota has been found in the penis, coronal sulcus, and male urethra, in the male partners of infected females. Those uncircumcised partners may act as a 'reservoir' increasing the likelihood of acquiring an infection after sexual intercourse. Another mode of transmission of the BV-associated microbiota is to a female sexual partner via the skin-to-skin transfer. BV may be transmitted via the perineal enteric bacteria from the microbiota of the female and male genitalia.

Diagnosis

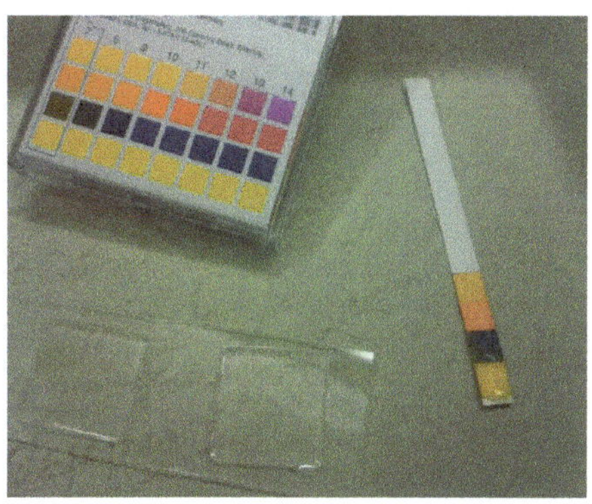

A pH indicator to detect vaginal alkalinization (here showing approximately pH 8), and a microscope slide to microscopically detect clue cells.

To make a diagnosis of bacterial vaginosis, a swab from inside the vagina should be obtained. These swabs should be tested for:

- A characteristic "fishy" odor on wet mount. This test, called the *whiff test*, is performed by adding a small amount of potassium hydroxide to a microscopic slide containing the vaginal discharge. A characteristic fishy odor is considered a positive whiff test and is suggestive of bacterial vaginosis.
- Loss of acidity. To control bacterial growth, the vagina is normally slightly acidic with a

pH of 3.8–4.2. A swab of the discharge is put onto litmus paper to check its acidity. A pH greater than 4.5 is considered alkaline and is suggestive of bacterial vaginosis.

- The presence of *clue cells* on wet mount. Similar to the whiff test, the test for clue cells is performed by placing a drop of sodium chloride solution on a slide containing vaginal discharge. If present, clue cells can be visualized under a microscope. They are so-named because they give a clue to the reason behind the discharge. These are epithelial cells that are coated with bacteria.

Two positive results in addition to the discharge itself are enough to diagnose BV. If there is no discharge, then all three criteria are needed. Differential diagnosis for bacterial vaginosis includes the following:

- Normal vaginal discharge.
- Candidiasis (thrush, or a yeast infection).
- Trichomoniasis, an infection caused by *Trichomonas vaginalis*.
- Aerobic vaginitis

The Center For Disease Control (CDC) defines STIs as "a variety of clinical syndromes and infections caused by pathogens that can be acquired and transmitted through sexual activity." But the CDC does not specifically identify BV as sexually transmitted infection.

Amsel Criteria

In clinical practice BV can be diagnosed using the Amsel criteria:

1. Thin, white, yellow, homogeneous discharge
2. Clue cells on microscopy
3. pH of vaginal fluid >4.5
4. Release of a fishy odor on adding alkali—10% potassium hydroxide (KOH) solution.

At least three of the four criteria should be present for a confirmed diagnosis. A modification of the Amsel criteria accepts the presence of two instead of three factors and is considered equally diagnostic.

Gram Stain

An alternative is to use a Gram-stained vaginal smear, with the Hay/Ison criteria or the Nugent criteria. The Hay/Ison criteria are defined as follows:

- Grade 1 (Normal): Lactobacillus morphotypes predominate.
- Grade 2 (Intermediate): Mixed flora with some Lactobacilli present, but Gardnerella or Mobiluncus morphotypes also present.
- Grade 3 (Bacterial Vaginosis): Predominantly Gardnerella and/or Mobiluncus morphotypes. Few or absent Lactobacilli. (Hay et al., 1994)

Gardnerella vaginalis is the main culprit in BV. Gardnerella vaginalis is a short rod (coccobacillus). Hence, the presence of clue cells and gram variable coccobacilli are indicative or diagnostic of bacterial vaginosis.

Nugent Score

The Nugent Score is now rarely used by physicians due to the time it takes to read the slides and requires the use of a trained microscopist. A score of 0-10 is generated from combining three other scores. The scores are as follows:

- 0–3 is considered negative for BV
- 4–6 is considered intermediate
- 7+ is considered indicative of BV.

At least 10–20 high power (1000× oil immersion) fields are counted and an average determined.

Lactobacillus morphotypes — average per high powered (1000× oil immersion) field. View multiple fields.	Gardnerella / Bacteroides morphotypes — average per high powered (1000× oil immersion) field. View multiple fields.	Curved Gram variable rods — average per high powered (1000× oil immersion) field. View multiple fields (note that this factor is less important — scores of only 0–2 are possible)
• Score 0 for >30 • Score 1 for 15–30 • Score 2 for 14 • Score 3 for <1 (this is an average, so results can be >0, yet <1) • Score 4 for 0	• Score 0 for 0 • Score 1 for <1 (this is an average, so results can be >0, yet <1) • Score 2 for 1–4 • Score 3 for 5–30 • Score 4 for >30	• Score 0 for 0 • Score 1 for <5 • Score 2 for 5+

DNA hybridization testing with Affirm VPIII was compared to the Gram stain using the Nugent criteria. The Affirm VPIII test may be used for the rapid diagnosis of BV in symptomatic women but uses expensive proprietary equipment to read results, and does not detect other pathogens that cause BV, including Prevotella spp, Bacteroides spp, & Mobiluncus spp.

Prevention

Some steps suggested to lower the risk include: not douching, avoiding sex, or limiting the number of sex partners.

One review concluded that probiotics may help prevent re-occurrence. Another review found that while there is tentative evidence it is not strong enough to recommend their use for this purpose.

Early evidence suggested that antibiotic treatment of male partners could re-establish the normal microbiota of the male urogenital tract and prevent the recurrence of infection. However a 2016 Cochrane review found high quality evidence that treating the sexual partners of women with bacterial vaginosis had no effect on symptoms, clinical outcomes, or recurrence in the affected women. It also found that such treatment may lead treated sexual partners to report increased adverse events.

Treatment

Antibiotics

Treatment is typically with the antibioticsmetronidazole or clindamycin. They can be either given by mouth or applied inside the vagina. About 10% to 15% of people; however, do not improve with the first course of antibiotics and recurrence rates of up to 80% have been documented. Recurrence rates are increased with sexual activity with the same pre-/posttreatment partner and inconsistent condom use although estrogen-containing contraceptives decrease recurrence. When clindamycin is given to pregnant women symptomatic with BV before 22 weeks of gestation the risk of pre-term birth before 37 weeks of gestation is lower.

Other antibiotics that may work include macrolides, lincosamides, nitroimidazoles, and penicillins.

Bacterial vaginosis is not considered a sexually transmitted infection, and treatment of a male sexual partner of a woman with bacterial vaginosis is not recommended.

Probiotics

A 2009 Cochrane review found tentative but insufficient evidence for probiotics as a treatment for BV. A 2014 review reached the same conclusion. A 2013 review found some evidence supporting the use of probiotics during pregnancy.

Complications

Although previously considered a mere nuisance infection, untreated bacterial vaginosis may cause complications, such as increased susceptibility to sexually transmitted infections including HIV and pregnancy complications.

It has been shown that HIV-infected women with bacterial vaginosis (BV) are more likely to transmit HIV to their sexual partners than those without BV. Diagnostic criteria for BV have also been associated with a female genital tract factor that induces expression of HIV.

There is evidence of an association between BV and increased rates of sexually transmitted infections such as HIV/AIDS. BV is associated with up to a six-fold increase in HIV shedding. There is also a correlation between the absence of vaginal lactobacilli and infection by Neisseria gonorrhoeae and Chlamydia trachomatis. BV is a risk factor for viral shedding and herpes simplex virus type 2 infection. BV may increase the risk infection or reactivation of human papillomavirus (HPV).

In addition, bacterial vaginosis an intercurrent disease in pregnancy may increase the risk of pregnancy complications, most notably premature birth or miscarriage. Pregnant women with BV have a higher risk of chorioamnionitis, miscarriage, preterm birth, premature rupture of membranes, and postpartum endometritis. BV is associated with gynaecological and obstetric complications. Data suggest an association between BV, tubal factor infertility, and pelvic inflammatory disease. Women with BV who are treated with vitro fertilization have a lower implantation rate and higher rates of early pregnancy loss.

Epidemiology

BV is the most common infection of the vagina in women of reproductive age. The percentage of women affected at any given time varies between 5% and 70%. BV is most common in parts of Africa, and least common in Asia and Europe. In the United States, about 30% of those between the ages of 14 and 49 are affected. Rates vary considerably between ethnic groups within a country.

Bifidobacterium Longum

Bifidobacterium longum is a Gram-positive, catalase-negative, rod-shaped bacterium present in the human gastrointestinal tract and one of the 32 species that belong to the genus *Bifidobacterium*. It is a microaerotolerant anaerobe and considered to be one of the earliest colonizers of the gastrointestinal tract of infants. When grown on general anaerobic medium, *B. longum* forms white, glossy colonies with a convex shape. While *B. longum* is not significantly present in the adult gastrointestinal tract, it is considered part of the gut flora and its production of lactic acid is believed to prevent growth of pathogenic organisms.*B. longum* is non-pathogenic and is often added to food products.

Classification

In 2002, three previously distinct species of *Bifidobacterium*, *B. infantis*, *B. longum*, and *B. suis*, were unified into a single species named *B. longum* with the biotypes *infantis, longum,* and *suis*, respectively. This occurred as the three species had extensive DNA similarity including a 16s rRNA gene sequence similarity greater than 97%. In addition, the three original species were phenotypically difficult to distinguish due to different carbohydrate fermentation patterns among strains of the same species. As probiotic activity varies among strains of *B. longum*, interest exists in the exact classification of new strains, although this is made difficult by the high gene similarity between the three biotypes. Currently, strain identification is done through polymerase chain reaction (PCR) on the subtly different 16s rRNA gene sequences.

Environment

B. longum colonizes the human gastrointestinal tract, where it, along with other *Bifidobacterium* species, represents up to 90% of the bacteria of an infant's gastrointestinal tract. This number gradually drops to 3% in an adult's gastrointestinal tract as other enteric bacteria such as *Bacteroides* and *Eubacterium* begin to dominate. Some strains of *B. longum* were found to have high tolerance for gastric acid and bile, suggesting that these strains would be able to survive the gastrointestinal tract to colonize the lower small and large intestines. The persistence of *B. longum* in the gut is attributed to the glycoprotein-binding fimbriae structures and bacterial polysaccharides, the latter of which possess strong electrostatic charges that aid in the adhesion of *B. longum* to intestinal endothelial cells. This adhesion is also enhanced by the fatty acids in the lipoteichoic acid of the *B. longum* cell wall.

Metabolism

B. longum is considered to be a scavenger, possessing multiple catabolic pathways to use a large variety of nutrients to increase its competitiveness among the gut flora. Up to 19 types of permease exist to transport various carbohydrates with 13 being ATP-binding cassette transporters.*B. longum* has several glycosyl hydrolases to metabolise complex oligosaccharides for carbon and energy. This is necessary as mono- and disaccharides have usually been consumed by the time they reach the lower gastrointestinal tract where *B. longum* resides. In addition, *B. longum* can uniquely ferment galactomannan-rich natural gum using glucosaminidases and alpha-mannosidases that participate in the fermentation of glucosamine and mannose, respectively. The high number of genes associated with oligosaccharide metabolism is a result of gene duplication and horizontal gene transfer, indicating that *B. longum* is under selective pressure to increase its capability to compete for various substrates in the gastrointestinal tract. Furthermore, *B. longum* possesses hydrolases, deaminases, and dehydratases to ferment amino acids.*B. longum* also has bile salt hydrolases to hydrolyze bile salts into amino acids and bile acids. The function of this is not clear, although *B. longum* could use the amino acids products or better tolerate bile salts.

Research

B. longum is a constituent in VSL#3. This proprietary, standardized, formulation of live bacteria may be used in combination with conventional therapies to treat ulcerative colitis, and requires a prescription.

Immune System Regulation

The use of *B. longum* was shown to shorten the duration and minimize the severity of symptoms associated with the common cold with a similar effect to that of neuraminidase inhibitors for influenza.

References

- Klich MA (2002). Identification of common Aspergillus species. Utrecht, The Netherlands, Centraalbureau voor Schimmelcultures. ISBN 90-70351-46-3.
- Singh, Harbhajan (2006). Mycoremediation: Fungal Bioremediation. John Wiley & Sons. p. 509. ISBN 0470050586.
- Appendix 2: Challenge test of water miscible cosmetic products". Challenge Organisms. Danish Environmental Protection Agency. 2010. ISBN 978-87-92668-66-0.
- K. Fugelsang, C. Edwards Wine Microbiology Second Edition, pp. 3–28 Springer Science and Business Media , New York (2010) ISBN 0387333495
- Enfert C, Hube B (editors) (2007). Candida: Comparative and Functional Genomics. Caister Academic Press. ISBN 978-1-904455-13-4.
- Goehring, Richard V. (2008). Mims' medical microbiology. (4th ed.). Philadelphia, PA: Mosby Elsevier. p. 656. ISBN 9780323044752.
- Turnbull PCB (1996). Baron S; et al., eds. Bacillus. In: Baron's Medical Microbiology (4th ed.). Univ of Texas Medical Branch. ISBN 0-9631172-1-1. (via NCBI Bookshelf).
- Roberts, T. A.; Baird-Parker, A. C.; Tompkin, R. B. (1996). Characteristics of microbial pathogens. London: Blackie Academic & Professional. p. 24. ISBN 0-412-47350-X. Retrieved 25 November 2010.
- Davis, Judi Ratliff; Lawley, Richard; Davis, Judy; Laurie Curtis (2008). The food safety hazard guidebook. Cambridge, UK: RSC Pub. p. 17. ISBN 0-85404-460-4. Retrieved 25 November 2010.
- Madigan, M; Martinko, J, eds. (2005). Brock Biology of Microorganisms (11th ed.). Prentice Hall. ISBN 0-13-144329-1.

- Noirot P (2007). "Replication of the Bacillus subtilis chromosome". In Graumann P. Bacillus: Cellular and Molecular Biology. Caister Academic Press. ISBN 978-1-904455-12-7.
- Bennett, John (2015). Mandell, Douglas, and Bennett's principles and practice of infectious diseases. Philadelphia, PA: Elsevier/Saunders. ISBN 9781455748013; Access provided by the University of Pittsburgh
- Borchardt, Kenneth A. (1997). Sexually transmitted diseases : epidemiology, pathology, diagnosis, and treatment. Boca Raton [u.a.]: CRC Press. p. 4. ISBN 9780849394768.
- Bennett, John (2015). Mandell, Douglas, and Bennett's principles and practice of infectious diseases. Philadelphia, PA: Elsevier/Saunders. ISBN 9781455748013.

3

Probiotic: A Comprehensive Study

Probiotics are types of microorganisms whose presence actually benefits the consumer's health. There are a number of benefits of using probiotics, such as the reduction of gastrointestinal discomfort and even in the strengthening of the immune system. The major components of probiotics are discussed in this chapter.

Probiotic

Élie Metchnikoff first suggested the possibility of colonizing the gut with beneficial flora in the early 20th century.

Probiotics are microorganisms that are believed to provide health benefits when consumed. The term probiotic is currently used to name ingested microorganisms associated with benefits for humans and animals. The term came into more common use after 1980. The introduction of the concept is generally attributed to Nobel recipient Élie Metchnikoff, who postulated that yogurt-consuming Bulgarian peasants lived longer lives because of this custom. He suggested in 1907 that "the dependence of the intestinal microbes on the food makes it possible to adopt measures to modify the flora in our bodies and to replace the harmful microbes by useful microbes". A significant expansion of the potential market for probiotics has led to higher requirements for scientific substantiation of putative benefits conferred by the microorganisms.

Although there are numerous claimed benefits of using commercial probiotics, such as reduction of gastrointestinal discomfort or strengthening of the immune system, such claims are not backed by scientific evidence.Yet one systematic review of 15 human randomized controlled trials from July 2016 found that certain commercially available strains of probiotic bacteria from the *Bifidobacterium* and *Lactobacillus*genera (*B. longum, B. breve, B. infantis, L. helveticus, L. rhamnosus, L. plantarum*, and *L. casei*), when taken by mouth in daily doses of 10^9–10^{10} colony forming units (CFU) for 1–2 months, possess treatment efficacy (i.e., improved behavioral outcomes) in certain psychological disorders, e.g. anxiety, depression, autism spectrum disorder, and obsessive-compulsive disorder – and improved certain aspects of memory.

Probiotics are considered to be generally safe, but they may cause bacteria-host interactions and unwanted side effects in certain cases.

Definition

The World Health Organization's (WHO) 2001 definition of probiotics is "live micro-organisms which, when administered in adequate amounts, confer a health benefit on the host". Following this definition, a working group convened by the FAO/WHO in May 2002 issued the "Guidelines for the Evaluation of Probiotics in Food". This first global effort was further developed in 2010; two expert groups of academic scientists and industry representatives made recommendations for the evaluation and validation of probiotic health claims. The same principles emerged from those groups as the ones expressed in the Guidelines of FAO/WHO in 2002. This definition, although widely adopted, is not acceptable to the European Food Safety Authority because it embeds a health claim which is not measurable.

A consensus definition of the term "probiotics", based on the available information and scientific evidence, was adopted after a joint Food and Agricultural Organization of the United Nations and World Health Organization expert consultation. In October 2001, this expert consultation defined probiotics as: "live micro-organisms which, when administered in adequate amounts, confer a health benefit on the host". The FAO/WHO consultation was also a first effort towards the assessment of probiotics efficacy and resulted in May 2002 in a document named "Guidelines for the Evaluation of Probiotics in Food". This effort was accompanied by local governmental and supra-governmental regulatory bodies requirements to better characterize health claims substantiations.

A group of scientific experts assembled in London, UK, on October 23, 2013, to discuss the scope and appropriate use of the term probiotic. The meeting was motivated by developments in the field since 2001. The panel's conclusions were published in June 2014.

Probiotics have to be alive when administered. One of the concerns throughout the scientific literature resides in the viability and reproducibility on a large scale of the observed results, as well as the viability and stability during use and storage, and finally the ability to survive in stomach acids and then in the intestinal ecosystem. Probiotics must have undergone controlled evaluation to document health benefits in the target host. Only products containing live organisms shown in reproducible human studies to confer a health benefit can actually claim to be a probiotic. The correct definition of health benefit, backed with solid scientific evidence, is a strong element for the proper identification and assessment of the effect of a probiotic. This aspect represents a major challenge for scientific and industrial investigations because several difficulties arise, such as variability in the site for probiotic use (oral, vaginal, intestinal) and mode of application.

The probiotic candidate must be a taxonomically defined microbe or combination of microbes (genus, species, and strain level). It is commonly admitted that most effects of probiotics are strain-specific and cannot be extended to other probiotics of the same genus or species. This calls for a precise identification of the strain, i.e. genotypic and phenotypic characterization of the tested microorganism.

Probiotics must be safe for their intended use. The 2002 FAO/WHO guidelines recommend that, though bacteria may be generally recognized as safe (GRAS), the safety of the potential probiotic should be assessed by the minimum required tests:

- Determination of antibiotic resistance patterns
- Assessment of certain metabolic activities (e.g., D-lactate production, bile salt deconjugation)
- Assessment of side effects during human studies
- Epidemiological surveillance of adverse incidents in consumers (after market)
- If the strain under evaluation belongs to a species that is a known mammalian toxin producer, it must be tested for toxin production. One possible scheme for testing toxin production has been recommended by the EU Scientific Committee on Animal Nutrition
- If the strain under evaluation belongs to a species with known hemolytic potential, determination of hemolytic activity is required

In Europe, EFSA has adopted a premarket system for safety assessment of microbial species used in food and feed productions, to set priorities for the need of risk assessment. The assessment is made for a selected group of microorganisms, which if favorable, leads to the "Qualified Presumption of Safety" status.

Finally, probiotics must be supplied in adequate numbers, which may be defined as the number able to trigger the targeted effect on the host. It depends on strain specificity, process, and matrix, as well as the targeted effect. Most of reported benefits demonstrated with the traditional probiotics have been observed after ingestion of a concentration around 10^7 to 10^8 probiotic cells per gram, with a serving size around 100 to 200 mg per day.

History

Probiotics have received renewed attention recently from product manufacturers, research studies, and consumers. The history of probiotics can be traced to the first use of cheese and fermented products, that were well known to the Greeks and Romans who recommended their consumption. The fermentation of dairy foods represents one of the oldest techniques for food preservation.

The original modern hypothesis of the positive role played by certain bacteria was first introduced by Russian scientist and Nobel laureateÉlie Metchnikoff, who in 1907 suggested that it would be possible to modify the gut flora and to replace harmful microbes with useful microbes. Metchnikoff, at that time a professor at the Pasteur Institute in Paris, proposed the hypothesis that the aging process results from the activity of putrefactive (proteolytic) microbes producing toxic substances in the large bowel. Proteolytic bacteria such as clostridia, which are part of the normal gut flora, produce toxic substances including phenols, indols, and ammonia from the digestion of proteins. According to Metchnikoff, these compounds were responsible for what he called "intestinal autointoxication", which would cause the physical changes associated with old age.

It was at that time known that milk fermented with lactic-acid bacteria inhibits the growth of proteolytic bacteria because of the low pH produced by the fermentation of lactose. Metchnikoff had also observed that certain rural populations in Europe, for example in Bulgaria and the Russian steppes, who lived largely on milk fermented by lactic-acid bacteria were exceptionally long lived. Based on these observations, Metchnikoff proposed that consumption of fermented milk would "seed" the intestine with harmless lactic-acid bacteria and decrease the intestinal pH, and that this would suppress the growth of proteolytic bacteria. Metchnikoff himself introduced in his diet sour milk fermented with the bacteria he called "Bulgarian Bacillus" and believed his health benefited. Friends in Paris soon followed his example and physicians began prescribing the sour-milk diet for their patients.

Bifidobacteria were first isolated from a breast-fed infant by Henry Tissier, who also worked at the Pasteur Institute. The isolated bacterium named *Bacillus bifidus communis* was later renamed to the genus *Bifidobacterium*. Tissier found that bifidobacteria are dominant in the gut flora of breast-fed babies and he observed clinical benefits from treating diarrhea in infants with bifidobacteria. The claimed effect was bifidobacterial displacement of proteolytic bacteria causing the disease.

During an outbreak of shigellosis in 1917, German professor Alfred Nissle isolated a strain of *Escherichia coli* from the feces of a soldier who was not affected by the disease. Methods of treating infectious diseases were needed at that time when antibiotics were not yet available, and Nissle used the *E. coli* Nissle 1917 strain in acute gastrointestinal infectious salmonellosis and shigellosis.

In 1920, Rettger and Cheplin reported that Metchnikoff's "Bulgarian Bacillus", later called *Lactobacillus delbrueckii* subsp. *bulgaricus,* could not live in the human intestine. They conducted experiments involving rats and humans volunteers, by feeding them with *Lactobacillus acidophilus.* They observed changes in composition of fecal microbiota, which they described as "transformation of the intestinal flora". Rettger further explored the possibilities of *L. acidophilus* and reasoned that bacteria originating from the gut were more likely to produce the desired effect in this environment. In 1935, certain strains of *L. acidophilus* were found to be very active when implanted in the human digestive tract. Trials were carried out using this organism, and encouraging results were obtained, especially in the relief of chronic constipation.

Contrasting antibiotics, probiotics were defined as microbially derived factors that stimulate the growth of other microorganisms. In 1989, Roy Fuller suggested a definition of probiotics that has been widely used: "*A live microbial feed supplement which beneficially affects the host animal by improving its intestinal microbial balance*". Fuller's definition emphasizes the requirement of viability for probiotics and introduces the aspect of a beneficial effect on the host.

The term "probiotic" originally referred to microorganisms that have effects on other microorganisms. The conception of probiotics involved the notion that substances secreted by one microorganism stimulated the growth of another microorganism. The term was used again to describe tissue extracts which stimulated microbial growth. The term probiotics was taken up by Parker, who defined the concept as, "organisms and substances that have a beneficial effect on the host animal by contributing to its intestinal microbial balance". Later, the definition was greatly improved by Fuller, whose explanation was very close to the definition used today. Fuller 89 described probiotics as a "live microbial feed supplement which beneficially affects the host animal by improving its intestinal microbial balance". He stressed two important facts of probiotics: the viable nature of probiotics and the capacity to help with intestinal balance.

In the following decades, intestinal lactic acid bacterial species with alleged health beneficial properties have been introduced as probiotics, including *Lactobacillus rhamnosus*, *Lactobacillus casei*, and *Lactobacillus johnsonii*.

EFSA and FDA Scientific Reviews

The European Food Safety Authority has rejected all petitions by commercial manufacturers for health claims on probiotic products in Europe due to insufficient research and thus inconclusive proof of effectiveness. Occurring over many years, the scientific reviews established that a cause-and-effect relationship had not been sufficiently proven in the products submitted.

In the United States where food product labeling requires language approval by the FDA, probiotic manufacturers have received warning letters of impending legal actions for using exaggerated claims of health benefits not supported by clinical evidence of efficacy.

Research

Probiotics have been the subject of research to see whether the health claims made for them have any supporting evidence. Overall scientific demonstration of probiotic effects requires defining a healthy microbiota and interactions between microbiota and host, and the difficulty to characterize probiotic effectiveness in health and disease. Recent developments of high-throughput sequencing technology and the consequent progresses of metagenomics represent a new approach for the future of probiotics research.

Studies are examining whether probiotics affect mechanisms of intestinal inflammation,diarrhea, or urogenital infections. Through 2012, however, in all cases proposed as health claims to the European Food Safety Authority, the scientific evidence remains insufficient to prove a cause-and-effect relationship between consumption of probiotic products and any health benefit.

Research into the potential health effects of supplemental probiotics has included the molecular biology and genomics of *Lactobacillus* in immune function, cancer, and antibiotic-associated diarrhea, travellers' diarrhea, pediatric diarrhea, inflammatory bowel disease, and irritable bowel syndrome. Testing of a probiotic applies to a specific strain under study. The scientific community cautions against extrapolating an effect from a tested strain to an untested strain.

Although research does suggest that the relationship between gut flora and humans is a mutualistic relationship, very little evidence supports claims that probiotic dietary supplements have any health benefits. Improved health through gut flora modulation appears to be directly related to long-term dietary changes.

Claims that some lactobacilli may contribute to weight gain in some humans remain controversial.

Allergies

Probiotics are ineffective in preventing allergies in children, with the possible exception of eczema.

Antibiotic-associated Diarrhea (AAD)

Antibiotics are a common treatment for children, and 20% of antibiotic-treated children develop diarrhea. Antibiotic-associated diarrhea results from an imbalance in the colonic microbiota caused by antibiotic therapy. Microbiota alteration changes carbohydrate metabolism, with decreased short-chain fatty acid absorption and osmotic diarrhea as a result. A review assessing the work of 16 different studies representing the evaluation of more than 3,400 patients concluded that the evidence gathered suggested a protective effect of some probiotics in this condition. In adults, some probiotics showed a beneficial role in reducing the occurrence of AAD.

Probiotic treatment might reduce the incidence and severity of AAD as indicated in several meta-analyses. For example, treatment with probiotic formulations including *L. rhamnosus* may reduce the risk of AAD, improve stool consistency during antibiotic therapy, and enhance the immune response after vaccination. However, further documentation of these findings through randomized, double-blind, placebo-controlled trials is required to confirm specific effects and obtain regulatory approval, which currently does not exist.

The potential efficacy of probiotic AAD prevention is dependent on the probiotic strain(s) used and on the dosage. A Cochrane Collaboration systematic review, in which 16 randomized clinical trials (n=3432 participants) were analyzed, concluded that treatments with less than 5×10^9CFUs/day did not show a significant decrease of AAD. However, patients treated with 5×10^9 CFUs/day (including *L. rhamnosus* and *Saccharomyces boulardii*) had 60% lower relative risk (95%CI: 44–71%) of experiencing AAD than untreated patients.

Bacterial Vaginosis

Probiotic treatment of bacterial vaginosis is the application or ingestion of bacterial species found in the healthy vagina to cure the infection of bacteria causing bacterial vaginosis. This treatment is based on the observation that 70% of healthy females have a group of bacteria in the genus *Lactobacillus* that dominate the population of organisms in the vagina. Currently, the success of such treatment has been mixed since the use of probiotics to restore healthy populations of *Lactobacillus* has not been standardized. Often, standard antibiotic treatment is used at the same time that probiotics are being tested. In addition, some groups of women respond to treatment based upon ethnicity, age, number of sexual partners, pregnancy, and the pathogens causing bacterial vaginosis. In 2013, researchers found that administration of hydrogen peroxide producing strains, such as *L. acidophilus* and *L. rhamnosus*, were able to normalize vaginal pH and rebalance vaginal flora, preventing and alleviating bacterial vaginosis.

Blood Pressure

The consumption of probiotics may modestly help to control high blood pressure.

Central Nervous System Function and Disorders

A systematic review of 15 human randomized controlled trials from July 2016 found that certain commercially available strains of probiotic bacteria from the Bifidobacterium and Lactobacillus-genera (*B. longum*, *B. breve*, *B. infantis*, *L. helveticus*, *L. rhamnosus*, *L. plantarum*, and *L. casei*), when taken by mouth in daily doses of 10^9–10^{10} colony forming units (CFU) for 1–2 months, possess treatment efficacy (i.e., improved behavioral outcomes) in certain central nervous system disorders – including anxiety, depression, autism spectrum disorder, and obsessive-compulsive disorder – and improved certain aspects of memory.

Cholesterol

Preliminary human and animal studies have demonstrated the efficacy of some strains of lactic acid bacteria (LAB) for reducing serum cholesterol levels, presumably by breaking down bile in the gut, thus inhibiting its reabsorption (where it enters the blood as cholesterol).

A meta-analysis that included five double-blind trials examining the short-term (2–8 weeks) effects of a yogurt with probiotic strains on serum cholesterol levels found a minor change of 8.5 mg/dL (0.22 mmol/L) (4% decrease) in total cholesterol concentration, and a decrease of 7.7 mg/dL (0.2 mmol/L) (5% decrease) in serum LDL concentration.

A slightly longer study evaluating the effect of a yogurt with probiotic strains on 29 subjects over six months found no statistically significant differences in total serum cholesterol or LDL values. However, the study did note a significant increase in serum HDL from, 50 to 62 mg/dL (1.28 to 1.6 mmol/L) following treatment. This corresponds to a possible improvement of LDL/HDL ratio.

Studies specifically on hyperlipidemic subjects are still needed.

Diarrhea

Some probiotics are suggested as a possible treatment for various forms of gastroenteritis, and a Cochrane Collaboration meta-analysis on the use of probiotics to treat acute infectious diarrhea based on a comprehensive review of medical literature through 2010 (35 relevant studies, >4500 participants) reported that use of any of the various tested probiotic formulations appeared to reduce the duration of diarrhea by a mean of 25 hours (vs. control groups, 95% confidence interval, 16–34 hours), also noting, however, that "the differences between the studies may be related to other unmeasured and unexplored environmental and host factors" and that further research was needed to confirm reported benefits.

Eczema

Probiotics are commonly given to breast-feeding mothers and their young children to prevent eczema, but some doubt exists over the strength of evidence supporting this practice.

Helicobacter Pylori

Some strains of lactic acid bacteria may affect *Helicobacter pylori* infections (which may cause peptic ulcers) in adults when used in combination with standard medical treatments, but no standard in medical practice or regulatory approval exists for such treatment.

Immune Function and Infections

Some strains of lactic acid bacteria (LAB) may affect pathogens by means of competitive inhibition (i.e., by competing for growth) and some evidence suggests they may improve immune function by increasing the number of IgA-producing plasma cells and increasing or improving phagocytosis, as well as increasing the proportion of T lymphocytes and natural killer cells.Clinical trials have demonstrated that probiotics may decrease the incidence of respiratory-tract infections and dental caries in children. LAB products might aid in the treatment of acute diarrhea, and possibly affect rotavirus infections in children and travelers' diarrhea in adults, but no products are approved for such indications.

Inflammation

Some strains of LAB may modulate inflammatory and hypersensitivity responses, an observation thought to be at least in part due to the regulation of cytokine function. Clinical studies are assessing whether they can prevent recurrences of inflammatory bowel disease in adults, as well as affect milk allergies. How probiotics may influence the immune system remains unclear.

Inflammatory Bowel Disease

Probiotics are being studied for their potential to influence inflammatory bowel disease. There is some evidence to support their use in conjunction with standard medications in treating ulcerative colitis and no evidence of their efficacy in treating Crohn's disease.

A live formulation of lyophilized *Bifidobacterium breve*, *Bifidobacterium longum*, *Bifidobacterium infantis*, *Lactobacillus acidophilus*, *Lactobacillus plantarum*, *Lactobacillus paracasei*, *Lactobacillus bulgaricus*, and *Streptococcus thermophilus* (VSL#3) has shown effectiveness in the small clinical trials, some of which were not randomized nor double-blinded, that had been done as of 2015; more high quality clinical trials are needed to determine safety and effectiveness.

Irritable Bowel Syndrome

Probiotics are under study for their potential to affect irritable bowel syndrome, although uncertainty remains around which type of probiotic works best, and around the size of possible effect.

Lactose Intolerance

Ingestion of certain active strains may help lactose-intolerant individuals tolerate more lactose than they would otherwise have tolerated.

Necrotizing Enterocolitis

Several clinical studies provide evidence for the potential of probiotics to lower the risk of necrotizing enterocolitis and mortality in premature infants. One meta-analysis indicated that probiotics reduce these risks by more than 50% compared with controls.

Urinary Tract

There is no good evidence that probiotics are of benefit in the management of infection or inflammation of the urinary tract.

Vitamin Production

Probiotic treatment has been studied as a means of addressing disorders associated with vitamin deficiencies including those of vitamin K,folic acid, and vitamin B12.

Side Effects

The manipulation of the gut microbiota is complex and may cause bacteria-host interactions. Although probiotics are considered to be safe, there are concerns about their safety in certain cases. Some people, such as those with immunodeficiency, short bowel syndrome, central venous catheters, cardiac valve disease and premature infants, may be at higher risk for adverse events. In severely ill people with inflammatory bowel disease there is a risk of the passage of viable bacteria from the gastrointestinal tract to the internal organs (bacterial translocation) as a consequence of bacteremia, which can cause adverse health consequences. Rarely, consumption of probiotics by children with lowered immune system function or who are already critically ill may result in bacteremia or fungemia (i.e., bacteria or fungi in the blood), which can lead to sepsis, a potentially fatal disease.

Although it has been suggested that *Lactobacillus* contributes to obesity in humans, no evidence of this relationship has been found.

Strains

Live probiotic cultures are available in fermented dairy products and probiotic fortified foods. However, tablets, capsules, powders, and sachets containing the bacteria in freeze-dried form are also available. Probiotics taken orally can be destroyed by the acidic conditions of the stomach. A number of microencapsulation techniques are being developed to address this problem.

Only preliminary evidence exists for most probiotic health claims. Even for the most studied strains, few have been sufficiently developed in basic and clinical research to warrant approval for health claim status by a regulatory agency such as the Food and Drug Administration or European Food Safety Authority, and as of 2010, no claims have been approved by those two agencies. Some experts are skeptical about the efficacy of different strains and believe not all subjects will benefit from the use of probiotics.

Some fermented products are reported to contain lactic acid bacteria, including pickled vegetables,fermented bean paste such as tempeh,miso,kefir,buttermilk or karnemelk,kimchi,pao cai,-sauerkraut, and soy sauce.

Commercial Probiotics

Labeling

The US National Yogurt Association gives a *Live & Active Cultures Seal* to refrigerated yogurt products which contain 100 million cultures per gram or frozen yogurt products containing 10 million cultures per gram at the time of manufacture. In 2002, the US Food and Drug Administration (FDA) and World Health Organization recommended that "the minimum viable numbers of each probiotic strain at the end of the shelf-life" be reported on labeling, but most companies that give a number report the viable cell count at the date of manufacture, a number probably much higher than existing at the moment of consumption. Because of variability in storage conditions and time before eating, it is difficult to tell exactly how much active culture remains at the time of consumption.

Due to these ambiguities, the European Commission placed a ban on putting the word "probiotic" on the packaging of products because such labeling misleads consumers to believe a health benefit is provided by the product when no scientific proof exists to demonstrate that health effect.

In the United States, the FDA has issued warning letters to various manufacturers of probiotic products whose labels claim unproven nutritional benefits or anti-disease activity.

History and Modern Products

The first commercially sold dairy-based probiotic was Yakult, a fermented milk with added *Lactobacillus casei*Shirota, in 1935. Since then, many more probiotic foods have come on the market, mostly in the form of dairy products. Recently, nondairy and unfermented probiotics have been produced, including breakfast cereal and snack bars, whereas other probiotic products include kefir, yogurt, kombucha, kimchi, sauerkraut, and other fermented foods and beverages.

Global Consumption

Sales of probiotic products have a rising trend from 2010 to 2014, increasing globally by 35% from US$23.1 billion to $31.3 billion. Some regions have increased their use by even more than the average, including Eastern Europe (67%), Asia Pacific (67%), and Latin America (47%), comprising nearly half of probiotics sold globally in 2014. By geographic region, the leading consumers of probiotics in 2014 were Western Europe ($8.3 billion), Asia Pacific ($7 billion), Japan ($5.4 billion), Latin America ($4.8 billion), North America ($3.5 billion), and Eastern Europe ($2.3 billion).

Multiple Probiotics

Preliminary research is evaluating the potential physiological effects of multiple probiotic strains, as opposed to a single strain. As the human gut may contain several hundred microbial species, one theory indicates that this diverse environment may benefit from consuming multiple probiotic strains, an effect that remains scientifically unconfirmed.

Lactobacillus Paracasei

Lactobacillus paracasei (commonly abbreviated as *L. paracasei*) is a gram-positive, facultatively-heterofermentative species of lactic acid bacteria that are commonly used in dairy product fermentation and probiotics. *L. paracasei* is a bacterium that operates by commensalism. It is commonly found in many human habitats such as our intestinal tracts and mouths as well as sewages, silages, and previously mentioned dairy products. The name includes morphology, a rod-shaped (bacillus shape) bacterium with a width of 2.0 to 4.0μm and length of 0.8 to 1.0μm.

So far, thirty four different strains of *L. paracasei* have been isolated from a variety of environments. Sixteen of those strains have been isolated from dairy, ten from plants, and eight from human and animal gastrointestinal tracts.*L. paracasei* is genotypically and phenotypically indistinguishable from other members of its genus such as *Lactobacillus casei* and *Lactobacillus rhamnosus*. However, they are easily differentiated from each other by their fermentation profiles. Its fermentative properties allows it to be used as biological food processors and supplements for diets and medical disorders, especially in the gastrointestinal tract.

Although probiotics are considered safe, they may cause bacteria-host interactions and adverse health consequences. In certain cases there is a risk of bacteremia when probiotics are used. Currently, the probiotic strain, frequency, dose and duration of the probiotic therapies are not established.

Physiology

Lactobacillus paracasei is a gram-positive, facultatively heterofermentative, non-spore forming microorganism. The cells of *L. paracasei* are typically rod shaped, with a size range of 2.0μm to 4.0μm in width, and 0.8 to 1.0μm in length. The organism is nonmotile. *L. Paracasei* cells often have square ends, and may exist either in single form or in chains.

L. paracasei grows optimally in a temperature range between 10 and 37 °C. No growth takes place above 40 °C. The organism is able to survive for approximately 40 seconds in a maximum temperature of 72 °C. As *L. Paracasei* is facultativelyheterofermentative: lactic acid is produced from most strains.

L. paracasei exist as a common inhabitant of the human gastrointestinal tract as part of the normal flora. Naturally fermented vegetables, milk, and meat may also contain strains of *L. paracasei*.

Phylogeny

Lactobacillus paracasei belongs to kingdom Bacteria. *L. paracasei* is part of the division Firmicutes, and also belongs to the class Bacilli. The order and family are Lactobacillales and Lactobacillaceae respectively. The argument on the nomenclature of *L. paracasei* versus *L. casei* has been one of intense debate. Most of the species profiled as *L. casei* or *L. paracasei* have been found to be part of the same species. In 1989, it was proposed that *L. paracasei* be designated a subspecies (paracasei) to account for the species that it shares DNA homology with. It has been shown their names have been used interchangeably in scientific literature.16S RNA sequence homology has confirmed the relatedness between these species.

Historically, the difference between *Lactobacillus paracasei* and other *lactobacilli* has been based on biochemical characteristics. There is an approximately 90% sequence identity between *casei*, *paracasei*, and *rhamnosus*. However, there are some differential criteria that are commonly used to differentiate between them. These differential criteria include nutritional requirements and growth environment.*L. Paracasei* has been found to show specific differences with other *Lactobacillus* spp. in that it is somewhat heat resistant, grows well in ripening cheese, and it has high proteolytic activity.

L. paracasei has been found to have 34 strains. These strains have been isolated from various countries around the world. Although there is a small correlation between phylogenetic relatedness and origin of isolation, currently there is not enough evidence to support direct proof of the relationship.

Genomics

L. paracasei's genome contains circular DNA and varies slightly among the different strains isolated. On average, the genomes are 2.9 to 3.0 millions of base pairs (commonly abbreviated Mb). It has a GC-content between 46.2 and 46.6% and is predicted to encode about 2800 to 3100 proteins. The difference in the genomes of these strains lies in variant cell envelopes, secretory proteins, and polysaccharides. Many of the commonly coded proteins are cell-surface associated cell-wall hydrolases that protect the cell against apoptosis. These enzymes have been shown to provide cellular protection to human epithelial cells.

Genetic diversity for the different *L. paracasei*genomes was assessed using multilocus sequence typing (MLST) and amplified fragment length polymorphism (AFLP). MLST is a technique used for classifying microbes by the use of DNA fragments from essential genes of the organism. AFLP is a Polymerase Chain Reaction (PCR) tool used in DNA profiling to amplify a desired DNA fragment with the use of restriction enzymes and ligands.

Clinical and Research Applications

Lactobacillus paracasei has been identified as a bacterium that has probiotic properties.*L. paracasei* is a part of the normal human gut microbiota.*L. paracasei* IMPC2.1 may be a chemoprophylactic in gastrointestinal cells.Gastrointestinal cells are susceptible to apoptosis and cell growth from both heat-killed and viable IMPC2.1 strains.*L. paracasei* 8700:2 has been isolated from healthy human gastrointestinal mucosa and human feces. Strain 8700:2 was also found to inhibit *Salmonella enterica* and *Heliobacter pylori*, two pathogens commonly found in the gastrointestinal tract. Strain 8700:2 breaks down oligofructose and inulin, while also growing rapidly on both and producing lactic acid as the end product.

L. paracasei is a constituent in the therapeutic, nutritional treatment VSL#3. This proprietary, standardized, formulation of live bacteria may be used in combination with conventional therapies to treat ulcerative colitis.

Health Concerns

The manipulation of the gut microbiota is complex and may cause bacteria-host interactions. Although probiotics are considered safe, when they are used by oral administration there is a risk of passage of viable bacteria from the gastrointestinal tract to the internal organs (bacterial translocation) and subsequent bacteremia, which can cause adverse health consequences. Some people, such as those with immune compromise, short bowel syndrome, central venous catheters, cardiac valve disease and premature infants, may be at higher risk for adverse events.

Currently, the probiotic strain, frequency, dose and duration of the probiotic therapy are not established. Live bacteria might not be essential because of beneficial effects of probiotics seems to be mediated by their DNA and by secreted soluble factors, and their therapeutic effects may be obtained by systemic administration rather than oral administration.

History

LAB (Lactic Acid Bacteria) were classified and grouped in the early 1900s after gaining scientists' attention after observing the bacteria's interactions in different foods, especially dairy products. In 1991, Martinus Beijerinck, a Dutch microbiologist, separated *Lactobacillus* as gram positive bacteria from the previously known LAB group.*L. paracasei* has been recently classified as a part of the *Lactobacillus casei* group of probiotics. However, it has no clear taxonomic position. The name *L. paracasei* was proposed for rejection in 1996 by Dicks, Duplessis, Dellaglio, and Lauer.

Probiotics in Pediatrics

Yakult, Mexico

Probiotics are defined to be non-pathogenic strains of organism that are incorporated into the diet to modify gut microbial ecology, leading to beneficial structural and functional changes in the gut. Some probiotics may serve as a barrier for the colonization of pathogens to prevent disease and enhance the immune system. In addition, some may carry out metabolic functions such as helping the fermentation of nondigestible fibres, and storing energy in the form of short-chain fatty acids. Of all the types of gut microbiota, *Bifidobacteria* and *Lactobacilli* are considered the two most essential bacteria beneficial to human health while *Staphylococci* and *Clostridia* are considered pathogenic to human health. Organisms that are considered as probiotics remain viable even after culturing, modification, and storage before ingestion. They are not killed during digestion, and ultimately adhere to the epithelium in the gut. As a consequence, the host may gain clinical health benefits, such as the reduction in antibiotic-associated diarrhea and allergies in infants. The microbiome, which is composed of ecosystems within and on the human body, is necessary for infants in the development of the gut-associated lymphoid tissue (GALT), an essential prerequisite for a normal immune system.

Intestinal Microflora in Infancy

Determinants of the Gut Microbiota

At birth, the infant's intestines are free of bacteria, but bacteria starts to colonize within the intestine quickly after birth. Facultative aerobes are the first to colonize the intestine in infants. However, as time goes by, these facultative aerobes start to deplete the oxygen store within the infant's intestine, making the intestine a favorable environment to harbor anaerobes. Overall, an infant's intestine is mainly colonized by enterobacteria during the first two days after birth.

Baby

Bifidobacteria, lactobacilli, Escherichia coli and *Clostridium difficile*bacteroids are the major types of bacteria that are commonly found in infants' GI tract but the proportion of each bacterium differs depending on several determinants.

The composition of the gut microbiota during the first few days is contributed mainly by the method of delivery, meaning that the composition is generally influenced by the mother and the surroundings. Other than these, the environment during birth, prematurity, hygiene measures, and the type of infant feeding also contribute to the composition of an infant's intestinal microflora.

The mode of delivery is a first major determinant of an infant's intestinal flora in early life.

Cesarean Section Versus Vaginal Delivery

Infants that were given birth through cesarean section tend to contain fewer *Bifidobacteria* and *Bacteroides* and a higher amount of *Clostridium difficile* when compared with infants that were given birth naturally. This is because infants that are delivered through cesarean section are exposed initially to bacteria originating from the hospital environment and health care workers while vaginally born infants are colonized at first by fecal and vaginal bacteria of the mother. Lab results show that *C. difficile* is commonly found in hospital environments but is uncommon in healthy individuals. This result is relevant to the higher count of *C. difficile* in infants born through cesarean section. This is also because infants born through cesarean section on average stay in the hospital longer than vaginally born infants. There are other contributors to the gut microbiota, including prematurity, hygiene level, and the method used to feed the infant.

The next major determinant of an infant's intestinal flora is the type of feeding.

Formula-fed Infants Versus Breast-fed Infants

When comparing formula-fed and breast-fed infants, formula-fed infants' intestinal culture mostly consist of coliforms such as *Escherichia coli, Clostridium difficile*, bacteroids, and *Lactobacilli* bacteria. Hence, formula-fed infants have a more complex intestinal microflora than breast-fed infants. In breast-fed infants, *Bifidobacteria* contribute 80-90% of the microflora, while *Lactobacilli* and bacteroids constitutes a small amount in the intestine. However, *Bacteroides, Veillonella*, and *Fusobacterium* in the intestines start to increase in number after weaning. At this point, the child's intestinal microflora starts to develop in line with that in adults.

Other Determinants

Premature babies are found to have higher counts of *C. difficile* in their intestine. Furthermore, infants with older siblings tend to be colonized by more bifidobacteria in their intestines than infants with no siblings. Hence, the mode of delivery, type of infant feeding, gestational age, infant hospitalization and antibiotic use by the infant all greatly contribute to the composition of microflora in infants.

Oral supplementation of *Bifidobacteria* and *Lactobacilli* are also useful in modifying the intestinal microflora in infants.

Immunity

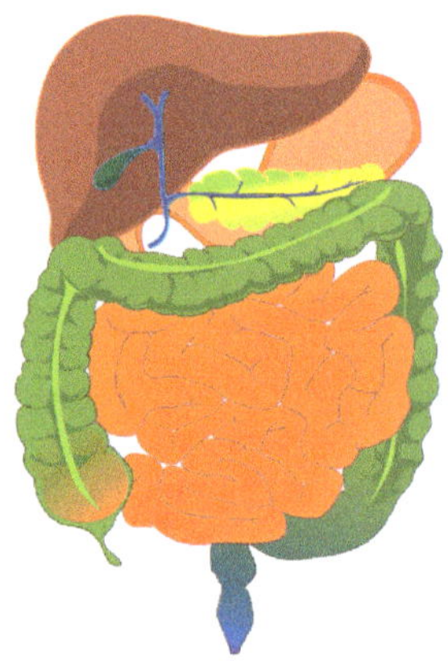

Complete GI tract

About 80% of all the immunologically active cells are located in the GALT. This implies that microbes in the gut are crucial in contributing to the immunological development of the GALT. It is necessary for the body to establish an efficient microbe-gut immune system interaction because there are about 10^{10} to 1012prokaryotic organisms that are harbored in an infant's intestine and some are potentially pathogenic.

Studies have been done to examine the gut-tissues of germ-free mice and found that the absence of microbes gave rise to an underdeveloped and sparse mucosal immune system. Characteristics of this underdeveloped mucosal immune system include a lamina propria that does not have IgA plasma cells or CD4 cells, and very small numbers of intraepithelial lymphocytes. This indicates that the mucosal immune system is dependent on gut bateria. Since new born infants do not show an underdeveloped and sparse mucosal immune system, it explains how microbes harbored in an infant's intestine are crucial for the development of a proper mucosal immune system.

Furthermore, intestinal bacteria affect epithelium cell functions through signaling receptors such as Toll-like receptors. The function of the epithelium cell is to regulate immune gut response by controlling T cell differentiation and antibody responses to T cell-dependent antigens. One example is the IgA responses to luminal antigens. When pathogen antigens and gut lumen protein are present, one antibody response to these molecules would be IgA secretion. Hence, microbial colonization has an effect on the IgA responses to antigens in the lumen. Another immune function that microbial colonization gives is the ability to reduce the possibility of immune hyperreactivity such as allergies through inducing modulation of the T-helper type 2 and T-helper type 1 ratio.

Importance of Intestinal Bacteria on Immune Function

Lactobacilli and *Bifidobacteria* are the two of the main types of bacteria that reside in the small intestine. Some commercial strains of these groups of bacteria are used as probiotics. Both bacterial groups may play a role in both local and systemic immunity of the intestine. Aside from providing a gut defense barrier, some strains of these two types of bacteria may secrete antimicrobial substates to inhibit the growth of pathogens.

Intestinal bacteria are also shown to improve the secretory immune function and the intestinal flora such as influencing secretory immunoglobulin A (sIgA) synthesis and sIgA precursors. sIgA is an immunoglobulin that is present in mucosal surfaces. Its function is mainly to protect against the attack of antigens, pathogens, toxins and viruses.

Researches have shown that high bacteroid and *Clostridium* counts accompanied by low *Bifidobacterium* counts can lead to an altered immune responses in infants. Premature infants are expected to have intestines that contain non-bifidobacterium species because of their frequent exposure to microbial environments and antibiotic use.

Health Benefits

Acute Diarrhea

Many studies show that probiotics exhibit a protective effect against different types of acute diarrhea, such as antibiotic-associated diarrhea, traveller's diarrhea and viral diarrhea. In the United States, more than 15 million children develop acute diarrhea per year, with the treatment typically being rehydration and recuperative procedures. As many prevention methods for diarrhea have adverse effects (e.g. intestinal intussusception in the usage of rotavirus vaccine), scientists are now turning to probiotics in hope of using it as a supplement to treat acute diarrhea. In a review that covered 34 masked, randomized, placebo-controlled trials related to diarrhea and probiotics, it was concluded that there was an overall reduction of 52% in antibiotic associated diarrhea, an 8% reduction in traveller's diarrhea, and a 34% reduction in other types of acute diarrhea. These numbers reflect a protective effect against diarrhea in subjects including both adults and children. In terms of children (age < 18 years), seven out of the 12 trials conducted in a health care setting showed a reduction in acute diarrhea of 57%. There was no significant difference in the protective effect of strains, namely *Saccharomyces boulardii, Lactobacillus rhamnosus GG, Lactobacillus acidophilus*, and *Lactobacillus bulgaricus*.

Probiotics were found to be effective in treating acute, infectious diarrhea in children when a review was completed in 2001, involving > 700 children aged from one to 48 months with acute diarrhea. A 2002 review showed that *Lactobacillus* reduced the average duration of acute diarrhea by 0.7 days and stool frequency by 1.6 on the second day of treatment. Dosage was most effective at more than ten billion colony-forming units of *Lactobacillus* during the first 48 hours of diarrhea. Moreover, there has been evidence showing that probiotics can prevent children and infants from gastroenteritis (also known as stomach flu), which may also cause diarrhea to occur.

The general mechanism in which probiotics protect against diarrhea is not known, but studies have shown that a dose of *Lactobacillus rhamnosus GG* of approximately 10^10 to 10^11 colony-forming units can inhibit the attachment of pathogenic organism in the intestine. Specifically for *Lactobacillus*, scientists have proposed that the organism fights diarrhea by increasing antimicrobial substances, competing for adherence sites with pathogenic organisms, and boosting the immune response to pathogens. Probiotics may also help in maintaining the integrity of the gastrointestinal epithelium by preventing the production of cytokines, which can cause injury to intestinal epithelial cells with resulting diarrhea. There have been no reports in where diarrhea has increased in severity due to patients receiving probiotic therapy.

Antibiotic-associated Diarrhea

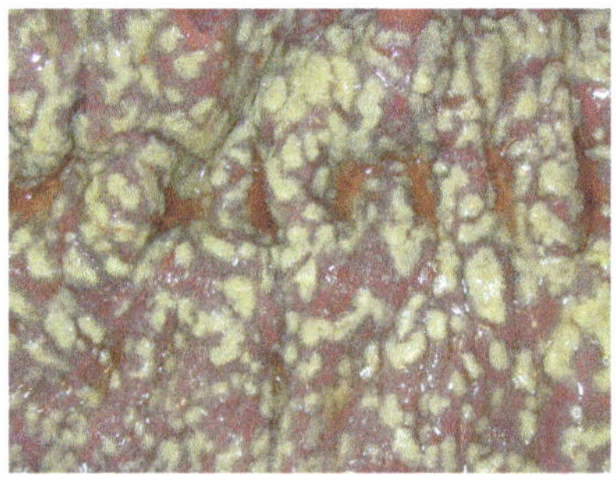

Pseudomembranous colitis

Diarrhea can occur in 11–40% of children treated with antibiotics, and is commonly known as antibiotic-associated diarrhea or AAD. The duration of the illness can range from the start of the therapy to two months after its termination. This illness can generally be accounted for by *Clostridium difficile*, a bacterium that can sometimes cause severe diarrhea known as pseudomembranous colitis. In a review of six trials related to antibiotic-associated diarrhea in 766 children aged one month to six years, there was an overall reduction in AAD when children were fed probiotics. The reduction, when compared to placebo, was seen to be from 28.5% to 11.9%; in fact, when patients were fed probiotics along with their dose of antibiotics, one fewer patient would develop AAD out of every seven. Antibiotics fed to the children in the trials included amoxicillin, penicillin and erythromycin. Evidence suggests that *S. boulardii* is best at preventing AAD in children, *Lactobacillus GG* ranking second, and a mixture of *S. thermophilus* and *B. lactis* relatively ranking last. No conclusion can be made on how probiotics can reduce the risk of AAD from any specific types of antibiotic.

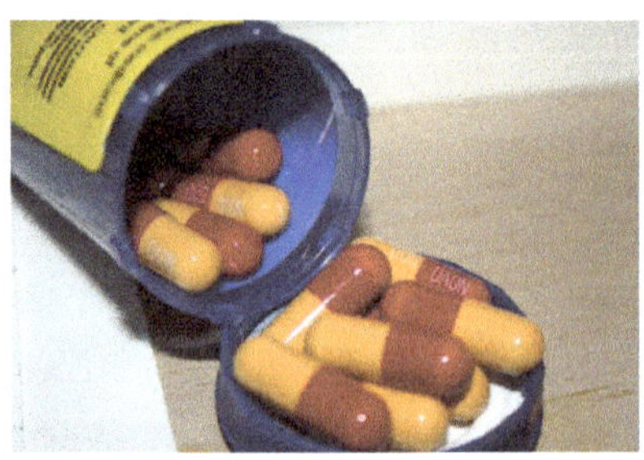

Novamoxin antibiotic

One of the proposed mechanisms of how probiotics protect from AAD is by regulating the composition of organisms in the intestines. Studies involving *L. acidophilus* and *Bifidobacterium* suggest that these microbes inhibit the growth of facultative anaerobic bacteria, which tends to increase during antibiotic treatment. As a result, it is seen that probiotics can alter and prevent changes in the intestinal microflora caused by antibiotics.

Viral Diarrhea

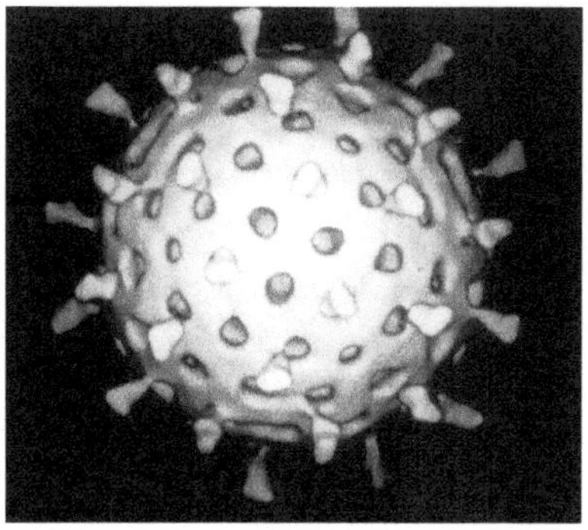

Rotavirus

Viral diarrhea refers to diarrhea caused by a rotavirus, a virus often affecting toddlers and infants. A meta-analysis of nine previously conducted studies has shown that probiotics are also effective in treating viral diarrhea in hospitalized children. Countries of where these studies originated from include Canada, Thailand, and Finland, with subjects ranging from one to above 36 months old. On average, the duration of the illness was shortened by 0.7 days when placing the children on *Lactobacillus* therapy. Although it is still largely unknown as to how probiotics do this, two mechanisms have been currently proposed. The first mechanism suggests that *Lactobacilli* augment the development of intestinal mucins (glyosylated proteins), which consequently protect the body from intestinal infections. For the second mechanism, scientists suggest that probiotics prevent the growth of urease-producing bacteria, which typically occurs in the second stage of rotavirus diarrhea.

Persistent Diarrhea

A 2013 review suggested probiotics are effective in treating persistent diarrhea in children, though more research is needed. Persistent diarrhea is an episode that starts acutely but then lasts for 14 days or more; In developing countries it is an important cause of morbidity and mortality in children under five years old. The study showed average duration reduced by four days and shorter hospital stay; stool frequency was reduced on the fifth day.

Allergy

Since the 1990s the incidence of allergic disease has increased in industrialized countries. Although current research is inconclusive, it suggests that probiotics may be beneficial in modulating host immune response and control allergic disease at an early age.Atopic infants have shown to have a lower count of *Bifidobacteria* when compared to non-atopic infants and are more susceptible to allergen sensitization.

Atopic eczema is a common allergic inflammation of skin among infants. One randomized double-blind placebo-controlled study was conducted in infants who manifested atopic eczema during exclusive breast-feeding and who had no exposure to any infant or substitute formula. These infants were designated to wean to either hydrolyzed whey formula supplemented with probiotic strains *Bifidobacterium lactis* or *Lactobacillus rhamnosus* GG, or receive the hydrolyzed formula alone. The purpose of the study was to investigate incidence and severity of allergic response, growth and nutrition of studied infants, as well as to evaluate whether previous documented immunomodulatory effects of probiotics provide clinical benefit. Results were shown to improve the severity of skin manifestations by decreasing allergic-type immune response (T-cell activation) while promoting the development of suppressive helper cells and immunologic tolerance to antigens. Therefore, chances for immune hyperactivity (i.e. allergic disease) can be reduced. Another study that measured inflammatory mediator levels in infants' stool showed that infants with atopic dermititis (one type of eczema that is highly associated with cow's milk protein) and who were fed with GG supplemented-hydrolyzed whey formula excreted smaller amounts of inflammatory mediators. This suggested that *Lactobacillus rhamnosus* GG is one of the probiotics that improves integrity of the intestinal mucosal barrier and decreases gut inflammation.

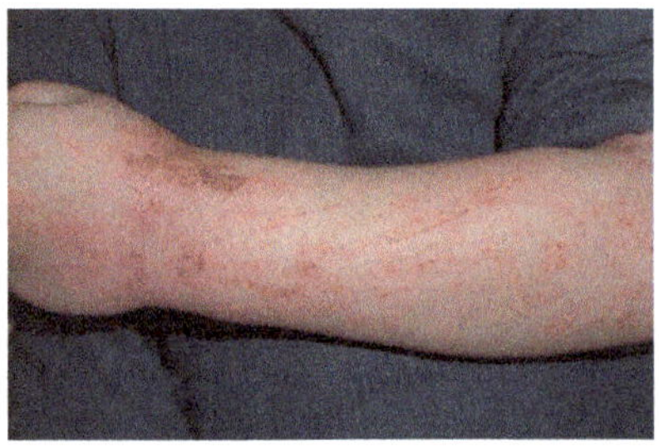

Atopic eczema

Numerous studies that used probiotics in different forms or sources—such as supplemented formula, supplemented water, and capsule in which contents were mixed with food—for treatment and management of atopy have associated with positive anti-inflammatory responses. These suggested that regular probiotic supplementation may stabilize intestinal mucosal barrier function and play a role in modulating allergic responses and reduce severity of atopic symptoms.

Necrotizing Enterocolitis

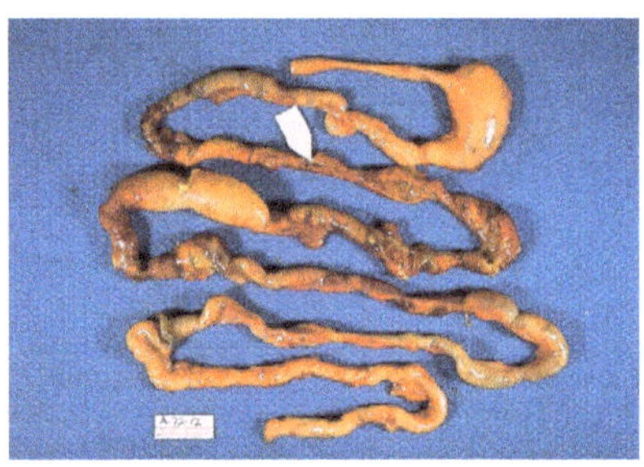

Neonatal necrotizing enterocolitis, gross pathology 20G0022 lores

Necrotizing enterocolitis (NEC) is a medical condition primarily seen in premature infants where a portion of bowel undergoes necrosis (tissue death). A variety of factors has shown negatively to affect premature infants in obtaining adequate or optimal colonization of gut microflora: lack of exposure to normal maternal flora and immature mucosa are two reasons. Thus microflora establishment and composition in premature infants is a major determinant in the pathophysiology of NEC.

Using probiotics to prevent NEC is considered one of the most exciting probiotics applications today. Data in this area of research have shown reduction of NEC risk. Probiotics are beneficial[*clarification needed*] in reducing enteric pathogens, improving gut structure and function, facilitating enteral nutrition, reducing dependence on parenteral nutrition, maintaining gut permeability, promoting intestinal mucosal barrier function, enhancing tolerance and reducing side-effects of antibiotic use. All of these lead to the clinical outcome of NEC prevention in infants. The mechanisms of action of probiotics are that colonization of favourable types of microflora is increased, while the amount of pathogens is reduced. Probiotics are also capable of modulating gut immune response and improving humoral immunity, and thus host response changes.

There have been many studies conducted on premature infants who weigh less than normal healthy infants and are fed with *Bifidobacteria* and/or *Lactobacillus* at an early stage of life. Results showed significant decrease in expression of NEC and reduced severity when NEC did occur. Other research conducted in more than 12,000 premature infants showed similar results and also showed reduction in NEC-associated mortality in the probiotic group. Clinical trials also demonstrated increased bifidobacterium count significantly improves gut flora balancing and is associated with lower intestinal permeability in preterm infants.Very low birth weight (VLBW) infants that are breast fed with probiotics supplemented have lower incidence of NEC when compared to VLBW infants that take breast milk alone.

Safety

It is reported that most of the lactic acid-producing bacteria (LAB) strains used in the food supply are nonpathogenic and nontoxigenic. They have been incorporated in food products for many years and are generally accepted as traditional food-grade organisms. Over 70 clinical studies to investigate potential probiotic-side effects have been conducted on more than 4000 children and infants that consume food containing microbial ingredients. None has shown any adverse effects. Other studies that focused on long-term intake of probiotic-supplemented formula in infants have also shown no toxicity effect and incorporation well tolerated. According to the 2002 report *Guidelines for the Evaluation of Probiotics in Food*, jointly published by the FAO and WHO, "documented correlations between systemetic infections and probiotic consumption are few and all occurred in patients with underlying medical conditions." Rarely, consumption of probiotics may cause bacteremia, fungemia and sepsis in children with lowered immune systems who are already critically ill. A 2013 systematic review found no adverse effects from *Lactobacillus* ingestion.

Dose and Canadian Regulations

Probiotics within Canadian regulations can be separated into two categories: probiotics in foods and in natural health products (NHP). For probiotics in foods, the bacterial cultures are considered as food ingredients and should be explicitly labeled on the product. Food products with microorganisms are regulated under the Food and Drugs Act, although there are no specific regulations for probiotics. In the case that the probiotic strain has been contained in the past in the food product with no dangers observed, Health Canada does not require a premarket safety review of the product. Probiotics that are sold as natural health products (e.g. in pharmaceutical form as in capsules) are regulated under the Natural Health Products Regulation.

Health Canada defines health claims to be "any representation in labeling or advertising that states, suggests, or implies that a relationship exists between consumption of a food or food constituent (including an ingredient in the food) and a person's health". The following is a part of the regulation for health claims pertaining to probiotics:

- (a) The term "probiotic(s)" and similar terms or representations should be accompanied by a statement of the demonstrated effect of the probiotic;

- (b) The claimed effect of the probiotic microorganism in a food should be clearly stated in a manner that is not false, misleading, deceptive, or likely to create an erroneous impression with respect to the effect of or benefit from the probiotic microorganism(s) in the food;
- (c) Where a health claim is made, the Latin name (i.e., genus and species) and the strain of the probiotic microorganism or mixed culture that is the subject of the claim should be declared;
- (d) The level of the probiotic strain expressed in colony forming units (cfu) in a serving of stated size of the food should be declared; and
- (e) If more than one probiotic strain is added to a food, the above recommendations apply to the mixed culture.

The probiotic products' manufacturer is responsible for having the labeled level of microorganism delivered to the consumer. Thus, they are responsible for maintaining the stability and viability of the bacterial culture throughout the product's shelf life. For example, when applying for a product to be an NHP, applicants must prove that the probiotic strain to be resistant to gastric and bile acid, safe for human use, and can benefit humans in one or more ways. Details on strains and quantity allowed for in bacterial cultures in NHP are outlined by the Natural Health Product Monograph.

Bifidobacterium Bifidum

Bifidobacterium bifidum is a bacterial species of the Bifidobacterium genus. B. bifidum is one of the most common probiotic bacteria that can be found in the body of mammals, including humans.

Structure and Characteristics

Since *B. bifidum* is a Gram-positive bacterium that is not motile, anaerobic, and not spore-forming. The bacterium is rod-shaped and can be found living in clusters, pairs, or even independently. The majority of the population of *B. bifidum* is found in the colon, lower small intestine, breast milk, and often in the vagina.

Benefits

The use of *B. bifidum* in probiotic applications may reduce the chances of acute diarrhea and the risk of *E. coli* infections, and contributes to the maintenance of vaginal homeostasis.

Health Concerns

The manipulation of the gut flora is complex and may cause bacteria-host interactions. Although probiotics, in general, are considered safe, there are concerns about their use in certain cases. Some people, such as those with compromised immune systems, short bowel syndrome, central venous catheters, heart valve disease and premature infants, may be at higher risk for adverse events. Rarely, consumption of probiotics may cause bacteremia, and sepsis, potentially fatal infections in children with lowered immune systems or who are already critically ill.

Transmission

B. bifidum can be found in breast milk. Breast feeding is one way to transmit the bacteria from mother to child. *B. bifidum* is found in the vagina; some studies show that vaginal births transmit more *B.bifidum* from mother to child than caesarean births. Transmission of *B. bifidium* allows a child to begin production of microflora which helps to colonize the child's intestines after birth.

Bifidobacterium Breve

Bifidobacterium breve is a bacterial species of the Bifidobacterium genus which has probiotic properties.

History and Clinical Uses

B. breve strains were originally isolated from the feces of human infants. In 1971, *Bifidobacterium parvulorum* and *Bifidobacterium breve* were merged under the name *Bifidobacterium breve.Bifidobacterim breve* administered in combination with prebiotics or other probiotics and standard therapy has showed some beneficial effect.*B. breve* is a constituent in the therapeutic, nutritional treatment VSL#3. This proprietary, standardized, formulation of live bacteria is used to treat ulcerative colitis and requires a prescription.

Characteristics

B. breve strains can fermentmannitol and sorbitol, but not arabinose or xylose.

Bacillus Coagulans

Bacillus coagulans is a lactic acid-forming bacterial species within the genus *Bacillus*. The organism was first isolated and described as *Bacillus coagulans* in 1915 by B.W. Hammer at the Iowa Agricultural Experiment Station as a cause of an outbreak of coagulation in evaporated milk packed by an Iowa condensary. Separately isolated in 1935 and described as *Lactobacillus sporogenes* in the fifth edition of *Bergey's Manual*, it exhibits characteristics typical of both genera *Lactobacillus* and *Bacillus*, its taxonomic position between the families Lactobacillaceae and Bacillaceae was often debated. However, in the seventh edition of Bergey's, it was finally transferred to the genus *Bacillus*. DNA-based technology was used in distinguishing between the two genera of bacteria which are morphologically similar and possess similar physiological and biochemical characteristics.

B. coagulans is a Gram-positive rod (0.9 by 3.0 to 5.0 μm in size), catalase positive, spore-forming, motile, and a facultative anaerobe. It may appear Gram-negative when entering the stationary phase of growth. The optimum temperature for growth is 50 °C (122 °F) ; range of temperatures tolerated are 30-55 °C (86-131 °F). IMViC tests VP and MR (methyl-red) tests are positive.

Uses

Bacillus coagulans has been added by the EFSA to their Qualified Presumption of Safety list and has been approved for veterinary purposes as GRAS by the U.S. Food and Drug Administration's Center for Veterinary Medicine, as well as by the European Union, and is listed by AAFCO for use as a direct-fed microbial in livestock production. It is often used in veterinary applications, especially as a probiotic in pigs, cattle, poultry, and shrimp. Many references to use of this bacterium in humans exist, especially in improving the vaginal flora, improving abdominal pain and bloating in irritable bowel syndrome patients, and increasing immune response to viral challenges. There is evidence from animal research that suggests that bacillus coagulans is effective in both treating as well as preventing recurrence of clostridium difficile associated diarrhea. One strain of this bacterium has also been assessed for safety as a food ingredient. Spores are activated in the acidic environment of the stomach and begin germinating and proliferating in the intestine. Sporeforming *B. coagulans* strains are used in some countries as probiotics for patients on antibiotics.

Marketing

B. coagulans is often marketed as *Lactobacillus sporogenes* or a 'sporeforming lactic acid bacterium' probiotic, but this is an outdated name due to taxonomic changes in 1939. Although *B. coagulans* does produce L+lactic acid, the bacterium used in these products is not a lactic-acid bacterium, as *Bacillus* species do not belong to the lactic acid bacteria. By definition, lactic acid bacteria (*Lactobacillus*, *Bifidobacterium*) do not form spores. Therefore, using the name *Lactobacillus sporogenes* is scientifically incorrect.

Bacillus Clausii

Bacillus clausii is a rod-shaped, Gram-positive, motile and spore-forming bacterium that lives in the soil. It is classified as probiotic; microrganisms that maintain a symbiotic relationship with the host organism. It is currently being studied in respiratory infections and some gastrointestinal disorders.*Bacillus clausii*, has been found to produce antimicrobial substances that are active against gram positive bacteria including *Staphylococcus aureus*, *Enterococcus faecium*, and *Clostridium difficile*.

Genome Structure

Bacillus clausii has relatively small genome that contains 4.30 Mbp with 4,108 protein coding genes.

Lactobacillus Fermentum

Lactobacillus fermentum is a Gram-positive species of bacterium in the genus *Lactobacillus*. It is associated with active dental caries lesions. It is also commonly found in fermenting animal and plant material. It has been found in sourdough. A few strains are considered probiotic or "friendly" bacteria in animals and at least one strain has been applied to treat urogenital infections in women. Some strains of lactobacilli formerly classified as *Lactobacillus fermentum* (such as RC-14) have since been reclassified as *Lactobacillus reuteri*. Commercialized strains of *L. fermentum* used as probiotics include PCC, ME-3 and CECT5716

Characteristics

Lactobacillus fermentum belongs to the genus *Lactobacillus*. Species in this genus are used for a wide variety of applications. These applications include food and feed fermentation. It has been found that some strains for *Lactobacillus fermentum* have natural resistances to certain antibiotics and chemotherapeutics. They are considered potential vectors of antibiotic resistance genes from the environment to humans or animals to humans.

Lactobacillus fermentum can also be a normal inhabitant of the human intestinal tract and some strains have been associated with cholesterol metabolism.

Probiotic

A microorganism is considered a probiotic by meeting certain characteristics, such as being of human origin, non-pathogenic, having high resistance to passing through the intestine, and being beneficial to the immune system. In general, they are seen as beneficial to the host's body and the human health. *Lactobacillus fermentum* has been identified as potential probiotic. The use of gut microbes as probiotics in food is aimed towards preventing and treating various health problems. Among these health problems allergies, neoplastic growth, and inflammatory bowel disease are included. Recent areas of study have focused on the influence of probiotics on metabolic functions of their host. One area has been the metabolism of cholesterol by LABs acting as probiotics. Research has shown that *Lactobacillus* species have been proven to remove cholesterol in vitro through various ways such as assimilation, binding to the surface cells, and incorporation into cellular membranes.

pH and Bile Tolerance

Testing of *Lactobacillus fermentum* against different pH concentration solutions revealed that it has a strong pH tolerance by its ability to grow and survive a few hours after being incubated in a 3-pH level solution. Strains of *Lactobacillus fermentum* have also been tested in different bileconcentrations and demonstrated to have good bile tolerance when incubated with 3 g L-1 of bile salt. The pH and bile tolerance that *L. fermentum* demonstrates is significant in terms of its consideration as a probiotic. It has to be strong enough to survive the stresses of the digestive system. The stomach has a pH between 1.5 and 3, and the upper intestine contains 3-5 g L-1 of bile. *Lactobacillus fermentum* has been found to survive in these conditions further supporting the idea that it can act as a probiotic.

Cholesterol Reduction

One of the ways in which *Lactobacillus fermentum* has been seen as a probiotic is by its ability to reduce cholesterol levels. Tests conducted using several strains of *Lactobacillus* and cholesterol broths demonstrated that *lactobacillus fermentum* had the largest removal of cholesterol. One of the mechanisms by which *L. fermentum* may remove cholesterol through in vivo is by the absorption of cholesterol, which as a result accelerates cholesterol metabolism. Another method is by the incorporation of cholesterol in the host body into its cell membrane or walls. This would also increase resistance of the bacterial cell membranes to environmental challenge. A third mechanism is by causing the body to consume more cholesterol. *L. fermentum* would interfere with the recycling of bile salt and facilitate its elimination, which as a result would increase the demand for bile salt made from cholesterol.

Lactobacillus Fermentum ME-3

The strain *Lactobacillus fermentum* ME-3 has recently been discovered and identified as an antimicrobial and antioxidative probiotic. This strain of *Lactobacillus fermentum* was discovered from the analysis of human fecal samples in 1994. One of the important characteristics of a probiotic microbe is the tolerance to conditions in the digestive tract. Tests conducted on the ME-3 strain in different bileconcentrations found that it was able to survive without large loss in numbers. It has also been found that *Lactobacillus fermentum* ME-3 has a tolerance to survive drops of pH levels. It can withstand a drop in values from 4.0 to 2.5 without decreasing in numbers. These characteristics of tolerance to bile concentrations and pH levels serve to classify ME-3 as a probiotic.

Lactobacillus fermentum ME-3 has also been found to have the capability to suppress mainly gram-negative bacteria. To a lesser extent, ME-3 has also been observed to be able to suppress *Enterococci* and *Staphylococcus aureus*. This would serve a beneficial purpose to the host. ME-3 has several antimicrobial characteristics. These include acetic, lactic and succinic acids, and putrescine. Research on the antioxidant properties of strain ME-3 in soft cheese products revealed that it prevented spoilage. Experimentation has also been conducted on the consumption of the ME-3 strain. The consumption had a positive influence on the microbiota of the gut. Volunteers were given goat milk fermented by strain ME-3 and capsulated ME-3. After three weeks analysis of fecal samples revealed that the ME-3 strain increased the number of beneficial *Lactobacilli* in comparison to those who were given non-fermented milk. Several human clinical studies performed on ME-3 focused on parameters related to cardiovascular disease development. Consumption of ME-3 indeed results in a reduction of oxidized LDL cholesterol, which is a major contributor to atherosclerosis development. Several mechanisms may contribute to the antioxidant effect of ME-3: the strain modulates the ratio of reduced glutathione/oxidized glutathione in the blood, and increases the levels of paraoxonase, an antioxidant enzyme which protects LDL particles from oxidative modifications.

Properties of the strain ME-3 can serve to classify it as a probiotic that has the ability to protect its host against food-derived infections and also help in the prevention of oxidative damage of food. Its multi-abilities have been tested and proven. Mice treated with a combination of ofloxacin and ME-3 revealed a reduction in liver and spleen granulomas of *Salmonella* thphimurium. ME-3 is commercialized in the USA, in Europe and in Asia in dietary supplement products for cardiovascular health, immune support or detoxification, under the brandname Reg'Activ.

Safety

In general, strains of *Lactobacillus* have been considered safe because of their association with food and because they are normal inhabitants of the human microflora. They have also been identified to have a low pathogenic potential further reinforcing the idea that they are safe microbes.

Recent research in regards to the safety of *Lactobacillus fermentum* has been carried out on mice. Mice were fed (intragastrically) different concentrations of *Lactobacillus fermentum* while a control group was also observed. After twenty-eight days blood samples were taken from the mice and analyzed. There was no health difference observed between the control mice and those fed *Lactobacillus fermentum* in terms of blood biochemistry, protein, albumin, glucose, and cholesterol. Also no negative side effects during the experiment such as change in body weight, feed intake, or clinical signs such as diarrhea and ruffled fur, were observed. The ingestion of *Lactobacillus fermentum* in mice appeared safe which led to further support that the use of *lactobacillus fermentum* in food is also safe.

Transferable Resistance Genes

One important consideration to determine the safety of *Lactobacillus fermentum* is transferable resistant genes. In order for *L. fermentum* to be considered as a potential probiotic, it must not contain any transferable resistant genes. If a resistance gene is transferable, it could lessen the effect of the use of antibiotics. Out of ten common antibiotic genes that were tested (gatamicin, cefazolin, penicillin, trimethoprim/sulfmethoxazole, ampicillin, carbenicillin, erythromycin, amikacin, chloramphenicol, and norfloxacin), *Lactobacillus fermentum* was found to only be resistant to amikacin and norfloxacin. Others studies have reported that most LABs are also resistant to these antibiotics, which led to the conclusion that it was a common characteristic of LABs. The resistance to these antibiotics can be considered natural or intrinsic. So far no observed *Lactobacillus fermentum* strains have been observed to have transferable resistance or acquired resistance genes

Dairy Products

Experiments conducted by introducing the strain ME-3 of *Lactobacillus fermentum* into dairy products as a probiotic ingredient revealed that it was able to suppress the reputed contaminants of food such as pathogenic *Salmonella* spp., *Shigella* spp., and urinary tract infections that are caused by *E. coli* and *Staphylococcus* spp. Also the introduction of *Lactobacillus fermentum* strains such as ME-3 in goat milk revealed that it was actually favorable to the host, resulting in an increase in number of beneficial *Lactobacilli.*

Heat Resistance

Although LABs have been associated with potential health advantages, they are also responsible for negative outcomes. They are the main organisms involved in the spoilage of tomato products. Species in the *Lactobacillus* genus have been identified to be the causative organisms. Research was carried out to observe the chemical constituents of tomato juice that stimulate the growth of bacteria that are responsible for the spoilage. These bacteria can resist high temperatures. A strain of *Lactobacillus fermentum* was extracted from a tomato juice concentrate. Meanwhile, eight different tomato juice mixtures were heated and the survival rate of *Lactobacillus fermentum* was measured. It was concluded that pectins are the main tomato juice constituents that protect the bacteria cells against destruction from heating. The breakdown of pectin from enzymic action would make the bacteria cells more susceptible to heat. However, it was found in previous research that heating had inactivated natural pectolytic enzymes and therefore *Lactobacillus fermentum* remained heat resistant. Heat resistance has also been found to correlate with the medium in which the bacteria are cultured, the better the medium used will result in a higher resistance to heat.

Antibiotic Resistance

Studies have shown that *L. fermentum* has antibiotic resistances. DNA was isolated from *Lactobacillus fermentum* and tested for antibiotic resistance against clinically important agents by using broth dilution tests. Different strains of *Lactobacillus fermentum* demonstrated uniform resistance patterns demonstrating resistance to glycopeptide vancomycin and to tetracycline.

Drug Resistance Plasmids

Research done on *Lactobacillus fermentum* strains has revealed the existence of tetracycline and erythromycin resistance plasmids.

Sensitivity to Antibiotics

While *Lactobacillus fermentum* has been found to have antibiotic resistant properties, other studies have demonstrated that lactobacillus fermentum is sensitive to some common antibiotics such as gentamicin, cefazolin, penicillin, trimethoprim/sulfamethoxazole, ampicillin, carbenicillin, erythromycin, amikacin, and cholorampehnicol.

Clostridium Butyricum

Clostridium butyricum is a strictly anaerobicendospore-forming Gram-positivebutyric acid producing bacillus subsisting by means of fermentation using an intracellularly accumulated amylopectin-like α-polyglucan (granulose) as a substrate. It is uncommonly reported as a human pathogen and is widely used as a probiotic in Asia (particularly in Japan, Korea and China).*C. butyricum* is a soil inhabitant in various parts of the world, has been cultured from the stool of healthy children and adults, and is common in soured milk and cheeses.

Therapeutic Uses

The first *C. butyricum* MIYAIRI strain was isolated from the feces of Dr. Chikaji Miyairi in Japan in 1933, and CBM 588 is the 588th MIYAIRI strain, isolated from a soil sample in Nagano, Japan, in 1963. Preparations based on CBM 588 have a long history of safe use in human populations in Asia, especially Japan, where such products are variously classed as pharmaceutical drugs, “quasi drugs”, and OTC (Over The Counter) probiotics. The safe history of CBM 588 in human Asian populations is supported by various peer-reviewed publications and case studies dating back to 1963, including reports of CBM 588 use in severely-ill, immune-compromised and hospitalized patients, whose ages range from infants to elderly people, and include pregnant women.

Its usefulness stems primarily from its ability to interfere with the growth of highly pathogenic *Clostridium difficile* by antagonizing its multiplication. It is often used in Japanese hospitals for *C. difficile* prophylaxis among in-patients and, particularly, during administration of certain powerful antibiotics (i.e.: Levofloxacin) associated with opportunistic *C. difficile* infection.

CBM 588 was approved for clinical use in humans by the Japanese Ministry of Health and Welfare in 1970. The standard preparation as marketed by Miyarisan Pharmaceutical (Tokyo, Japan) consists of white, marked tablets each containing 0.35×10^6 colony forming units (CFU) of *C. butyricum* MIYAIRI 588 (as active agent). CBM 588 does not establish permanently in the gut, in common with other orally administered probiotic bacteria. CBM 588 for clinical use is produced by submerged anaerobic fermentation followed by centrifugation, drying, blending and packaging.

The MIYAIRI 588 strain of *C. butyricum* does not carry any genes encoding any toxins and virulence factors associated with *Clostridium* or other enteropathogens. Absence of neurotoxin production has been demonstrated by polymerase chain reaction (PCR) and Southern blot hybridisation for type E botulinum toxin gene. The absence of genes encoding botulinum neurotoxin A,B,F and genes encoding non-toxic haemagglutinin (NTNH) and genes encoding *Clostridium perfringens* toxins (alpha, beta, epsilon and iota) has been demonstrated by PCR assay.

This strain is deposited at the Fermentation Research Institute, Agency of Industrial Science and Technology, Japan under the strain name *Clostridium butyricum* MIYAIRI 588 strain, deposit number FERM BP-2789. Recent European Food Safety Authority opinions confirm the official strain nomenclature as *Clostridium butyricum* FERM BP-2789.

Yeast in Winemaking

The process of fermentation at work on Pinot noir. As yeast consume the sugar in the must it releases alcohol and carbon dioxide (seen here as the foaming bubbles) as byproducts.

The role of yeast in winemaking is the most important element that distinguishes wine from grape-juice. In the absence of oxygen, yeast converts the sugars of wine grapes into alcohol and carbon dioxide through the process of fermentation. The more sugars in the grapes, the higher the potential alcohol level of the wine if the yeast are allowed to carry out fermentation to dryness. Sometimes winemakers will stop fermentation early in order to leave some residual sugars and sweetness in the wine such as with dessert wines. This can be achieved by dropping fermentation temperatures to the point where the yeast are inactive, sterile filtering the wine to remove the yeast or fortification with brandy to kill off the yeast cells. If fermentation is unintentionally stopped, such as when the yeasts become exhausted of available nutrients, and the wine has not yet reached dryness this is considered a stuck fermentation.

The most common yeast associated with winemaking is *Saccharomyces cerevisiae* which has been favored due to its predictable and vigorous fermentation capabilities, tolerance of relatively high levels of alcohol and sulfur dioxide as well as its ability to thrive in normal wine pH between 2.8 and 4. Despite its widespread use which often includes deliberate inoculation from cultured stock, *S.cerevisiae* is rarely the only yeast species involved in a fermentation. Grapes brought in from harvest are usually teeming with a variety of "wild yeast" from the *Kloeckera* and *Candida*genera. These yeasts often begin the fermentation process almost as soon as the grapes are picked when the weight of the clusters in the harvest bins begin to crush the grapes, releasing the sugar-rich must. While additions of sulfur dioxide (often added at the crusher) may limit some of the wild yeast activities, these yeasts will usually die out once the alcohol level reaches about 5% due to the toxicity of alcohol on the yeast cells physiology while the more alcohol tolerant *Saccharomyces* species take over. In addition to *S. cerevisiae*, Saccharomyces bayanus is a species of yeast that can tolerate alcohol levels of 17–20% and is often used in fortified wine production such as ports and varieties such as Zinfandel and Syrah harvested at high Brix sugar levels. Another common yeast involved in wine production is *Brettanomyces* whose presence in a wine may be viewed by different winemakers as either a wine fault or in limited quantities as an added note of complexity.

History

French scientist Louis Pasteur discovered the connection between microscopic yeast and the process of fermentation.

For most of the history of wine, winemakers did not know the mechanism that somehow converted sugary grape juice into alcoholic wine. They could observe the fermentation process which was often described as "boiling", "seething" or the wine being "troubled" due to release of carbon dioxide that gave the wine a frothy, bubbling appearance. This history is preserved in the etymology of the word "yeast" itself which essentially means "to boil".

In the mid-19th century, the French scientist Louis Pasteur was tasked by the French government to study what made some wines spoil. His work, which would later lead to Pasteur being considered one of the "Fathers of Microbiology", would uncover the connection between microscopic yeast cells and the process of the fermentation. It was Pasteur who discovered that yeast converted sugars in the must into alcohol and carbon dioxide, though the exact mechanisms of how the yeast would accomplish this task was not discovered till the 20th century with the Embden–Meyerhof–Parnas pathway.

The yeast species commonly known as *Saccharomyces cerevisiae* was first identified in late 19th century enology text as *Saccharomyces ellipsoideus* due to the elliptical (as opposed to circular) shape of the cells. Throughout the 20th century, more than 700 different strains of *Saccharomyces cerevisiae* were identified. The difference between the vast majority of these strains are mostly minor, though individual winemakers will develop a preference for particular strains when making certain wines or working with particular grape varieties. Some of these difference include the "vigor" or speed of fermentation, temperature tolerance, the production of volatile sulfur compounds (such as hydrogen sulfide) and other compounds that may influence the aroma of the wine.

Role in Winemaking

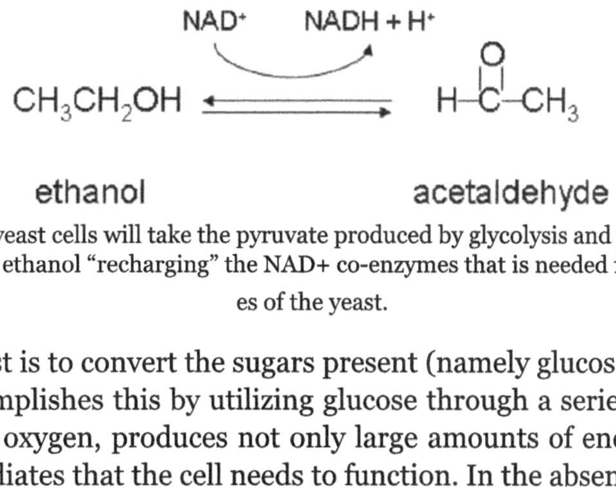

In the absence of oxygen, yeast cells will take the pyruvate produced by glycolysis and reduce it into acetaldehyde which is further reduced into ethanol "recharging" the NAD+ co-enzymes that is needed for various metabolic processes of the yeast.

The primary role of yeast is to convert the sugars present (namely glucose) in the grape must into alcohol. The yeast accomplishes this by utilizing glucose through a series of metabolic pathways that, in the presence of oxygen, produces not only large amounts of energy for the cell but also many different intermediates that the cell needs to function. In the absence of oxygen (and sometimes even in the presence oxygen), the cell will continue some metabolic functions (such as glycolysis) but will rely on other pathways such as reduction of acetaldehyde into ethanol (fermentation) to "recharge" the co-enzymes needed to keep metabolism going. It is through this process of fermentation that ethanol is released by the yeast cells as a waste product. Eventually, if the yeast cells are healthy and fermentation is allowed to run to the completion, all fermentable sugars will be used up by the yeast with only the unfermentable pentose leaving behind a negligible amount of residual sugar.

Other Compounds in Wine Produced by Yeast

While the production of alcohol is the most noteworthy by-product of yeast metabolism from a winemaking perspective, there are a number of other products that yeast produce that can be also influence the resulting wine. This includes glycerol which is produced when an intermediate of the glycolysis cycle (dihydroxyacetone) is reduced to "recharge" the NADH enzyme needed to continue other metabolic activities. This is usually produced early in the fermentation process before the mechanisms to reduce acetaldehyde into ethanol to recharge NADH becomes the cell's primary means of maintaining redox balance. As glycerol contributes increased body and a slightly sweet taste without increasing the alcohol level of the wine, some winemakers try to intentionally favor conditions that would promote glycerol production in wine. This includes selecting yeast strains that favor glycerol production (or allowing some wild yeast like *Kloeckera* and *Metschnikowia* to ferment), increased oxygen exposure and aeration as well as fermenting at higher temperatures. Glycerol production is also encouraged if most available acetaldehyde is made unavailable by binding with bisulfite molecules in the wine, but it would take a substantial amount of sulfur dioxide addition (far beyond legal limits) to prolong glycerol production beyond just these very nascent stages of fermentation.

If a Chardonnay has too much "buttery" diacetyl notes, winemakers may add fresh yeast to the wine to consume the diacetyl and reduce it to the more neutral-smelling fusel oil 2,3-Butanediol.

Other by-products of yeast include:

- Methanol – Caused by the demethylation of pectins in the must by enzymes of the yeast. More commonly found in red wines than white but only in very small amounts between 20–200 mg/L.

- Fusel oils – Formed by the decomposition of amino acids by the yeast. This includes 2,3-Butanediol which is formed by yeast-consuming diacetyl, the compound that gives Chardonnay and other wines a "buttery" aroma, reducing it first to acetoin and then to the more neutral-smelling 2,3-Butanediol. Many beer and winemakers who have a wine with too much "butteriness" will often "pitch" fresh yeast cultures into the no longer fermenting tank so that the yeast will consume the diacetyl and reduce the aroma.

- Succinic acid – Like glycerol, this is often formed early in fermentation. Usually found in concentrations of 500–1200 mg/L, it is a minor acid in the overall acidity of wine.

- Acetic acid – Considered a main component of volatile acidity that can make a wine taste unbalanced and overly acidic. While acetic acid is the main volatile acid produced by yeast, trace amounts of butyric, formic and propionic acids can also be formed depending on the yeast strain. Most countries have wine laws setting the legal limit of volatile acidity, usually expressed as acetic acid, to 1200–2000 mg/L. Acetic acid can also lead to the development of the wine fault ethyl acetate which is characterized by a "nail polish remover" smell. However, small amounts of acetic acid are actually beneficial for the yeast as they use them to synthesis lipids in the cell membrane.

The distinctive "aldehydic" notes of Sherry wines are caused by special yeast native to the Jerez wine region.

- Acetaldehyde – While most of the acetaldehyde produce gets reduced to ethanol or is bound by sulfur dioxide, concentrations between 50–100 mg/L can remain in the wine. The flor yeast strains that produce the Spanish wineSherry will produce higher amounts that contributes to the characterized "aldehydic" aromas of Sherries. In the presence of oxygen, yeast can convert some of the ethanol presence in the wine back into acetaldehyde creating oxidized aromas.

- Hydrogen sulfide – Often produced by yeast during fermentation because of a nitrogen deficiency in the must. This can be done by a reduction of sulfates or sulfites available in the must or by the decomposition of dead yeast cells by other yeast that releases sulfur-containing amino acids that are further broken down by the yeast. The latter often happens with wines that sit in contact with their lees for long periods of time between rackings. In the presence of alcohol, hydrogen sulfide can react with ethanol to form ethyl mercaptans and disulfides that contribute to off aromas and wine faults. Some commercial yeast strains, such as *Montrachet 522* are known to produce higher levels of hydrogen sulfides than other strains, particularly if the must has some nutrient deficiencies.

- Pyruvic acid – Along with acetaldehyde, this compound can react with anthocyanins extracted from contact with grape skins to create a more stable color pigment (pyranoanthocyanin) that can enhance the color of some red wines.

- Various esters, ketones, lactones, phenols and acetals.

Lees

When yeast cells die, they sink to the bottom of the fermentation vessel where they combine with insoluble tartrates, grape seeds, skin and pulp fragments to form the lees. During fermentation, the first significant racking which removes the bulk of dead yeast cells is often referred to as the *gross lees* as opposed to the less coarse *fine lees* that come as the wine continues to settle and age. During the time that the wine spends in contact with the lees, a number of changes can impact the wine due to both the autolysis (or self-metabolize) of the dead yeast cells as well as the reductive conditions that can develop if the lees are not aerated or stirred (a process that the French call *bâtonnage*). The length of time that a wine spends on its lees (called *sur lie*) will depend on the winemaking style and type of wine.

The lees leftover from the secondary fermentation of sparkling wine can be seen on the bottom side of this bottle being inspected. Eventually this wine will go through riddling to collect the lees in the neck where it will be removed prior to corking.

The process of leaving the wine to spend some contact with the lees has a long history in winemaking, being known to the Ancient Romans and described by Cato the Elder in the 2nd century BC. Today the practice is widely associated with any red wines that are barrel fermented, Muscadet, sparkling wine Champagne as well as Chardonnay produced in many wine regions across the globe. Typically when wines are left in contact with their lees, they are regularly stirred in order to release the mannoproteins, polysaccharides and other compounds that were present in the yeast cell walls and membranes. This stirring also helps avoid the development of reductive sulfur compounds like mercaptans and hydrogen sulfide that can appear if the lees layer is more than 10 cm (4 inches) thick and undisturbed for more than a week.

Most of the benefits associated with lees contact deals with the influence on the wine of the mannoproteins released during the autolysis of the yeast cells. Composed primarily of mannose and proteins, with some glucose, mannoproteins are often bound in the cell wall of yeast with hydrophobic aroma compounds that become volatilized as the cell wall breaks down. Not only does the release of mannoproteins impart sensory changes in the wine but they can contribute to tartrate and protein stability, help enhance the body and mouthfeel of the wine as well as decrease the perception of bitterness and astringency of tannins.

Secondary Fermentation

The production of Champagne and many sparkling wines requires a second fermentation to occur in the bottle in order to produce the carbonation necessary for the style. A small amount of sugared liquid is added to individual bottles, and the yeast is allowed to convert this to more alcohol and carbon dioxide. The lees are then ricked into the neck of the bottle, frozen, and expelled via pressure of the carbonated wine.

Types of Yeasts Used in Winemaking

Yeast taxonomy includes classification of yeast species depending on the presence or absence of a sexual phase. Therefore, some winemaking yeasts are classified by their asexual anamorph (or "imperfect" form) while others may be classified by their sexual teleomorph (or "perfect" form).

A common example of this is *Brettanomyces* (or "Brett") that is usually referenced in wine and viticulture text under its asexual classification though some scientific and winemaking texts may describe specific species (such as *Dekkera bruxellensis*) under its sporulating sexual classification of *Dekkera*. Unless otherwise noted, this article will commonly refer to the asexual form of wine yeast.

Film yeast on the surface of wine in a barrel of Vin jaune from the Jura wine region of France.

The most common yeast generally associated with winemaking is *Saccharomyces cerevisiae* which is also used in bread making and brewing. Other genera of yeast that can be involved in winemaking (either beneficially or as the cause of potential wine faults) include:

- *Brettanomyces* (Teleomorph *Dekkera*)
- *Candida* (Teleomorphs for different species from several genera including *Pichia*, *Metschnikowia*, *Issatchenkia*, *Torulaspora* and *Kluyveromyces*)
- *Kloeckera* (Teleomorph *Hanseniaspora*), usually the most common "wild yeast" found in the vineyard. Some species are known "killer yeast" that produce inhibitory levels of ethyl acetate and acetic acid that can kill off sensitive strains of *Saccharomyces cerevisiae*
- *Saccharomycodes*
- *Schizosaccharomyces*, the only wine yeast that reproduced by fission whereas most wine yeast reproduce by budding.
- *Zygosaccharomyces*, very alcohol-tolerant and can grow in wines up to 18% v/v. Additionally this yeast can survive in extremely high sugar levels (as much as 60% w/w or 60 Brix) and is very resistant to sulfur dioxide.
- *Aureobasidium*, particularly the "black yeast" species of *Aureobasidium pullulans* found in moist cellars that can contaminate aging wine in barrels.

Saccharomyces

The *Saccharomyces* (sugar mold) yeast genus is favored for winemaking (for both grapes as well as other fruit wines in addition also to being used in brewing and breadmaking) because of the generally reliable and positive attributes it can bring to the wine. These yeasts will usually readily fermentglucose, sucrose and raffinose and metabolize glucose, sucrose, raffinose, maltose and ethanol. However, *Saccharomyces* cannot ferment or utilize pentoses (such as arabinose) which is usually present in small amount in wines as residual sugars.

In addition to *Saccharomyces cerevisiae*, other species within the *Saccharomyces* genus that are involved with winemaking include:

- *Saccharomyces bayanus*
- *Saccharomyces beticus*
- *Saccharomyces fermentati*
- *Saccharomyces paradoxus*
- *Saccharomyces pastorianus*
- *Saccharomyces uvarum*

Influences of Different Strains on Fermentation

While some strains of yeast may influence the sensory characteristics and aromas of young wine, these differences seem to fade as the wine ages.

In 1996, *Saccharomyces cerevisiae* was the first single-celled, eukaryotic organism to have its entire genomesequenced. This sequencing helped confirm the nearly century of work by mycologists and enologists in identifying different strains of *Saccharomyces cerevisiae* that are used in beer, bread and winemaking. Today there are several hundred different strains of *S. cerevisiae* identified. Not all of the strains are suitable for winemaking and even among the strains that are, there is debate among winemakers and scientists about the actual magnitude of differences between the various strains and their potential impact on the wine. Even among strains that have demonstrated distinctive difference when compared among young wines, these differences seem to fade and become less distinctive as the wines age.

Some distinct difference among various strains include the production of certain "off-flavor" and aromas that may be temporary (but producing a "stinky fermentation") or could stay with the wine and either have to be dealt with through other winemaking means (such as the presence of volatile sulfur compounds like hydrogen sulfide) or leave a faulty wine. Another difference includes the "vigor" or speed of fermentation (which can also be influenced by other factors beyond yeast selection) with some yeast strains having the tendency to do "fast ferments" while others may take longer to get going.

Another less measurable difference that are subject to more debate and questions of winemakers preference is the influence of strain selection on the varietal flavors of certainly grape varieties such as Sauvignon blanc and Sémillon. It is believed that these wines can be influenced by thiols produced by the hydrolysis of certain cysteine-linked compounds by enzymes that are more prevalent in particular strains. Other aromatic varieties such as Gewürztraminer, Riesling and Muscat may also be influenced by yeast strains containing high levels of glycosidases enzymes that can modify monoterpenes. Similarly, though potentially to a much smaller extent, other varieties could be influenced by hydrolytic enzymes working on aliphatics, norisoprenoids, and benzene derivatives such as polyphenols in the must.

In sparkling wine production some winemakers select strains (such as one known as *Épernay* named after the town in the Champagne wine region of France and *California Champagne*, also known as UC-Davis strain 505) that are known to flocculate well, allowing the dead yeast cells to be removed easily by riddling and disgorgement. In Sherry production, the surface film of yeast known as flor used to make the distinctive style of fino and manzanilla sherries comes from different strains of *Saccharomyces cerevisiae*, though the commercial flor yeast available for inoculation is often from different species of *Saccharomyces*, *Saccharomyces beticus*, *Saccharomyces fermentati* and *Saccharomyces bayanus*.

Wild Yeasts and Natural Fermentation

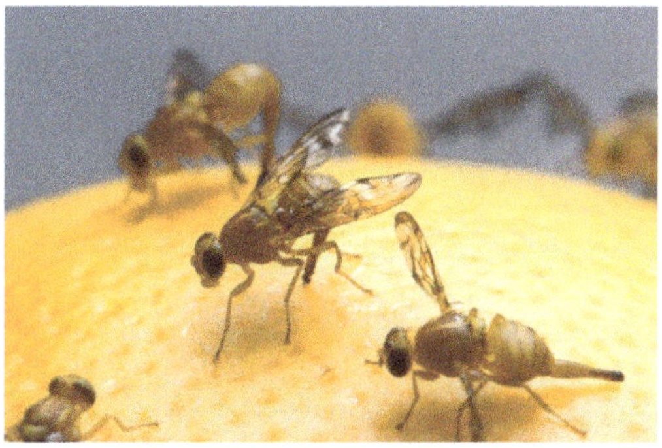

Fruit flies are a common vector that transports ambient or "wild" yeast strains within wineries.

In winemaking, the term "wild yeast" has multiple meanings. In its most basic context, it refers to yeast that has not been introduced to the must by intentional inoculation of a cultured strain. Instead, these "wild yeasts" often come into contact with the must through their presence on harvest equipment, transport bins, the surface winemaking equipment and as part of the natural flora of a winery. Very often these are strains of *Saccharomyces cerevisiae* that have taken residence in these places over the years, sometimes being previously introduced by inoculation of prior vintages. In this context, these wild yeasts are often referred to as ambient, indigenous or natural yeast as opposed to inoculated, selected or cultured yeast. Wineries that often solely rely on these "in-house" strains will sometimes market their wines as being the product of wild or *natural fermentations*. The (c. 304) *Nanfang Caomu Zhuang* has the earliest description of winemaking using "herb ferment" wild yeast with rice and various herbs, including the poisonous *Gelsemium elegans*.

Another use of the term "wild yeast" refers to the non-*Saccharomyces* genera of yeasts that are present in the vineyard, on the surface of grapevines and of the grapes themselves. Anywhere from 160 to 100,000 colony forming units of wild yeasts per berry could exist in a typical vineyard. These yeasts can be carried by air currents, birds and insects through the vineyard and even into the winery (such as by fruit flies). The most common wild yeasts found in the vineyard are from the genera *Kloeckera*, *Candida* and *Pichia* with the species *Kloeckera apiculata* being the most dominant species by far.*Saccharomyces cerevisiae*, itself, is actually quite rarely found in the vineyard or on the surface freshly harvested wine grapes unless the winery frequently reintroduced winery waste (such as lees and pomace) into the vineyard.

Unlike the "ambient" *Saccharomyces* wild yeast, these genera of wild yeasts have very low tolerance to both alcohol and sulfur dioxide. They are capable of starting a fermentation and often begin this process as early as the harvest bin when clusters of grapes get slightly crushed under their own weight. Some winemakers will try to "knock out" these yeasts with doses of sulfur dioxide, most often at the crusher before the grapes are pressed or allowed to macerate with skin contact. Other winemakers may allow the wild yeasts to continue fermenting until they succumb to the toxicity of the alcohol they produce which is often between 3–5% alcohol by volume and then letting either inoculated or "ambient" *Saccharomyces* strains finish the fermentation.

Wineries that wish to cultivate an "in-house" ambient yeast strain will often recycle the leftover pomace of previous vintages as compost in the vineyard.

The use of both "ambient" and non-*Saccharomyces* wild yeasts carries both potential benefits and risk. Some winemakers feel that the use of resident/indigenous yeast helps contribute to the unique expression of *terroir* in the wine. In wine regions such as Bordeaux, classified and highly regarded estates will often tout the quality of their resident "chateau" strains. To this extent, wineries will often take the leftover pomace and lees from winemaking and return them to the vineyard to be used as compost in order to encourage the sustained presence of favorable strains. But compared to inoculated yeast, these ambient yeasts hold the risk of having a more unpredictable fermentation. Not only could this unpredictability include the presence of off-flavors/aromas and higher volatile acidity but also the potential for a stuck fermentation if the indigenous yeast strains are not vigorous enough to fully convert all the sugars.

It is virtually inevitable that non-*Saccharomyces* wild yeast will have a role in beginning the fermentation of virtually every wine but for the wineries that choose to allow these yeasts to continue fermenting versus minimizing their influence do so with the intent of enhancing complexity through bio-diversity. While these non-*Saccharomyces* ferment glucose and fructose into alcohol, they also have the potential to create other intermediates that could influence the aroma and flavor profile of the wine. Some of these intermediates could be positive, such as phenylethanol, which can impart a rose-like aroma. However, as with ambient yeasts, the products of these yeasts can be very unpredictable – especially in terms of the types of flavors and aromas that these yeasts can produce.

Inoculated Yeast

When winemakers select a cultured yeast strain, it is largely done because the winemaker wants a predictable fermentation taken to completion by a strain that has a track record of dependability. Among the particular considerations that are often important to winemakers is a yeast's tendency to:

- Quickly begin fermentation, out-competing other "wild yeasts" for nutrients in the must
- Completely utilize all fermentable sugars with a predictable sugar-to-alcohol conversion rate
- Have an alcohol-tolerance up to 15% or even higher depending on the winemaking style
- Have a high sulfur dioxide tolerance but low production of sulfur compounds such as hydrogen sulfide or dimethyl sulfide
- Produce a minimum amount of residual pyruvate, acetic acid and acetaldehyde
- Produce minimum foaming during fermentation which may create difficulties for cap management during maceration or cause bungs to pop out during barrel fermentation.
- Have high levels of flocculation and lees compaction that makes racking, fining and filtering of the wine easier.

Inoculated (or *pure cultured*) yeasts are strains of *Saccharomyces cerevisiae* that have been identified and plated from wineries across the world (including notable producers from well-known wine regions such as Bordeaux, Burgundy, Napa Valley and the Barossa Valley). These strains are tested in laboratories to determine a strain's vigor, sulfur dioxide and alcohol tolerance, production levels of acetic acid and sulfur compounds, ability to re-ferment (positive for sparkling wine but a negative attribute for sweet late-harvest wines), development of surface film on the wine (positive for some Sherry styles but a negative attribute for many other wines), enhancement of a wine's color or certain varietal characteristics by enzymes in the yeast cells and other metabolic products produced by the yeast, foaming and flocculation tendencies, yeasticidal properties (a trait known as "Killer yeast") and tolerance for nutritional deficiencies in a must that may lead to a stuck fermentation.

Re-hydrating Freeze Dried Yeast Cultures

Preparing a yeast starter culture and gradually cooling the culture down to the must temperature by adding some wine.

Pure culture yeasts that are grown in a lab are often freeze dried and packaged for commercial use. Prior to their addition into must, these yeasts need to be re-hydrated in "starter cultures" that must be carefully monitored (particularly in regards to temperature) to insure that the yeast cells are not killed off by cold shock. Ideally winemakers want to add enough inoculum to have a viable cell population density of 5 million cells per milliliter. The exact amount of freeze-dried culture varies by manufacturer and strain of yeast but it is often around 1 gram per gallon (or 25 grams per 100 liters). Wines that could have potentially problematic fermentation (such as high sugar level late harvest or botryized wines) may have more yeast added.

Similarly, re-hydration procedures will also vary depending on the manufacturer and winery. Yeast is often inoculated in a volume of water or grape must that is 5–10 times the weight of the dry yeast. This liquid is often brought to temperature of 40 °C (104 °F) prior to the introduction of the yeast (though some yeast strains may need temperatures below 38 °C (100 °F)) to allow the cells to disperse easily rather than clump and sink to the bottom of the container. The heat activation also allows the cells to quickly reestablish their membrane barrier before soluble cytoplasmic components escape the cell. Re-hydration at lower temperatures can greatly reduce the viability of the yeast with up to 60% cell death if the yeast is re-hydrated at 15 °C (60 °F). The culture is then stirred and aerated to incorporate oxygen into the culture which the yeast uses in the synthesis of needed survival factors.

The temperature of the starter culture is then slowly reduced, often by the graduated addition of must to get within 5–10 °C (41–50 °F) of the must that the culture will be added to. This is done to avoid the sudden cold shock that the yeast cells may experience if the starter culture was added directly to the must itself which can kill up 60% of the culture. Additionally, surviving cells exposed to cold shock tend to see an increase in hydrogen sulfide production.

Nutritional Needs of Wine Yeast

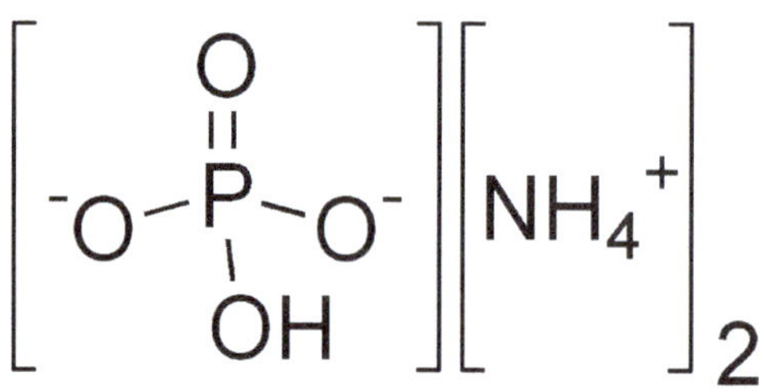

Diammonium phosphate (or DAP) is a common additive that provides two necessary nutrients for yeast to have a healthy and sustained fermentation—nitrogen and phosphate.

In order to successfully complete a fermentation with minimum to no negative attributes being added to the wine, yeast needs to have the full assortment of its nutritional needs met. These include not only an available energy source (carbon in the form of sugars such as glucose) and yeast assimilable nitrogen (ammonia and amino acids or YAN) but also minerals (such as magnesium) and vitamins (such as thiamin and riboflavin) that serve as important growth and survival factors. Among the other nutritional needs of wine yeast:

- Phosphate – used for the production of nucleic acids, phospholipids (an important component of the cell membrane) and ATP (Adenosine triphosphate which the cell uses for transferring energy for metabolism).

- Potassium – important for the uptake and utilization of phosphate
- Biotin – involved in the synthesis of proteins, fatty acids and nucleic acids.
- Pantothenic acid – involved in the metabolism of sugars and lipids. A deficiency of this vitamin could lead into increase hydrogen sulfide production with off-aromas in the resulting wine.
- Nicotinic acid – involved in the synthesis of Nicotinamide adenine dinucleotide (NAD+), a co-enzyme that is important in maintaining the redox balance of the cell as well as in the process of ethanol fermentation itself.
- Inositol – involved with the secondary messenger molecules that facilitate cell division.
- Trace amounts of calcium, chlorine, copper, iron, manganese and zinc for healthy cell function.

Many of these nutrients are available in the must and skins of the grapes themselves but sometimes are supplemented by winemakers with additions such as diammonium phosphate (DAP), freeze-dried micro-nutrients (such as *Go-Ferm* and *Ferm-K*) and even the remnant of dead or extracted yeast cells such that the fermenting yeast can break down to mine for available nitrogen and nutrients. One historical winemaking tradition that is still practiced in some Italian wine regions is the *ripasso* method of adding the leftover pomace from the pressing of other wines into a newly fermenting batch of wine as an additional food source for the yeast.

One traditional way of providing nutrients for the yeast is the *ripasso* method where the leftover grape skins and pomace *(pictured)* from a previous fermentation is added to a newly fermenting wine.

Saccharomyces cerevisiae can assimilate nitrogen from both inorganic (ammonia and ammonium) and organic forms (amino acids, particularly arginine). As yeast cells die, enzymes within the cells begin autolyzing by breaking down the cell, including the amino acids. This autolysis of the cell provides an available nitrogen source for the still-fermenting and viable yeast cells. However, this autolysis can also release sulfur-link compounds (such as the breakdown of amino acid cysteine) which can combine with other molecules and react with alcohol to create volatile thiols that can contribute to a "stinky fermentation" or later development into various wine faults.

The Role of Oxygen

Yeasts are facultative anaerobes meaning that they can exist in both the presence and absence of oxygen. While fermentation is traditionally thought of as an anaerobic process done in the absence of oxygen, early exposure of the yeast to oxygen can be a vital component in the successful completion of that fermentation. This is because oxygen is important in the synthesis of cell "survival factors" such as ergosterol and lanosterol. These sterols are important in maintaining the selective permeability of the yeast cell membrane which becomes critical as the yeast becomes exposed to increasing osmotic pressure and levels of alcohol in the wine. As a waste product of its own metabolism, alcohol is actually very toxic to yeast cells. Yeast with weak survival factors and lacking sterols may succumb to these conditions before fermenting a wine to complete dryness, leaving a stuck fermentation.

Cultured yeasts that are freeze-dried and available for inoculation of wine must are deliberately grown in commercial labs in high oxygen/low sugar conditions that favor the development of these survival factors. One of the reasons that some winemakers prefer using inoculated yeast is the predictability of fermentation due to the high level of survival factors that cultured yeast are assured of having without necessarily needing to expose the wine to additional levels of oxygen. Winemakers using "ambient" yeasts that are resident in their winery may not have this same assurance of survival factors and may need to compensate with other winemaking techniques.

Wild non-*Saccharomyces* yeasts often need a much greater exposure to oxygen in order to build up survival factors which is why many of these yeasts are often found living oxidatively as "film yeast" on the surface of wines in tanks or barrels.

Wine Faults Related to Yeast

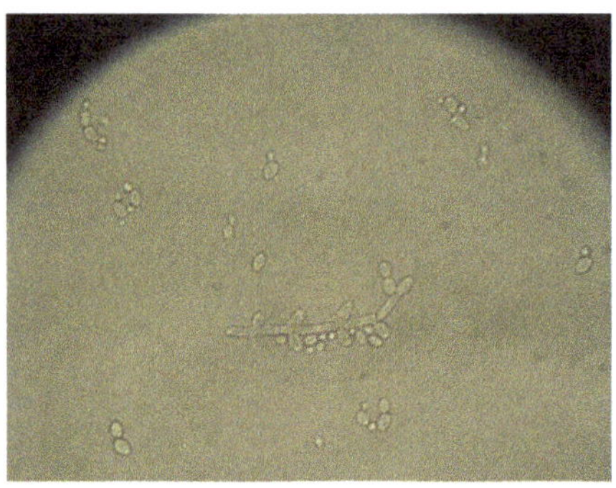

Film yeast like *Candida* (*pictured*) and *Pichia* can cover the surface of a wine with a film layer that not only consumed most of the free sulfur dioxide available to protect the wine but also produces high levels of acetic acid that will contribute to volatile acidity in a wine.

Either directly or indirectly, wine yeast can be a culprit behind a wide variety of wine faults. These can include the presence of "off flavors" and aromas that can be the by-product of some "wild yeast" fermentation such as those by species within the genera of *Kloeckera* and *Candida*. Even the common wine yeast *Saccharomyces cerevisiae* can be behind some wine faults with some strains of the yeast known to produce higher than ideal levels of acetic acid, acetaldehyde and volatile sulfur compounds such as thiols. Also any yeast can have a low tolerance to nutritional deficiencies, temperature fluctuation or extremes and excessive or low sugar levels that may lead to a stuck fermentation.

In the presence of oxygen several species of *Candida* and *Pichia* can create a film surface on top of the wine in the tank of barrel. Allowed to go unchecked, these yeasts can rapidly deplete the available free sulfur compounds that keeps a wine protected from oxidation and other microbial attack. The presence of these yeasts is often identified by elevated levels of volatile acidity, particularly acetic acid. Some strains of *Pichia* will metabolize acetic acid (as well as ethyl acetate and isoamyl acetate that may also be produced) with the side-effect of substantially decreasing the titratable acidity and shifting the pH of wine upwards to levels that make the wine prone to attack by other spoilage microbes. Commonly called "film yeast", these yeasts are distinguished from the flor sherry yeast that are usually welcomed by winemakers in producing the delicate fino-style wines.

Growth of many unfavorable wild yeasts is generally slowed at lower cellar temperatures, so many winemakers who wish to inhibit the activities of these yeasts before the more favorable *Saccharomyces* yeast kick in, will often chill their must, such as the practice of "cold soaking" the must during a pre-fermentation maceration at temperatures between 4–15 °C (39–50 °F). Though some species, such as *Brettanomyces*, will not be inhibited and may even thrive during an extended period of cold soaking.

Brettanomyces

The wine yeast *Brettanomyces* (or "Brett") produces very distinctive aroma compounds, 4-Ethylphenol (4-EP) and 4-Ethylguaiacol (4-EG), that can have a wine being described as smelling like a "barnyard", "wet saddle" or "band-aid". To some winemakers and with some wine styles (such as Pinot noir from Burgundy), a limited amount of these compounds could be considered a positive attribute that adds to the complexity of wine. To other winemakers and with other wine styles (such as Riesling from the Mosel), the presence of any Brett will be considered a fault.Fruit flies are common vector in the transfer of *Brettanomyces* between tanks and even nearby wineries.

While some wine regions view the influence of Brettanomyces on the wine, in limited amounts, as added complexity, many winemakers view the presence of Brettanomyces species such as Brettanomyces bruxellensis (pictured) in their wineries as a negative influence that needs to be controlled.

As a fermentation yeast, *Brettanomyces* can usually ferment a wine up to 10-11% alcohol levels before they die out. Sometimes *Brettanomyces* already present in a wine that has been inoculated with *Saccharomyces cerevisiae* will out compete the *Saccharomyces* strain for nutrients and even inhibit it due to the high levels of acetic acid, decanoic acid and octanoic acid that many strains of *Brettanomyces* can produce.

Once Brett is in a winery, it is very difficult to control even with strict hygiene and the discarding of barrels and equipment that has previously come into contact with "Bretty" wine. This is because many species of *Brettanomyces* can use a wide variety of carbon sources in wine and grape must, including ethanol, for metabolism. Additionally, Brett can produce a wide range of by-products that could influence the wine beyond just the 4-EP and 4-EG compounds previously discussed. Many of these compounds, such as the "footprints" of the 4-EP and 4-EG, will still remain in the wine even after yeast cells die and are removed by racking and sterile filtration.

Baker's Yeast

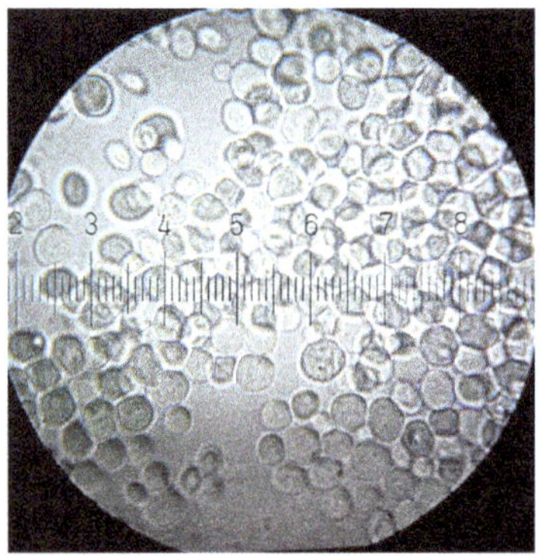

Saccharomyces cerevisiae, the yeast commonly used as baker's yeast. Gradation marks are 1 μm apart

Baker's yeast is the common name for the strains of yeast commonly used as a leavening agent in bakingbread and bakery products, where it converts the fermentablesugars present in the dough into carbon dioxide and ethanol. Baker's yeast is of the species *Saccharomyces cerevisiae*, which is the same species (but a different strain) commonly used in alcoholic fermentation, which is called brewer's yeast. Baker's yeast is also a single-cell microorganism found on and around the human body.

The use of steamed or boiled potatoes, water from potato boiling, or sugar in a bread dough provides food for the growth of yeasts; however, too much sugar will dehydrate them. Yeast growth is inhibited by both salt and sugar, but more so with salt than sugar.Fats, such as butter or eggs, slow down yeast growth; however, others say the effect of fat on dough remains unclear, presenting evidence that small amounts of fat are beneficial for baked bread volume.

Saccharomyces exiguus (also known as *S. minor*) is a wild yeast found on plants, fruits, and grains that is occasionally used for baking; however, in general it is not used in a pure form but comes from being propagated in a sourdough starter.

History

A block of compressed fresh yeast in its wrapper

It is not known when yeast was first used to bake bread; the earliest definite records come from Ancient Egypt. Researchers speculate that a mixture of flour meal and water was left longer than usual on a warm day and the yeasts that occur in natural contaminants of the flour caused it to ferment before baking. The resulting bread would have been lighter and tastier than the previous hard flatbreads. It is generally assumed that the earliest forms of leavening were likely very similar to modern sourdough; the leavening action of yeast would have been discovered from its action on flatbread doughs, and would have been either cultivated separately or transferred from batch to batch by means of previously mixed ("old") dough. Also, the development of leavened bread seems to have developed in close proximity to the development of beer brewing, and barm from the beer fermentation process can also be used in bread making.

Without an understanding of microbiology, early bakers would have had little ability to directly control yeast cultures, but still kept locally interesting cultures by reusing doughs and starters to leaven later batches. However, it became possible to isolate and propagate favored yeast strains in the same manner as was done in the beer industry, and it eventually became practical to propagate yeast in a slurry with a composition similar to beer wort, usually including malted barley and wheat flour. Such cultures (sometimes referred to in old American cookery as "emptins", from their origins as the dregs of beer or cider fermentation) would become the ancestors of modern baker's yeast, as, in general, they were carefully maintained to avoid what would later be discovered to be bacterial contamination, including using preservatives such as hops as well as boiling the growth medium.

In the 19th century, bread bakers obtained their yeast from beer brewers, and this led to sweet-fermented breads such as the Imperial "Kaiser-Semmel" roll, which in general lacked the sourness created by the acidification typical of *Lactobacillus*. However, beer brewers slowly switched from top-fermenting to bottom-fermenting yeast (both S. cerevisiae) and this created a shortage of yeast for making bread, so the Vienna Process was developed in 1846. While the innovation is often popularly credited for using steam in baking ovens, leading to a different crust characteristic, it is notable for including procedures for high milling of grains, cracking them incrementally instead of mashing them with one pass; as well as better processes for growing and harvesting top-fermenting yeasts, known as press-yeast.

Refinements in microbiology following the work of Louis Pasteur led to more advanced methods of culturing pure strains. In 1879, Great Britain introduced specialized growing vats for the production of *S. cerevisiae*, and in the United States around the turn of the century centrifuges were used for concentrating the yeast, making modern commercial yeast possible, and turning yeast production into a major industrial endeavor. The slurry yeast made by small bakers and grocery shops became cream yeast, a suspension of live yeast cells in growth medium, and then compressed yeast, the fresh cake yeast that became the standard leaven for bread bakers in much of the Westernized world during the early 20th century.

During World War II, Fleischmann's developed a granulated active dry yeast for the United States armed forces, which did not require refrigeration and had a longer shelf-life and better temperature tolerance than fresh yeast; it is still the standard yeast for US military recipes. The company created yeast that would rise twice as fast, cutting down on baking time. Lesaffre would later create instant yeast in the 1970s, which has gained considerable use and market share at the expense of both fresh and dry yeast in their various applications.

Types of Baker's Yeast

Active dried yeast, a granulated form in which yeast is commercially sold

Baker's yeast is available in a number of different forms, the main differences being the moisture contents. Though each version has certain advantages over the others, the choice of which form to use is largely a question of the requirements of the recipe at hand and the training of the cook preparing it. Dry yeast forms are good choices for longer-term storage, often lasting more than a year at room temperatures without significant loss of viability. In general, with occasional allowances for liquid content and temperature, the different forms of commercial yeast are considered interchangeable.

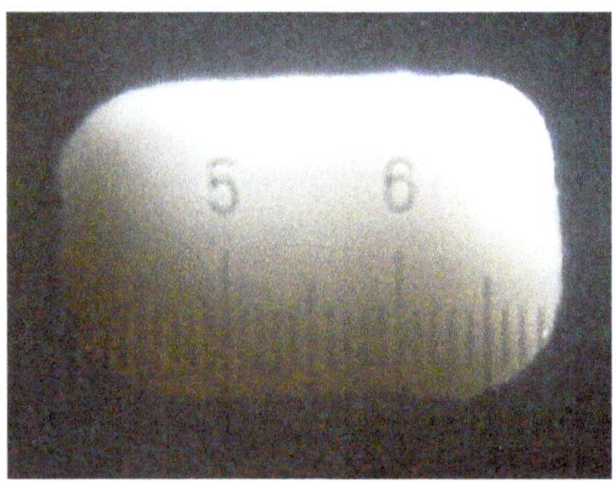

A single grain of active dry yeast. The numbered ticks on the scale are 230 μm apart

- Cream yeast is the closest form to the yeast slurries of the 19th century, in essence being a suspension of yeast cells in liquid, siphoned off from the growth medium. Its primary use is in industrial bakeries with special high-volume dispensing and mixing equipment, and it is not readily available to small bakeries or home cooks.
- Compressed yeast is, in essence, cream yeast with most of the liquid removed. It is a soft solid, beige in color, and best known in the consumer form as small, foil-wrapped cubes of cake yeast. It is also available in larger-block form for bulk usage. It is highly perishable; though formerly widely available for the consumer market, it has become less common in supermarkets in some countries due to its poor keeping properties, having been superseded in some such markets by active dry and instant yeast. It is still widely available for commercial use, and is somewhat more tolerant of low temperatures than other forms of commercial yeast; however, even there, instant yeast has made significant market inroads.
- Active dry yeast is the form of yeast most commonly available to non-commercial bakers in the United States. It consists of coarse oblong granules of yeast, with live yeast cells

encapsulated in a thick jacket of dry, dead cells with some growth medium. Under most conditions, active dry yeast must first be proofed or rehydrated. It can be stored at room temperature for a year, or frozen for more than a decade, which means that it has better keeping qualities than other forms, but it is generally considered more sensitive than other forms to thermal shock when actually used in recipes.

- Instant yeast appears similar to active dry yeast, but has smaller granules with substantially higher percentages of live cells per comparable unit volumes. It is more perishable than active dry yeast but also does not require rehydration, and can usually be added directly to all but the driest doughs. In general, instant yeast has a small amount of ascorbic acid added as a preservative. Some producers provide two or more forms of instant yeast in their product portfolio; for example, LeSaffre's "SAF Instant Gold" is designed specifically for doughs with high sugar contents, and such yeasts are more generally known as osmotolerant yeasts.
- Rapid-rise yeast is a variety of dried yeast (usually a form of instant yeast) that is of a smaller granular size, thus it dissolves faster in dough, and it provides greater carbon dioxide output to allow faster rising. There is considerable debate as to the value of such a product; while most baking experts believe it reduces the flavor potential of the finished product, *Cook's Illustrated* magazine, among others, feels that, at least for direct-rise recipes, it makes little difference. Rapid-rise yeast is often marketed specifically for use in bread machines.
- Deactivated yeast is dead yeast which has no leavening value and is not interchangeable with other yeast types. Typically used for pizza and pan bread doughs, it is used at a rate of 0.1% of the flour weight, though manufacturer specifications may vary. It is a powerful reducing agent used to increase the extensibility of a dough.

For most commercial uses, yeast of any form is packaged in bulk (blocks or freezer bags for fresh yeast; vacuum-packed brick bags for dry or instant); however, yeast for home use is often packaged in pre-measured doses, either small squares for compressed yeast or sealed packets for dry or instant. For active dry and instant yeast, in general a single dose (reckoned for the average bread recipe of between 500 g and 1000 g of dough) is about 2.5 tsp (~12 mL) or about 7 g ($\frac{1}{4}$ oz), though comparatively lesser amounts are used when the yeast is used in a pre-ferment. In general, a yeast flavor in the baked bread is not noticeable when the bakers' percent of added yeast is less than 2.5%.

Use in Research

Model Organism

Because it is readily available and easy to culture, baker's yeast has long been used in chemical, biological, and genetic research as a model organism. In 1996, after 6 years of work, *S. cerevisiae* became the first eukaryote to have its entire genomesequenced. It has over 12 million base pairs and around 6000 genes. Since then, it has remained in the forefront of genetic research. For example, most of our knowledge of the cell division cycle was worked out from experiments with yeast.

Organic Synthesis

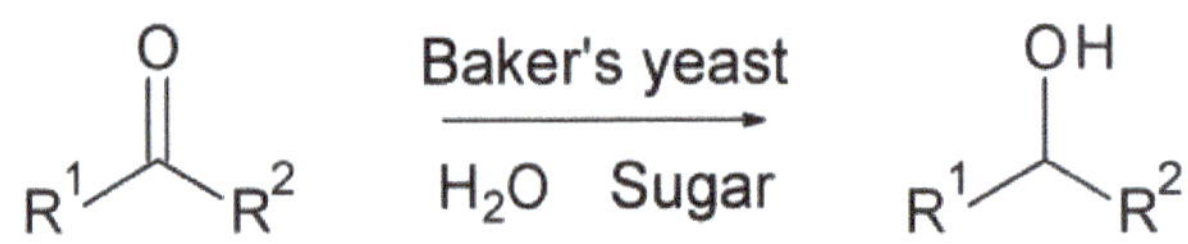

Reduction of a carbonyl to a hydroxyl with baker's yeast.

Baker's yeast contains enzymes that can reduce a carbonyl group into a hydroxyl group in fairly high yield, thus making it useful for biotransformations in organic syntheses. It is known to reduce organometallic carbonyl compounds in very high yield.

Baker's yeast can also be used to produce ethanol via fermentation for use in chemical synthesis, although doing so in some places requires permits.

Industrial Production

The baking industry relies on industrial production of its ingredients, including baking yeasts. Much effort has been put into developing and marketing yeasts that will perform reliably in mass production. Since the end of the nineteenth century, baker's yeast has been produced by companies that specialize in its production.

In 2012 the companies that dominated the US market included: Lesaffre Group, AB Vista, DSM, GB Plange and AB Mauri, producing hundreds of thousands of metric tons of yeast each year in industrial plants.

Aspergillus Oryzae

*Aspergillus oryzae, kn*own in English as koji, is a filamentousfungus (a mold) used in Chinese and other East Asian cuisines to fermentsoybeans for making soy sauce and fermented bean paste, and also to saccharify rice, other grains, and potatoes in the making of alcoholic beverages such as *huangjiu, sake, makgeolli*, and *shōchū*. The domestication of *A. oryzae* occurred at least 2000 years ago.*A. oryzae* is also used for the production of rice vinegars.

Dr. Eiji Ichishima of Tohoku University called the *kōji* fungus a "national fungus" (*kokkin*) in the journal of the Brewing Society of Japan, because of its importance not only for making the *koji* for *sake* brewing, but also for making the *koji* for *miso*, soy sauce, and a range of other traditional Japanese foods. His proposal was approved at the society's annual meeting in 2006.

"Red *kōji-kin*" is a separate species, *Monascus purpureus*.

History of Koji

300 BCE – A. *oryzae* is first mentioned in the *Zhouli* (Rites of the Zhou dynasty) in China. Its development is a milestone in Chinese food technology, for it provides the conceptual framework for three major fermented soy foods: soy sauce, *jiang / miso*, and *douchi*, not to mention grain-based wines (including Japanese *sake*) and *li* (the Chinese forerunner of Japanese *amazake*).

Properties Desirable in Sake Brewing and Testing

These properties of *A. oryzae* strains are important in rice saccharification for sake brewing:

- Growth: rapid mycelial growth on and into the rice kernels
- Enzymes: strong secretion of amylases (α-amylase and glucoamylase); some carboxypeptidase; low tyrosinase
- Aesthetics: pleasant fragrance; accumulation of flavoring compounds
- Color: low production of deferriferrichrome (a siderophore), flavins, and other colored substances

Varieties Used for Shōchū Making

Three varieties of kōji mold are used for making *shōchū*, each with distinct characteristics.

- White was discovered at the beginning of the Taishō period when natural mutation and separation of some black *kōji* to white was observed. This effect was researched and white *kōji* was successfully grown independently. White *kōji* is easy to cultivate and its enzymes promote rapid saccharization; as a result, it is used to produce most *shōchū* today. It gives rise to a drink with a refreshing, gentle, sweet taste.
- Black is mainly used in Okinawa to produce *awamori*. It produces plenty of citric acid which helps to prevent the souring of the *moromi*. Of all three *kōji*, it most effectively extracts the taste and character of the base ingredients, giving its *shōchū* a rich aroma with a slightly sweet, mellow taste. Its spores disperse easily, covering production facilities and workers' clothes in a layer of black. Such issues led to it falling out of favour, but due to the development of *new kuro-kōji* (NK-*kōji*) in the mid-1980s, interest in black *kōji* resurged amongst *honkaku shōchū* makers because of the depth and quality of the taste it produced. Several popular brands now explicitly state they use black *kōji* on their labels.
- Yellow is used to produce sake, and at one time all *honkaku shōchū*. However, yellow *kōji* is extremely sensitive to temperature; its *moromi* can easily sour during fermentation. This makes it difficult to use in warmer regions such as Kyūshū, and gradually black and white *kōji* became more common. Its strength is that it gives rise to a rich, fruity, refreshing taste, so despite the difficulties and great skill required, it is still used by some manufacturers. It is popular amongst young people and women who previously had no interest in typically strong potato *shōchū*, playing a role in its recent revival.

Genome

Initially kept secret, the *A. oryzae*genome was released by a consortium of Japanese biotechnology companies in late 2005. The eight chromosomes together comprise 37 million base pairs and 12 thousand predicted genes. The genome of *A. oryzae* is thus one-third larger than that of two related *Aspergillus* species, the genetics model organism *A. nidulans* and the potentially dangerous *A. fumigatus*. Many of the extra genes present in *A. oryzae* are predicted to be involved in secondary metabolism. The sequenced strain isolated in 1950 is called RIB40 or ATCC 42149; its morphology, growth, and enzyme production are typical of strains used for *sake* brewing.

Use in Biotechnology

Resveratrol can be produced from its glucoside piceid through the process of fermentation by *A. oryzae*.

In Fiction

A. oryzae is a supporting character (of sorts) in the manga series *Moyasimon: Tales of Agriculture* and its anime adaptation.

References

- Ljungh A, Wadstrom T, eds. (2009). Lactobacillus Molecular Biology: From Genomics to Probiotics. Caister Academic Press. ISBN 978-1904455417.[page needed]
- Golden, David M.; Jay, James M.; Martin J. Loessner (2005). Modern food microbiology. Berlin: Springer. p. 179. ISBN 0-387-23180-3.
- Jeff Cox "From Vines to Wines: The Complete Guide to Growing Grapes and Making Your Own Wine" pgs 133–136 Storey Publishing 1999 ISBN 1-58017-105-2
- J. Robinson (ed) "The Oxford Companion to Wine" Third Edition pgs 778–780 Oxford University Press 2006 ISBN 0198609906
- K. Fugelsang, C. Edwards Wine Microbiology Second Edition pgs 3–28 Springer Science and Business Media , New York (2010) ISBN 0387333495
- B. Zoecklein, K. Fugelsang, B. Gump, F. Nury Wine Analysis and Production pgs 281–290 Kluwer Academic Publishers, New York (1999) ISBN 0834217015
- J. Robinson (ed) "The Oxford Companion to Wine" Third Edition pgs 267 & 508 Oxford University Press 2006 ISBN 0198609906
- B. Zoecklein, K. Fugelsang, B. Gump, F. Nury Wine Analysis and Production pgs 97-114 Kluwer Academic Publishers, New York (1999) ISBN 0834217015
- Dr. Yair Margalit, Winery Technology & Operations A Handbook for Small Wineries pgs 67–74 The Wine Appreciation Guild (1996) ISBN 0-932664-66-0
- J. Robinson (ed) "The Oxford Companion to Wine" Third Edition pgs 398–399 Oxford University Press 2006 ISBN 0198609906
- Young, Linda; Cauvain, Stanley P. (2007). Technology of Breadmaking. Berlin: Springer. p. 79. ISBN 0-387-38563-0. The scientific name for baker's yeast is Saccharomyces cerevisiae, ...
- Christian, Elizabeth W.; Vaclavik, Vickie (2003). Essentials of food science. New York: Kluwer Academic/Plenum Publishers. p. 323. ISBN 0-306-47363-1.

4

Fermentation: An Overview

Fermentation is the process that is used in converting sugar to gases or to alcohol. It is also used to refer to the growth of microorganisms with the aim of producing a specific chemical product. This section is an overview of the subject matter incorporating all the major aspects of fermentation.

Fermentation

Fin progress: Bubbles of CO2 form a froth on top of the fermentation mixture.

Fermentation is a metabolic process that converts sugar to acids, gases, or alcohol. It occurs in yeast and bacteria, and also in oxygen-starved muscle cells, as in the case of lactic acid fermentation. Fermentation is also used more broadly to refer to the bulk growth of microorganisms on a growth medium, often with the goal of producing a specific chemical product. French microbiologist Louis Pasteur is often remembered for his insights into fermentation and its microbial causes. The science of fermentation is known as zymology.

Fermentation takes place when the electron transport chain is unusable (often due to lack of a final electron receptor, such as oxygen). In this case it becomes the cell's primary means of ATP (energy) production. It turns NADH and pyruvate produced in glycolysis into NAD^+ and an organic molecule (which varies depending on the type of fermentation). In the presence of O_2, NADH and pyruvate are used to generate ATP in respiration. This is called oxidative phosphorylation, and it generates much more ATP than glycolysis alone. For that reason, cells generally benefit from avoiding fermentation when oxygen is available, the exception being obligate anaerobes which cannot tolerate oxygen.

The first step, glycolysis, is common to all fermentation pathways:

$$C_6H_{12}O_6 + 2\ NAD^+ + 2\ ADP + 2\ P_i \rightarrow 2\ CH_3COCOO^- + 2\ NADH + 2\ ATP + 2\ H_2O + 2H^+$$

Pyruvate is CH_3COCOO^-. P_i is inorganic phosphate. Two ADP molecules and two P_i are converted to two ATP and two water molecules via substrate-level phosphorylation. Two molecules of NAD^+ are also reduced to NADH.

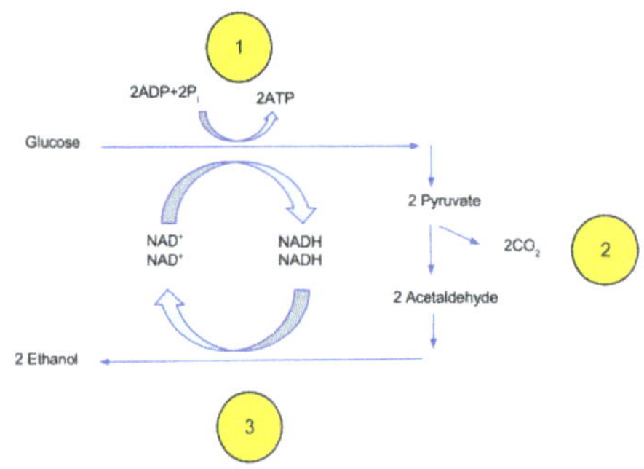

Overview of ethanol fermentation. One glucose molecule breaks down into two pyruvate molecules (1). The energy from this exothermic reaction is used to bind inorganic phosphates to ATP and convert NAD+ to NADH. The two pyruvates are then broken down into two acetaldehyde molecules and give off two CO2 molecules as a waste product (2). The acetaldehyde is then reduced into ethanol using the energy and hydrogen from NADH; in this process the NADH is oxidized into NAD+ so that the cycle may repeat (3).

In oxidative phosphorylation the energy for ATP formation is derived from an electrochemical proton gradient generated across the inner mitochondrial membrane (or, in the case of bacteria, the plasma membrane) via the electron transport chain. Glycolysis has substrate-level phosphorylation (ATP generated directly at the point of reaction).

Humans have used fermentation to produce drinks and beverages since the Neolithic age. For example, fermentation is used for preservation in a process that produces lactic acid as found in such sour foods as pickled cucumbers, kimchi and yogurt, as well as for producing alcoholic beverages such as wine and beer. Fermentation can even occur within the stomachs of animals, such as humans.

Definitions

To many people, fermentation simply means the production of alcohol: grains and fruits are fermented to produce beer and wine. If a food soured, one might say it was 'off' or fermented. Here are some definitions of fermentation. They range from informal, general usage to more scientific definitions.

1. Preservation methods for food via microorganisms (general use).
2. Any process that produces alcoholic beverages or acidic dairy products (general use).
3. Any large-scale microbial process occurring with or without air (common definition used in industry).
4. Any energy-releasing metabolic process that takes place only under anaerobic conditions (becoming more scientific).
5. Any metabolic process that releases energy from a sugar or other organic molecules, does not require oxygen or an electron transport system, and uses an organic molecule as the final electron acceptor (most scientific).

Examples

Fermentation does not necessarily have to be carried out in an anaerobic environment. For example, even in the presence of abundant oxygen, yeast cells greatly prefer fermentation to aerobic respiration, as long as sugars are readily available for consumption (a phenomenon known as the Crabtree effect). The antibiotic activity of hops also inhibits aerobic metabolism in yeast.

Fermentation reacts NADH with an endogenous, organic electron acceptor. Usually this is pyruvate formed from the sugar during the glycolysis step. During fermentation, pyruvate is metabolized to various compounds through several processes:

- ethanol fermentation, aka alcoholic fermentation, is the production of ethanol and carbon dioxide
- lactic acid fermentation refers to two means of producing lactic acid:

1. homolactic fermentation is the production of lactic acid exclusively
2. heterolactic fermentation is the production of lactic acid as well as other acids and alcohols.

Sugars are the most common substrate of fermentation, and typical examples of fermentation products are ethanol, lactic acid, carbon dioxide, and hydrogen gas (H_2). However, more exotic compounds can be produced by fermentation, such as butyric acid and acetone. Yeast carries out fermentation in the production of ethanol in beers, wines, and other alcoholic drinks, along with the production of large quantities of carbon dioxide. Fermentation occurs in mammalian muscle during periods of intense exercise where oxygen supply becomes limited, resulting in the creation of lactic acid.

Chemistry

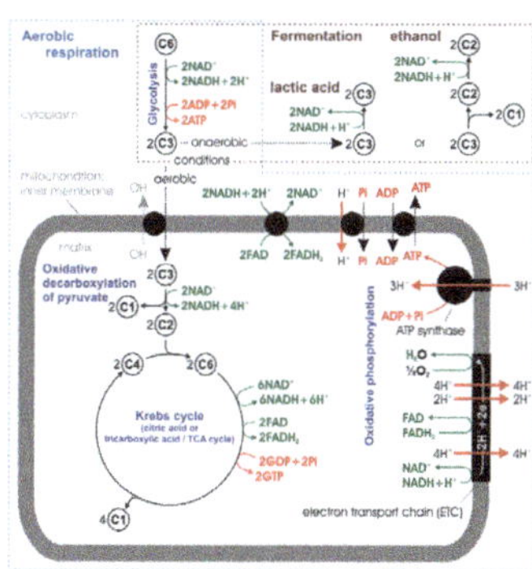

Comparison of aerobic respiration and most known fermentation types in eucaryotic cell. Numbers in circles indicate counts of carbon atoms in molecules, C6 is glucose $C_6H_{12}O_6$, C1 carbon dioxide CO_2. Mitochondrial outer membrane is omitted.

Fermentation products contain chemical energy (they are not fully oxidized), but are considered waste products, since they cannot be metabolized further without the use of oxygen.

Ethanol Fermentation

The chemical equation below shows the alcoholic fermentation of glucose, whose chemical formula is $C_6H_{12}O_6$. One glucose molecule is converted into two ethanol molecules and two carbon dioxide molecules:

$$C_6H_{12}O_6 \rightarrow 2\ C_2H_5OH + 2\ CO_2$$

C_2H_5OH is the chemical formula for ethanol.

Before fermentation takes place, one glucose molecule is broken down into two pyruvate molecules. This is known as glycolysis.

Lactic Acid Fermentation

Homolactic fermentation (producing only lactic acid) is the simplest type of fermentation. The pyruvate from glycolysis undergoes a simple redox reaction, forming lactic acid. It is unique because it is one of the only respiration processes to not produce a gas as a byproduct. Overall, one molecule of glucose (or any six-carbon sugar) is converted to two molecules of lactic acid: $C_6H_{12}O_6 \rightarrow 2\ CH_3CHOHCOOH$ It occurs in the muscles of animals when they need energy faster than the blood can supply oxygen. It also occurs in some kinds of bacteria (such as lactobacilli) and some fungi. It is this type of bacteria that converts lactose into lactic acid in yogurt, giving it its sour taste. These lactic acid bacteria can carry out either homolactic fermentation, where the end-product is mostly lactic acid, or

Heterolactic fermentation, where some lactate is further metabolized and results in ethanol and carbon dioxide (via the phosphoketolase pathway), acetate, or other metabolic products, e.g.: $C_6H_{12}O_6 \rightarrow CH_3CHOHCOOH + C_2H_5OH + CO_2$If lactose is fermented (as in yogurts and cheeses), it is first converted into glucose and galactose (both six-carbon sugars with the same atomic formula): $C_{12}H_{22}O_{11} + H_2O \rightarrow 2\ C_6H_{12}O_6$Heterolactic fermentation is in a sense intermediate between lactic acid fermentation, and other types, e.g. alcoholic fermentation. The reasons to go further and convert lactic acid into anything else are:

- The acidity of lactic acid impedes biological processes; this can be beneficial to the fermenting organism as it drives out competitors that are unadapted to the acidity; as a result, the food will have a longer shelf life (part of the reason foods are purposely fermented in the first place); however, beyond a certain point, the acidity starts affecting the organism that produces it.

- The high concentration of lactic acid (the final product of fermentation) drives the equilibrium backwards (Le Chatelier's principle), decreasing the rate at which fermentation can occur, and slowing down growth.

- Ethanol, into which lactic acid can be easily converted, is volatile and will readily escape, allowing the reaction to proceed easily. CO_2 is also produced, but it is only weakly acidic, and even more volatile than ethanol.

- Acetic acid (another conversion product) is acidic, and not as volatile as ethanol; however, in the presence of limited oxygen, its creation from lactic acid releases additional energy. It is a lighter molecule than lactic acid, that forms fewer hydrogen bonds with its surroundings (due to having fewer groups that can form such bonds), thus is more volatile and will also allow the reaction to move forward more quickly.

- If propionic acid, butyric acid, and longer monocarboxylic acids are produced, the amount of acidity produced per glucose consumed will decrease, as with ethanol, allowing faster growth.

Aerobic Respiration

In aerobic respiration, the pyruvate produced by glycolysis is oxidized completely, generating additional ATP and NADH in the citric acid cycle and by oxidative phosphorylation. However, this can occur only in the presence of oxygen. Oxygen is toxic to organisms that are obligate anaerobes, and is not required by facultative anaerobic organisms. In the absence of oxygen, one of the fermentation pathways occurs in order to regenerate NAD^+; lactic acid fermentation is one of these pathways.

Hydrogen Gas Production in Fermentation

Hydrogen gas is produced in many types of fermentation (mixed acid fermentation, butyric acid fermentation, caproate fermentation, butanol fermentation, glyoxylate fermentation), as a way to regenerate NAD^+ from NADH. Electrons are transferred to ferredoxin, which in turn is oxidized by hydrogenase, producing H_2. Hydrogen gas is a substrate for methanogens and sulfate reducers, which keep the concentration of hydrogen low and favor the production of such an energy-rich compound, but hydrogen gas at a fairly high concentration can nevertheless be formed, as in flatus.

As an example of mixed acid fermentation, bacteria such as *Clostridium pasteurianum* ferment glucose producing butyrate, acetate, carbon dioxide and hydrogen gas: The reaction leading to acetate is:

$$C_6H_{12}O_6 + 4\,H_2O \rightarrow 2\,CH_3COO^- + 2\,HCO_3^- + 4\,H^+ + 4\,H_2$$

Glucose could theoretically be converted into just CO_2 and H_2, but the global reaction releases little energy.

Methane Gas Production in Fermentation

Acetic acid can also undergo a dismutation reaction to produce methane and carbon dioxide:

$$CH_3COO^- + H^+ \rightarrow CH_4 + CO_2 \qquad \Delta G° = -36\ \text{kJ/reaction}$$

This disproportionation reaction is catalysed by methanogen archaea in their fermentative metabolism. One electron is transferred from the carbonyl function (e^- donor) of the carboxylic group to the methyl group (e^- acceptor) of acetic acid to respectively produce CO_2 and methane gas.

History of Human Use

The use of fermentation, particularly for beverages, has existed since the Neolithic and has been documented dating from 7000–6600 BCE in Jiahu, China, 5000 BCE in India, Ayurveda mentions many Medicated Wines, 6000 BCE in Georgia, 3150 BCE in ancient Egypt, 3000 BCE in Babylon, 2000 BCE in pre-Hispanic Mexico, and 1500 BC in Sudan. Fermented foods have a religious significance in Judaism and Christianity. The Baltic god Rugutis was worshiped as the agent of fermentation.

The first solid evidence of the living nature of yeast appeared between 1837 and 1838 when three publications appeared by C. Cagniard de la Tour, T. Swann, and F. Kuetzing, each of whom independently concluded as a result of microscopic investigations that yeast is a living organism that reproduces by budding. It is perhaps because wine, beer, and bread were each basic foods in Europe that most of the early studies on fermentation were done on yeasts, with which they were made. Soon, bacteria were also discovered; the term was first used in English in the late 1840s, but it did not come into general use until the 1870s, and then largely in connection with the new germ theory of disease.

Louis Pasteur in his laboratory

Louis Pasteur (1822–1895), during the 1850s and 1860s, showed that fermentation is initiated by living organisms in a series of investigations. In 1857, Pasteur showed that lactic acid fermentation is caused by living organisms. In 1860, he demonstrated that bacteria cause souring in milk, a process formerly thought to be merely a chemical change, and his work in identifying the role of microorganisms in food spoilage led to the process of pasteurization. In 1877, working to improve the French brewing industry, Pasteur published his famous paper on fermentation, "*Etudes sur la Bière*", which was translated into English in 1879 as "Studies on fermentation". He defined fermentation (incorrectly) as "Life without air", but correctly showed that specific types of microorganisms cause specific types of fermentations and specific end-products.

Although showing fermentation to be the result of the action of living microorganisms was a breakthrough, it did not explain the basic nature of the fermentation process, or prove that it is caused by the microorganisms that appear to be always present. Many scientists, including Pasteur, had unsuccessfully attempted to extract the fermentation enzyme from yeast. Success came in 1897 when the German chemist Eduard Buechner ground up yeast, extracted a juice from them, then found to his amazement that this "dead" liquid would ferment a sugar solution, forming carbon dioxide and alcohol much like living yeasts. Buechner's results are considered to mark the birth of biochemistry. The "unorganized ferments" behaved just like the organized ones. From that time on, the term enzyme came to be applied to all ferments. It was then understood that fermentation is caused by enzymes that are produced by microorganisms. In 1907, Buechner won the Nobel Prize in chemistry for his work.

Advances in microbiology and fermentation technology have continued steadily up until the present. For example, in the late 1970s, it was discovered that microorganisms could be mutated with physical and chemical treatments to be higher-yielding, faster-growing, tolerant of less oxygen, and able to use a more concentrated medium. Strain selection and hybridization developed as well, affecting most modern food fermentations. Other approaches to advancing the fermentation industry has been done by companies such as BioTork, a biotechnology company that naturally evolves microorganisms to improve fermentation processes. This approach differs from the more popular genetic modification, which has become the current industry standard.

Etymology

The word "ferment" is derived from the Latin verb *fervere,* which means to boil. It is thought to have been first used in the late 14th century in alchemy, but only in a broad sense. It was not used in the modern scientific sense until around 1600.

Fermentation in Food Processing

Grapes being trodden to extract the juice and made into wine in storage jars. Tomb of Nakht, 18th dynasty, Thebes, Ancient Egypt

Fermentation in food processing is the process of converting carbohydrates to alcohol or organic acids using microorganisms—yeasts or bacteria—under anaerobic conditions. Fermentation usually implies that the action of microorganisms is desired. The science of fermentation is known as zymology or zymurgy.

The term fermentation sometimes refers specifically to the chemical conversion of sugars into ethanol, producing alcoholic drinks such as wine, beer, and cider. However, similar processes take place in the leavening of bread (CO_2 produced by yeast activity), and in the preservation of sour foods with the production of lactic acid, such as in sauerkraut and yogurt.

Apart from alcohol, widely consumed fermented foods include vinegar, olives, yogurt, bread, and cheese. In various parts of the world, more localised foods prepared by fermentation may also be based on beans, dough, grain, vegetables, fruit, honey, dairy products, fish, meat, or tea.

History and Prehistory

Natural fermentation precedes human history. Since ancient times, humans have exploited the fermentation process. The earliest evidence of an alcoholic drink, made from fruit, rice, and honey, dates from 7000 to 6600 BC, in the Neolithic Chinese village of Jiahu, and winemaking dates from 6000 BC, in Georgia, in the Caucasus area. Seven-thousand-year-old jars containing the remains of wine, now on display at the University of Pennsylvania, were excavated in the Zagros Mountains in Iran. There is strong evidence that people were fermenting alcoholic drinks in Babylon c. 3000 BC, ancient Egypt c. 3150 BC, pre-Hispanic Mexico c. 2000 BC, and Sudan c. 1500 BC.

The French chemist Louis Pasteur founded zymology, when in 1856 he connected yeast to fermentation. When studying the fermentation of sugar to alcohol by yeast, Pasteur concluded that the fermentation was catalyzed by a vital force, called "ferments", within the yeast cells. The "ferments" were thought to function only within living organisms. "Alcoholic fermentation is an act correlated with the life and organization of the yeast cells, not with the death or putrefaction of the cells", he wrote.

Nevertheless, it was known that yeast extracts can ferment sugar even in the absence of living yeast cells. While studying this process in 1897, Eduard Buchner of Humboldt University of Berlin, Germany, found that sugar was fermented even when there were no living yeast cells in the mixture, by a yeast secreted enzyme complex that he termed *zymase*. In 1907 he received the Nobel Prize in Chemistry for his research and discovery of "cell-free fermentation".

One year earlier, in 1906, ethanol fermentation studies led to the early discovery of NAD^+.

Uses

Beer and bread, two major uses of fermentation in food

Food fermentation is the conversion of sugars and other carbohydrates into alcohol or preservative organic acids and carbon dioxide. All three products have found human uses. The production of alcohol is made use of when fruit juices are converted to wine, when grains are made into beer, and when foods rich in starch, such as potatoes, are fermented and then distilled to make spirits such as gin and vodka. The production of carbon dioxide is used to leaven bread. The production of organic acids is exploited to preserve and flavor vegetables and dairy products.

Food fermentation serves five main purposes: to enrich the diet through development of a diversity of flavors, aromas, and textures in food substrates; to preserve substantial amounts of food through lactic acid, alcohol, acetic acid, and alkaline fermentations; to enrich food substrates with protein, essential amino acids, and vitamins; to eliminate antinutrients; and to reduce cooking time and the associated use of fuel.

Fermented Foods by Region

Nattō, a Japanese fermented soybean food

- Worldwide: alcohol(beer, wine), vinegar, olives, yogurt, bread, cheese
- Asia
    - East and Southeast Asia: amazake, atchara, bai-ming, belacan, burong mangga, com ruou, dalok, doenjang, douchi, jeruk, lambanog, kimchi, kombucha, leppet-so, narezushi, miang, miso, nata de coco, nata de pina, natto, naw-mai-dong, oncom, pak-siam-dong, paw-tsaynob, prahok, ruou nep, sake, seokbakji, soju, soy sauce, stinky tofu, szechwan cabbage, tai-tan tsoi, chiraki, tape, tempeh, totkal kimchi, yen tsai, zha cai
    - Central Asia: kumis (mare milk), kefir, shubat (camel milk)
    - South Asia: achar, appam, dosa, dhokla, dahi (yogurt), idli, kaanji, mixed pickle, ngari, hawaichaar, jaand (rice beer), sinki, tongba, paneer
- Africa: fermented millet porridge, garri, hibiscus seed, hot pepper sauce, injera, lamoun makbouss, laxoox, mauoloh, msir, mslalla, oilseed, ogi, ogili, ogiri, iru
- Americas: sourdough bread, cultured milk, chicha, elderberry wine, kombucha, pickling (pickled vegetables), sauerkraut, lupin seed, oilseed, chocolate, vanilla, tabasco, tibicos, pulque, mikyuk (fermented bowhead whale)
- Middle East: kushuk, lamoun makbouss, mekhalel, torshi, boza
- Europe: rakfisk, sauerkraut, pickled cucumber, surströmming, mead, elderberry wine, salami, sucuk, prosciutto, cultured milk products such as quark, kefir, filmjölk, crème fraîche, smetana, skyr, rakı, tupí.
- Oceania: poi, kaanga pirau (rotten corn), sago

Fermented Foods by Type

Bean-based

Cheonggukjang, doenjang, miso, natto, soy sauce, stinky tofu, tempeh, oncom, soybean paste, Beijing mung bean milk, kinama, iru

Dough-based

Proofing (baking technique)

Grain-based

Batter made from rice and lentil (Vigna mungo) prepared and fermented for baking idlis and dosas

Amazake, beer, bread, choujiu, gamju, injera, kvass, makgeolli, murri, ogi, rejuvelac, sake, sikhye, sourdough, sowans, rice wine, malt whisky, grain whisky, idli, dosa, vodka, boza

Vegetable-based

Kimchi, mixed pickle, sauerkraut, Indian pickle, gundruk, tursu

Fermenting cocoa beans

Fruit-based

Wine, vinegar, cider, perry, brandy, atchara, nata de coco, burong mangga, asinan, pickling, vişinată, chocolate, rakı

Honey-based

Mead, metheglin

Dairy-based

Some kinds of cheese also, kefir, kumis (mare milk), shubat (camel milk), cultured milk products such as quark, filmjölk, crème fraîche, smetana, skyr, and yogurt

Cheeses in art: Still Life with Cheeses, Almonds and Pretzels, Clara Peeters, c. 1615

Fish-based

Bagoong, faseekh, fish sauce, Garum, Hákarl, jeotgal, rakfisk, shrimp paste, surströmming, shidal

Meat-based

Chin som mok is a northern Thai speciality made with grilled, banana leaf-wrapped pork (both skin and meat) that has been fermented with glutinous rice

Chorizo, salami, sucuk, pepperoni, nem chua, som moo, saucisson

Tea-based

Pu-erh tea, Kombucha

Risks

Alaska has witnessed a steady increase of cases of botulism since 1985. It has more cases of botulism than any other state in the United States of America. This is caused by the traditional Eskimo practice of allowing animal products such as whole fish, fish heads, walrus, sea lion, and whale flippers, beaver tails, seal oil, and birds, to ferment for an extended period of time before being consumed. The risk is exacerbated when a plastic container is used for this purpose instead of the old-fashioned, traditional method, a grass-lined hole, as the *Clostridium botulinum* bacteria thrive in the anaerobic conditions created by the air-tight enclosure in plastic.

The World Health Organization has classified pickled foods as possibly carcinogenic, based on epidemiological studies. Other research found that fermented food contains a carcinogenic by-product, ethyl carbamate (urethane). "A 2009 review of the existing studies conducted across Asia concluded that regularly eating pickled vegetables roughly doubles a person's risk for esophageal squamous cell carcinoma."

Fermentation in Winemaking

Fermenting must

The process of fermentation in winemaking turns grape juice into an alcoholic beverage. During fermentation, yeasts transform sugars present in the juice into ethanol and carbon dioxide (as a by-product). In winemaking, the temperature and speed of fermentation are important considerations as well as the levels of oxygen present in the must at the start of the fermentation. The risk of stuck fermentation and the development of several wine faults can also occur during this stage, which can last anywhere from 5 to 14 days for *primary fermentation* and potentially another 5 to 10 days for a *secondary fermentation*. Fermentation may be done in stainless steel tanks, which is common with many white wines like Riesling, in an open wooden vat, inside a wine barrel and inside the wine bottle itself as in the production of many sparkling wines.

History

The natural occurrence of fermentation means it was probably first observed long ago by humans. The earliest uses of the word "fermentation" in relation to winemaking was in reference to the apparent "boiling" within the must that came from the anaerobic reaction of the yeast to the sugars in the grape juice and the release of carbon dioxide. The Latin *fervere* means, literally, *to boil*. In the mid-19th century, Louis Pasteur noted the connection between yeast and the process of the fermentation in which the yeast act as catalyst and mediator through a series of a reaction that convert sugar into alcohol. The discovery of the Embden–Meyerhof–Parnas pathway by Gustav Embden, Otto Fritz Meyerhof and Jakub Karol Parnas in the early 20th century contributed more to the understanding of the complex chemical processes involved in the conversion of sugar to alcohol.

Process

"Bloom", visible as a dusting on the berries

In winemaking, there are distinctions made between *ambient yeasts* which are naturally present in wine cellars, vineyards and on the grapes themselves (sometimes known as a grape's "bloom" or "blush") and *cultured yeast* which are specifically isolated and inoculated for use in winemaking. The most common genera of wild yeasts found in winemaking include *Candida, Klöckera/ Hanseniaspora, Metschnikowiaceae, Pichia* and *Zygosaccharomyces*. Wild yeasts can produce high-quality, unique-flavored wines; however, they are often unpredictable and may introduce less desirable traits to the wine, and can even contribute to spoilage. It should be noted that few yeast, and lactic and acetic acid bacterial colonies naturally live on the surface of grapes, but traditional wine makers, particularly in Europe, advocate use of ambient yeast as a characteristic of the region's *terroir*; nevertheless, many winemakers prefer to control fermentation with predictable cultured yeast. The cultured yeasts most commonly used in winemaking belong to the *Saccharomyces cerevisiae* (also known as "sugar yeast") species. Within this species are several hundred different strains of yeast that can be used during fermentation to affect the heat or vigor of the process and enhance or suppress certain flavor characteristics of the varietal. The use of different strains of yeasts is a major contributor to the diversity of wine, even among the same grape variety. Alternative, non-*Saccharomyces cerevisiae*, yeasts are being used more prevalently in the industry to add greater complexity to wine. After a winery has been in operation for a number of years, few yeast strains are actively involved in the fermentation process. The use of active dry yeasts reduces the variety of strains that appear in spontaneous fermentation by outcompeting those strains that are naturally present.

The addition of cultured yeast normally occurs with the yeast first in a dried or "inactive" state and is reactivated in warm water or diluted grape juice prior to being added to the must. To thrive and be active in fermentation, the yeast needs access to a continuous supply of carbon, nitrogen, sulfur, phosphorus as well as access to various vitamins and minerals. These components are naturally present in the grape must but their amount may be corrected by adding nutrients to the wine, in order to foster a more encouraging environment for the yeast. Newly formulated time-release nutrients, specifically manufactured for wine fermentations, offer the most advantageous conditions for yeast. Oxygen is needed as well, but in wine making, the risk of oxidation and the lack of alcohol production from oxygenated yeast requires the exposure of oxygen to be kept at a minimum.

Dry winemaking yeast (left) and yeast nutrients used in the rehydration process to stimulate yeast cells.

Upon the introduction of active yeasts to the grape must, phosphates are attached to the sugar and the six-carbon sugar molecules begin to be split into three-carbon pieces and go through a series of rearrangement reactions. During this process, the carboxylic carbon atom is released in the form of carbon dioxide with the remaining components becoming acetaldehyde. The absence of oxygen in this anaerobic process allows the acetaldehyde to be eventually converted, by reduction, to ethanol. During the conversion of acetaldehyde, a small amount is converted, by oxidation, to acetic acid which, in excess, can contribute to the wine fault known as volatile acidity (vinegar taint).

After the yeast has exhausted its life cycle, they fall to the bottom of the fermentation tank as sediment known as lees. Yeast ceases its activity whenever all of the sugar in must has been converted into other chemicals or whenever the alcohol content has reached 15% alcohol per unit volume; a concentration strong enough to halt the enzymatic activity of almost all strains of yeast.

Other Compounds Involved

The metabolism of amino acids and breakdown of sugars by yeasts has the effect of creating other biochemical compounds that can contribute to the flavor and aroma of wine. These compounds can be considered "volatile" like aldehydes, ethyl acetate, ester, fatty acids, fusel oils, hydrogen sulfide, ketones and mercaptans or "non-volatile" like glycerol, acetic acid and succinic acid. Yeast also has the effect during fermentation of releasing glycoside hydrolase which can hydrolyse the flavor precursors of aliphatics (a flavor component that reacts with oak), benzene derivatives, monoterpenes (responsible for floral aromas from grapes like Muscat and Traminer), norisoprenoids (responsible for some of the spice notes in Chardonnay), and phenols.

Some strains of yeasts can generate volatile thiols which contribute to the fruity aromas in many wines such as the gooseberry scent commonly associated with Sauvignon blanc. *Brettanomyces* yeasts are responsible for the "barnyard aroma" characteristic in some red wines like Burgundy and Pinot noir.

Methanol is not a major constituent of wine. The usual concentration range is between 0.1 g/liter and 0.2 g/liter. These small traces have no adverse effect on people and no direct effect on the senses.

Winemaking Considerations

During fermentation, there are several factors that winemakers take into consideration, with the most influential to ethanol production being sugar content in the must, the yeast strain used, and the fermentation temperature. The biochemical process of fermentation itself creates a lot of residual heat which can take the must out of the ideal temperature range for the wine. Typically, white wine is fermented between 18-20 °C (64-68 °F) though a wine maker may choose to use a higher temperature to bring out some of the complexity of the wine. Red wine is typically fermented at higher temperatures up to 29 °C (85 °F). Fermentation at higher temperatures may have adverse effect on the wine in stunning the yeast to inactivity and even "boiling off" some of the flavors of the wines. Some winemakers may ferment their red wines at cooler temperatures, more typical of white wines, in order to bring out more fruit flavors.

To control the heat generated during fermentation, the winemaker must choose a suitable vessel size or else use a cooling device. Various kinds of cooling devices are available, ranging from the ancient Bordeaux practice of placing the fermentation vat atop blocks of ice to sophisticated fermentation tanks that have built-in cooling rings.

A risk factor involved with fermentation is the development of chemical residue and spoilage which can be corrected with the addition of sulfur dioxide (SO_2), although excess SO_2 can lead to a wine fault. A winemaker who wishes to make a wine with high levels of residual sugar (like a dessert wine) may stop fermentation early either by dropping the temperature of the must to stun the yeast or by adding a high level of alcohol (like brandy) to the must to kill off the yeast and create a fortified wine.

The ethanol produced through fermentation acts as an important co-solvent to the non-polar compound that water cannot dissolve, such as pigments from grape skins, giving wine varieties their distinct color, and other aromatics. Ethanol and the acidity of wine act as an inhibitor to bacterial growth, allowing wine to be safely kept for years in the absence of air.

Other Types of Fermentation

A California Chardonnay that shows it has been barrel fermented.

In winemaking, there are different processes that fall under the title of "Fermentation" but might not follow the same procedure commonly associated with wine fermentation.

Bottle Fermentation

Bottle fermentation is a method of sparkling wine production, originating in the Champagne region where after the cuvee has gone through a primary yeast fermentation the wine is then bottled and goes through a secondary fermentation where sugar and additional yeast known as *liqueur de tirage* is added to the wine. This secondary fermentation is what creates the carbon dioxide bubbles that sparkling wine is known for.

Carbonic Maceration

The process of carbonic maceration is also known as whole grape fermentation where instead of yeast being added, the grapes fermentation is encouraged to take place inside the individual grape berries. This method is common in the creation of Beaujolais wine and involves whole clusters of grapes being stored in a closed container with the oxygen in the container being replaced with carbon dioxide. Unlike normal fermentation where yeast converts sugar into alcohol, carbonic maceration works by enzymes within the grape breaking down the cellular matter to form ethanol and other chemical properties. The resulting wines are typically soft and fruity.

Malolactic Fermentation

Instead of yeast, bacteria play a fundamental role in malolactic fermentation which is essentially the conversion of malic acid into lactic acid. This has the benefit of reducing some of the tartness and making the resulting wine taste softer. Depending on the style of wine that the winemaker is trying to produce, malolactic fermentation may take place at the very same time as the yeast fermentation. Alternatively, some strains of yeast may be developed that can convert L-malate to L-lactate during alcohol fermentation. For example, *Saccharomyces cerevisiae* strain ML01 (*S.*

cerevisiae strain ML01), which carries a gene encoding malolactic enzyme from *Oenococcus oeni* and a gene encoding malate permease from *Schizosaccharomyces pombe. S. cerevisiae strain* ML01 has received regulatory approval in both Canada and the United States.

- Co-fermentation

Yeast in Winemaking

The role of yeast in winemaking is the most important element that distinguishes wine from grape juice. In the absence of oxygen, yeast converts the sugars of wine grapes into alcohol and carbon dioxide through the process of fermentation. The more sugars in the grapes, the higher the potential alcohol level of the wine if the yeast are allowed to carry out fermentation to dryness. Sometimes winemakers will stop fermentation early in order to leave some residual sugars and sweetness in the wine such as with dessert wines. This can be achieved by dropping fermentation temperatures to the point where the yeast are inactive, sterile filtering the wine to remove the yeast or fortification with brandy to kill off the yeast cells. If fermentation is unintentionally stopped, such as when the yeasts become exhausted of available nutrients, and the wine has not yet reached dryness this is considered a stuck fermentation.

The most common yeast associated with winemaking is *Saccharomyces cerevisiae* which has been favored due to its predictable and vigorous fermentation capabilities, tolerance of relatively high levels of alcohol and sulfur dioxide as well as its ability to thrive in normal wine pH between 2.8 and 4. Despite its widespread use which often includes deliberate inoculation from cultured stock, *S.cerevisiae* is rarely the only yeast species involved in a fermentation. Grapes brought in from harvest are usually teeming with a variety of "wild yeast" from the *Kloeckera* and *Candida* genera. These yeasts often begin the fermentation process almost as soon as the grapes are picked when the weight of the clusters in the harvest bins begin to crush the grapes, releasing the sugar-rich must. While additions of sulfur dioxide (often added at the crusher) may limit some of the wild yeast activities, these yeasts will usually die out once the alcohol level reaches about 5% due to the toxicity of alcohol on the yeast cells physiology while the more alcohol tolerant *Saccharomyces* species take over. In addition to *S. cerevisiae*, Saccharomyces bayanus is a species of yeast that can tolerate alcohol levels of 17–20% and is often used in fortified wine production such as ports and varieties such as Zinfandel and Syrah harvested at high Brix sugar levels. Another common yeast involved in wine production is *Brettanomyces* whose presence in a wine may be viewed by different winemakers as either a wine fault or in limited quantities as an added note of complexity.

History

For most of the history of wine, winemakers did not know the mechanism that somehow converted sugary grape juice into alcoholic wine. They could observe the fermentation process which was often described as "boiling", "seething" or the wine being "troubled" due to release of carbon dioxide that gave the wine a frothy, bubbling appearance. This history is preserved in the etymology of the word "yeast" itself which essentially means "to boil".

In the mid-19th century, the French scientist Louis Pasteur was tasked by the French government to study what made some wines spoil. His work, which would later lead to Pasteur being considered one of the "Fathers of Microbiology", would uncover the connection between microscopic yeast cells and the process of the fermentation. It was Pasteur who discovered that yeast converted sugars in the must into alcohol and carbon dioxide, though the exact mechanisms of how the yeast would accomplish this task was not discovered till the 20th century with the Embden–Meyerhof–Parnas pathway.

The yeast species commonly known as *Saccharomyces cerevisiae* was first identified in late 19th century enology text as *Saccharomyces ellipsoideus* due to the elliptical (as opposed to circular) shape of the cells. Throughout the 20th century, more than 700 different strains of *Saccharomyces cerevisiae* were identified. The difference between the vast majority of these strains are mostly minor, though individual winemakers will develop a preference for particular strains when making certain wines or working with particular grape varieties. Some of these difference include the "vigor" or speed of fermentation, temperature tolerance, the production of volatile sulfur compounds (such as hydrogen sulfide) and other compounds that may influence the aroma of the wine.

Role in Winemaking

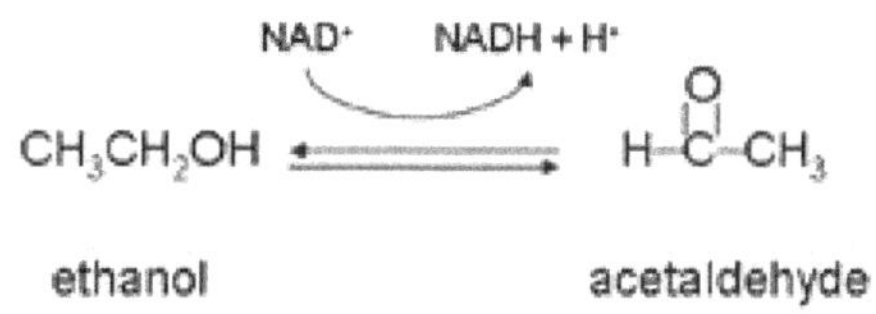

In the absence of oxygen, yeast cells will take the pyruvate produced by glycolysis and reduce it into acetaldehyde which is further reduced into ethanol "recharging" the NAD+ co-enzymes that is needed for various metabolic processes of the yeast.

The primary role of yeast is to convert the sugars present (namely glucose) in the grape must into alcohol. The yeast accomplishes this by utilizing glucose through a series of metabolic pathways that, in the presence of oxygen, produces not only large amounts of energy for the cell but also many different intermediates that the cell needs to function. In the absence of oxygen (and sometimes even in the presence oxygen), the cell will continue some metabolic functions (such as glycolysis) but will rely on other pathways such as reduction of acetaldehyde into ethanol (fermentation) to "recharge" the co-enzymes needed to keep metabolism going. It is through this process of fermentation that ethanol is released by the yeast cells as a waste product. Eventually, if the yeast cells are healthy and fermentation is allowed to run to the completion, all fermentable sugars will be used up by the yeast with only the unfermentable pentose leaving behind a negligible amount of residual sugar.

Other Compounds in Wine Produced by Yeast

While the production of alcohol is the most noteworthy by-product of yeast metabolism from a winemaking perspective, there are a number of other products that yeast produce that can be also influence the resulting wine. This includes glycerol which is produced when an intermediate of the glycolysis cycle (dihydroxyacetone) is reduced to "recharge" the NADH enzyme needed to continue other metabolic activities. This is usually produced early in the fermentation process before the mechanisms to reduce acetaldehyde into ethanol to recharge NADH becomes the cell's primary means of maintaining redox balance. As glycerol contributes increased body and a slightly sweet taste without increasing the alcohol level of the wine, some winemakers try to intentionally favor conditions that would promote glycerol production in wine. This includes selecting yeast strains that favor glycerol production (or allowing some wild yeast like *Kloeckera* and *Metschnikowia* to ferment), increased oxygen exposure and aeration as well as fermenting at higher temperatures. Glycerol production is also encouraged if most available acetaldehyde is made unavailable by binding with bisulfite molecules in the wine, but it would take a substantial amount of sulfur dioxide addition (far beyond legal limits) to prolong glycerol production beyond just these very nascent stages of fermentation.

Other by-products of yeast include:

- Methanol – Caused by the demethylation of pectins in the must by enzymes of the yeast. More commonly found in red wines than white but only in very small amounts between 20–200 mg/L.
- Fusel oils – Formed by the decomposition of amino acids by the yeast. This includes 2,3-Butanediol which is formed by yeast-consuming diacetyl, the compound that gives Chardonnay and other wines a "buttery" aroma, reducing it first to acetoin and then to the more neutral-smelling 2,3-Butanediol. Many beer and winemakers who have a wine with too much "butteriness" will often "pitch" fresh yeast cultures into the no longer fermenting tank so that the yeast will consume the diacetyl and reduce the aroma.
- Succinic acid – Like glycerol, this is often formed early in fermentation. Usually found in concentrations of 500–1200 mg/L, it is a minor acid in the overall acidity of wine.
- Acetic acid – Considered a main component of volatile acidity that can make a wine taste unbalanced and overly acidic. While acetic acid is the main volatile acid produced by yeast, trace amounts of butyric, formic and propionic acids can also be formed depending on the yeast strain. Most countries have wine laws setting the legal limit of volatile acidity, usually expressed as acetic acid, to 1200–2000 mg/L. Acetic acid can also lead to the development of the wine fault ethyl acetate which is characterized by a "nail polish remover" smell. However, small amounts of acetic acid are actually beneficial for the yeast as they use them to synthesis lipids in the cell membrane.
- Acetaldehyde – While most of the acetaldehyde produce gets reduced to ethanol or is bound by sulfur dioxide, concentrations between 50–100 mg/L can remain in the wine. The flor yeast strains that produce the Spanish wine Sherry will produce higher amounts that contributes to the characterized "aldehydic" aromas of Sherries. In the presence of oxygen, yeast can convert some of the ethanol presence in the wine back into acetaldehyde creating oxidized aromas.
- Hydrogen sulfide – Often produced by yeast during fermentation because of a nitrogen deficiency in the must. This can be done by a reduction of sulfates or sulfites available in the must or by the decomposition of dead yeast cells by other yeast that releases sulfur-containing amino acids that are further broken down by the yeast. The latter often happens with wines that sit in contact with their lees for long periods of time between rackings. In the presence of alcohol, hydrogen sulfide can react with ethanol to form ethyl mercap-

tans and disulfides that contribute to off aromas and wine faults. Some commercial yeast strains, such as *Montrachet 522* are known to produce higher levels of hydrogen sulfides than other strains, particularly if the must has some nutrient deficiencies.

- Pyruvic acid – Along with acetaldehyde, this compound can react with anthocyanins extracted from contact with grape skins to create a more stable color pigment (pyranoanthocyanin) that can enhance the color of some red wines.
- Various esters, ketones, lactones, phenols and acetals.

Lees

When yeast cells die, they sink to the bottom of the fermentation vessel where they combine with insoluble tartrates, grape seeds, skin and pulp fragments to form the lees. During fermentation, the first significant racking which removes the bulk of dead yeast cells is often referred to as the *gross lees* as opposed to the less coarse *fine lees* that come as the wine continues to settle and age. During the time that the wine spends in contact with the lees, a number of changes can impact the wine due to both the autolysis (or self-metabolize) of the dead yeast cells as well as the reductive conditions that can develop if the lees are not aerated or stirred (a process that the French call *bâtonnage*). The length of time that a wine spends on its lees (called *sur lie*) will depend on the winemaking style and type of wine.

The process of leaving the wine to spend some contact with the lees has a long history in winemaking, being known to the Ancient Romans and described by Cato the Elder in the 2nd century BC. Today the practice is widely associated with any red wines that are barrel fermented, Muscadet, sparkling wine Champagne as well as Chardonnay produced in many wine regions across the globe. Typically when wines are left in contact with their lees, they are regularly stirred in order to release the mannoproteins, polysaccharides and other compounds that were present in the yeast cell walls and membranes. This stirring also helps avoid the development of reductive sulfur compounds like mercaptans and hydrogen sulfide that can appear if the lees layer is more than 10 cm (4 inches) thick and undisturbed for more than a week.

Most of the benefits associated with lees contact deals with the influence on the wine of the mannoproteins released during the autolysis of the yeast cells. Composed primarily of mannose and proteins, with some glucose, mannoproteins are often bound in the cell wall of yeast with hydrophobic aroma compounds that become volatilized as the cell wall breaks down. Not only does the release of mannoproteins impart sensory changes in the wine but they can contribute to tartrate and protein stability, help enhance the body and mouthfeel of the wine as well as decrease the perception of bitterness and astringency of tannins.

Secondary Fermentation

The production of Champagne and many sparkling wines requires a second fermentation to occur in the bottle in order to produce the carbonation necessary for the style. A small amount of sugared liquid is added to individual bottles, and the yeast is allowed to convert this to more alcohol and carbon dioxide. The lees are then ricked into the neck of the bottle, frozen, and expelled via pressure of the carbonated wine.

Types of Yeasts Used in Winemaking

Yeast taxonomy includes classification of yeast species depending on the presence or absence of a sexual phase. Therefore, some winemaking yeasts are classified by their asexual anamorph (or "imperfect" form) while others may be classified by their sexual teleomorph (or "perfect" form). A common example of this is *Brettanomyces* (or "Brett") that is usually referenced in wine and viticulture text under its asexual classification though some scientific and winemaking texts may describe specific species (such as *Dekkera bruxellensis*) under its sporulating sexual classification of *Dekkera*. Unless otherwise noted, this article will commonly refer to the asexual form of wine yeast.

The most common yeast generally associated with winemaking is *Saccharomyces cerevisiae* which is also used in bread making and brewing. Other genera of yeast that can be involved in winemaking (either beneficially or as the cause of potential wine faults) include:

- *Brettanomyces* (Teleomorph *Dekkera*)
- *Candida* (Teleomorphs for different species from several genera including *Pichia*, *Metschnikowia*, *Issatchenkia*, *Torulaspora* and *Kluyveromyces*)
- *Kloeckera* (Teleomorph *Hanseniaspora*), usually the most common "wild yeast" found in the vineyard. Some species are known "killer yeast" that produce inhibitory levels of ethyl acetate and acetic acid that can kill off sensitive strains of *Saccharomyces cerevisiae*
- *Saccharomycodes*
- *Schizosaccharomyces*, the only wine yeast that reproduced by fission whereas most wine yeast reproduce by budding.
- *Zygosaccharomyces*, very alcohol-tolerant and can grow in wines up to 18% v/v. Additionally this yeast can survive in extremely high sugar levels (as much as 60% w/w or 60 Brix) and is very resistant to sulfur dioxide.
- *Aureobasidium*, particularly the "black yeast" species of *Aureobasidium pullulans* found in moist cellars that can contaminate aging wine in barrels.

Saccharomyces

The *Saccharomyces* (sugar mold) yeast genus is favored for winemaking (for both grapes as well as other fruit wines in addition also to being used in brewing and breadmaking) because of the generally reliable and positive attributes it can bring to the wine. These yeasts will usually readily ferment glucose, sucrose and raffinose and metabolize glucose, sucrose, raffinose, maltose and ethanol. However, *Saccharomyces* cannot ferment or utilize pentoses (such as arabinose) which is usually present in small amount in wines as residual sugars.

In addition to *Saccharomyces cerevisiae*, other species within the *Saccharomyces* genus that are involved with winemaking include:

- *Saccharomyces bayanus*
- *Saccharomyces beticus*
- *Saccharomyces fermentati*

- *Saccharomyces paradoxus*
- *Saccharomyces pastorianus*
- *Saccharomyces uvarum*

Influences of Different Strains on Fermentation

In 1996, *Saccharomyces cerevisiae* was the first single-celled, eukaryotic organism to have its entire genome sequenced. This sequencing helped confirm the nearly century of work by mycologists and enologists in identifying different strains of *Saccharomyces cerevisiae* that are used in beer, bread and winemaking. Today there are several hundred different strains of *S. cerevisiae* identified. Not all of the strains are suitable for winemaking and even among the strains that are, there is debate among winemakers and scientists about the actual magnitude of differences between the various strains and their potential impact on the wine. Even among strains that have demonstrated distinctive difference when compared among young wines, these differences seem to fade and become less distinctive as the wines age.

Some distinct difference among various strains include the production of certain "off-flavor" and aromas that may be temporary (but producing a "stinky fermentation") or could stay with the wine and either have to be dealt with through other winemaking means (such as the presence of volatile sulfur compounds like hydrogen sulfide) or leave a faulty wine. Another difference includes the "vigor" or speed of fermentation (which can also be influenced by other factors beyond yeast selection) with some yeast strains having the tendency to do "fast ferments" while others may take longer to get going.

Another less measurable difference that are subject to more debate and questions of winemakers preference is the influence of strain selection on the varietal flavors of certainly grape varieties such as Sauvignon blanc and Sémillon. It is believed that these wines can be influenced by thiols produced by the hydrolysis of certain cysteine-linked compounds by enzymes that are more prevalent in particular strains. Other aromatic varieties such as Gewürztraminer, Riesling and Muscat may also be influenced by yeast strains containing high levels of glycosidases enzymes that can modify monoterpenes. Similarly, though potentially to a much smaller extent, other varieties could be influenced by hydrolytic enzymes working on aliphatics, norisoprenoids, and benzene derivatives such as polyphenols in the must.

In sparkling wine production some winemakers select strains (such as one known as *Épernay* named after the town in the Champagne wine region of France and *California Champagne*, also known as UC-Davis strain 505) that are known to flocculate well, allowing the dead yeast cells to be removed easily by riddling and disgorgement. In Sherry production, the surface film of yeast known as flor used to make the distinctive style of fino and manzanilla sherries comes from different strains of *Saccharomyces cerevisiae*, though the commercial flor yeast available for inoculation is often from different species of *Saccharomyces*, *Saccharomyces beticus*, *Saccharomyces fermentati* and *Saccharomyces bayanus*.

Wild Yeasts and Natural Fermentation

In winemaking, the term "wild yeast" has multiple meanings. In its most basic context, it refers to yeast that has not been introduced to the must by intentional inoculation of a cultured strain. Instead, these "wild yeasts" often come into contact with the must through their presence on harvest equipment, transport bins, the surface winemaking equipment and as part of the natural flora of a winery. Very often these are strains of *Saccharomyces cerevisiae* that have taken residence in these places over the years, sometimes being previously introduced by inoculation of prior vintages. In this context, these wild yeasts are often referred to as ambient, indigenous or natural yeast as opposed to inoculated, selected or cultured yeast. Wineries that often solely rely on these "in-house" strains will sometimes market their wines as being the product of wild or *natural fermentations*. The (c. 304) *Nanfang Caomu Zhuang* has the earliest description of winemaking using "herb ferment" wild yeast with rice and various herbs, including the poisonous *Gelsemium elegans*.

Another use of the term "wild yeast" refers to the non-*Saccharomyces* genera of yeasts that are present in the vineyard, on the surface of grapevines and of the grapes themselves. Anywhere from 160 to 100,000 colony forming units of wild yeasts per berry could exist in a typical vineyard. These yeasts can be carried by air currents, birds and insects through the vineyard and even into the winery (such as by fruit flies). The most common wild yeasts found in the vineyard are from the genera *Kloeckera*, *Candida* and *Pichia* with the species *Kloeckera apiculata* being the most dominant species by far. *Saccharomyces cerevisiae*, itself, is actually quite rarely found in the vineyard or on the surface freshly harvested wine grapes unless the winery frequently reintroduced winery waste (such as lees and pomace) into the vineyard.

Unlike the "ambient" *Saccharomyces* wild yeast, these genera of wild yeasts have very low tolerance to both alcohol and sulfur dioxide. They are capable of starting a fermentation and often begin this process as early as the harvest bin when clusters of grapes get slightly crushed under their own weight. Some winemakers will try to "knock out" these yeasts with doses of sulfur dioxide, most often at the crusher before the grapes are pressed or allowed to macerate with skin contact. Other winemakers may allow the wild yeasts to continue fermenting until they succumb to the toxicity of the alcohol they produce which is often between 3–5% alcohol by volume and then letting either inoculated or "ambient" *Saccharomyces* strains finish the fermentation.

The use of both "ambient" and non-*Saccharomyces* wild yeasts carries both potential benefits and risk. Some winemakers feel that the use of resident/indigenous yeast helps contribute to the unique expression of *terroir* in the wine. In wine regions such as Bordeaux, classified and highly regarded estates will often tout the quality of their resident "chateau" strains. To this extent, wineries will often take the leftover pomace and lees from winemaking and return them to the vineyard to be used as compost in order to encourage the sustained presence of favorable strains. But compared to inoculated yeast, these ambient yeasts hold the risk of having a more unpredictable fermentation. Not only could this unpredictability include the presence of off-flavors/aromas and higher volatile acidity but also the potential for a stuck fermentation if the indigenous yeast strains are not vigorous enough to fully convert all the sugars.

It is virtually inevitable that non-*Saccharomyces* wild yeast will have a role in beginning the fermentation of virtually every wine but for the wineries that choose to allow these yeasts to continue fermenting versus minimizing their influence do so with the intent of enhancing complexity through bio-diversity. While these non-*Saccharomyces* ferment glucose and fructose into alcohol, they also have the potential to create other intermediates that could influence the aroma and flavor profile of the wine. Some of these intermediates could be positive, such as phenylethanol, which can impart a rose-like aroma. However, as with ambient yeasts, the products of these yeasts can be very unpredictable – especially in terms of the types of flavors and aromas that these yeasts can produce.

Inoculated Yeast

When winemakers select a cultured yeast strain, it is largely done because the winemaker wants a predictable fermentation taken to completion by a strain that has a track record of dependability. Among the particular considerations that are often important to winemakers is a yeast's tendency to:

- Quickly begin fermentation, out-competing other "wild yeasts" for nutrients in the must
- Completely utilize all fermentable sugars with a predictable sugar-to-alcohol conversion rate
- Have an alcohol-tolerance up to 15% or even higher depending on the winemaking style
- Have a high sulfur dioxide tolerance but low production of sulfur compounds such as hydrogen sulfide or dimethyl sulfide
- Produce a minimum amount of residual pyruvate, acetic acid and acetaldehyde
- Produce minimum foaming during fermentation which may create difficulties for cap management during maceration or cause bungs to pop out during barrel fermentation.
- Have high levels of flocculation and lees compaction that makes racking, fining and filtering of the wine easier.

Inoculated (or *pure cultured*) yeasts are strains of *Saccharomyces cerevisiae* that have been identified and plated from wineries across the world (including notable producers from well-known wine regions such as Bordeaux, Burgundy, Napa Valley and the Barossa Valley). These strains are tested in laboratories to determine a strain's vigor, sulfur dioxide and alcohol tolerance, production levels of acetic acid and sulfur compounds, ability to re-ferment (positive for sparkling wine but a negative attribute for sweet late-harvest wines), development of surface film on the wine (positive for some Sherry styles but a negative attribute for many other wines), enhancement of a wine's color or certain varietal characteristics by enzymes in the yeast cells and other metabolic products produced by the yeast, foaming and flocculation tendencies, yeasticidal properties (a trait known as "Killer yeast") and tolerance for nutritional deficiencies in a must that may lead to a stuck fermentation.

Re-hydrating Freeze Dried Yeast Cultures

Pure culture yeasts that are grown in a lab are often freeze dried and packaged for commercial use. Prior to their addition into must, these yeasts need to be re-hydrated in "starter cultures" that must be carefully monitored (particularly in regards to temperature) to insure that the yeast cells are not killed off by cold shock. Ideally winemakers want to add enough inoculum to have a viable cell population density of 5 million cells per milliliter. The exact amount of freeze-dried culture varies by manufacturer and strain of yeast but it is often around 1 gram per gallon (or 25 grams per 100 liters). Wines that could have potentially problematic fermentation (such as high sugar level late harvest or botryized wines) may have more yeast added.

Similarly, re-hydration procedures will also vary depending on the manufacturer and winery. Yeast is often inoculated in a volume of water or grape must that is 5–10 times the weight of the dry yeast. This liquid is often brought to temperature of 40 °C (104 °F) prior to the introduction of the yeast (though some yeast strains may need temperatures below 38 °C (100 °F)) to allow the cells to disperse easily rather than clump and sink to the bottom of the container. The heat activation also allows the cells to quickly reestablish their membrane barrier before soluble cytoplasmic components escape the cell. Re-hydration at lower temperatures can greatly reduce the viability of the yeast with up to 60% cell death if the yeast is re-hydrated at 15 °C (60 °F). The culture is then stirred and aerated to incorporate oxygen into the culture which the yeast uses in the synthesis of needed survival factors.

The temperature of the starter culture is then slowly reduced, often by the graduated addition of must to get within 5–10 °C (41–50 °F) of the must that the culture will be added to. This is done to avoid the sudden cold shock that the yeast cells may experience if the starter culture was added directly to the must itself which can kill up 60% of the culture. Additionally, surviving cells exposed to cold shock tend to see an increase in hydrogen sulfide production.

Nutritional Needs of Wine Yeast

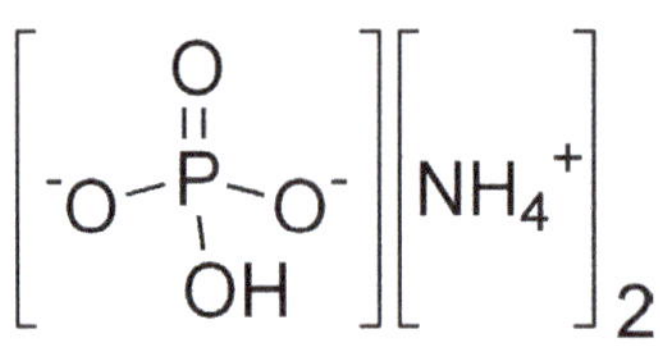

Diammonium phosphate (or DAP) is a common additive that provides two necessary nutrients for yeast to have a healthy and sustained fermentation—nitrogen and phosphate.

In order to successfully complete a fermentation with minimum to no negative attributes being added to the wine, yeast needs to have the full assortment of its nutritional needs met. These include not only an available energy source (carbon in the form of sugars such as glucose) and yeast assimilable nitrogen (ammonia and amino acids or YAN) but also minerals (such as magnesium) and vitamins (such as thiamin and riboflavin) that serve as important growth and survival factors. Among the other nutritional needs of wine yeast:

- Phosphate – used for the production of nucleic acids, phospholipids (an important component of the cell membrane) and ATP (Adenosine triphosphate which the cell uses for transferring energy for metabolism).
- Potassium – important for the uptake and utilization of phosphate
- Biotin – involved in the synthesis of proteins, fatty acids and nucleic acids.
- Pantothenic acid – involved in the metabolism of sugars and lipids. A deficiency of this vitamin could lead into increase hydrogen sulfide production with off-aromas in the resulting wine.
- Nicotinic acid – involved in the synthesis of Nicotinamide adenine dinucleotide (NAD+),

a co-enzyme that is important in maintaining the redox balance of the cell as well as in the process of ethanol fermentation itself.

- Inositol – involved with the secondary messenger molecules that facilitate cell division.
- Trace amounts of calcium, chlorine, copper, iron, manganese and zinc for healthy cell function.

Many of these nutrients are available in the must and skins of the grapes themselves but sometimes are supplemented by winemakers with additions such as diammonium phosphate (DAP), freeze-dried micro-nutrients (such as *Go-Ferm* and *Ferm-K*) and even the remnant of dead or extracted yeast cells such that the fermenting yeast can break down to mine for available nitrogen and nutrients. One historical winemaking tradition that is still practiced in some Italian wine regions is the *ripasso* method of adding the leftover pomace from the pressing of other wines into a newly fermenting batch of wine as an additional food source for the yeast.

Saccharomyces cerevisiae can assimilate nitrogen from both inorganic (ammonia and ammonium) and organic forms (amino acids, particularly arginine). As yeast cells die, enzymes within the cells begin autolyzing by breaking down the cell, including the amino acids. This autolysis of the cell provides an available nitrogen source for the still-fermenting and viable yeast cells. However, this autolysis can also release sulfur-link compounds (such as the breakdown of amino acid cysteine) which can combine with other molecules and react with alcohol to create volatile thiols that can contribute to a "stinky fermentation" or later development into various wine faults.

The Role of Oxygen

Yeasts are facultative anaerobes meaning that they can exist in both the presence and absence of oxygen. While fermentation is traditionally thought of as an anaerobic process done in the absence of oxygen, early exposure of the yeast to oxygen can be a vital component in the successful completion of that fermentation. This is because oxygen is important in the synthesis of cell "survival factors" such as ergosterol and lanosterol. These sterols are important in maintaining the selective permeability of the yeast cell membrane which becomes critical as the yeast becomes exposed to increasing osmotic pressure and levels of alcohol in the wine. As a waste product of its own metabolism, alcohol is actually very toxic to yeast cells. Yeast with weak survival factors and lacking sterols may succumb to these conditions before fermenting a wine to complete dryness, leaving a stuck fermentation.

Cultured yeasts that are freeze-dried and available for inoculation of wine must are deliberately grown in commercial labs in high oxygen/low sugar conditions that favor the development of these survival factors. One of the reasons that some winemakers prefer using inoculated yeast is the predictability of fermentation due to the high level of survival factors that cultured yeast are assured of having without necessarily needing to expose the wine to additional levels of oxygen. Winemakers using "ambient" yeasts that are resident in their winery may not have this same assurance of survival factors and may need to compensate with other winemaking techniques.

Wild non-*Saccharomyces* yeasts often need a much greater exposure to oxygen in order to build up survival factors which is why many of these yeasts are often found living oxidatively as "film yeast" on the surface of wines in tanks or barrels.

Wine Faults Related to Yeast

Either directly or indirectly, wine yeast can be a culprit behind a wide variety of wine faults. These can include the presence of "off flavors" and aromas that can be the by-product of some "wild yeast" fermentation such as those by species within the genera of *Kloeckera* and *Candida*. Even the common wine yeast *Saccharomyces cerevisiae* can be behind some wine faults with some strains of the yeast known to produce higher than ideal levels of acetic acid, acetaldehyde and volatile sulfur compounds such as thiols. Also any yeast can have a low tolerance to nutritional deficiencies, temperature fluctuation or extremes and excessive or low sugar levels that may lead to a stuck fermentation.

In the presence of oxygen several species of *Candida* and *Pichia* can create a film surface on top of the wine in the tank of barrel. Allowed to go unchecked, these yeasts can rapidly deplete the available free sulfur compounds that keeps a wine protected from oxidation and other microbial attack. The presence of these yeasts is often identified by elevated levels of volatile acidity, particularly acetic acid. Some strains of *Pichia* will metabolize acetic acid (as well as ethyl acetate and isoamyl acetate that may also be produced) with the side-effect of substantially decreasing the titratable acidity and shifting the pH of wine upwards to levels that make the wine prone to attack by other spoilage microbes. Commonly called "film yeast", these yeasts are distinguished from the flor sherry yeast that are usually welcomed by winemakers in producing the delicate fino-style wines.

Growth of many unfavorable wild yeasts is generally slowed at lower cellar temperatures, so many winemakers who wish to inhibit the activities of these yeasts before the more favorable *Saccharomyces* yeast kick in, will often chill their must, such as the practice of "cold soaking" the must during a pre-fermentation maceration at temperatures between 4–15 °C (39–50 °F). Though some species, such as *Brettanomyces*, will not be inhibited and may even thrive during an extended period of cold soaking.

Brettanomyces

The wine yeast *Brettanomyces* (or "Brett") produces very distinctive aroma compounds, 4-Ethylphenol (4-EP) and 4-Ethylguaiacol (4-EG), that can have a wine being described as smelling like a "barnyard", "wet saddle" or "band-aid". To some winemakers and with some wine styles (such as Pinot noir from Burgundy), a limited amount of these compounds could be considered a positive attribute that adds to the complexity of wine. To other winemakers and with other wine styles (such as Riesling from the Mosel), the presence of any Brett will be considered a fault. Fruit flies are common vector in the transfer of *Brettanomyces* between tanks and even nearby wineries.

As a fermentation yeast, *Brettanomyces* can usually ferment a wine up to 10-11% alcohol levels before they die out. Sometimes *Brettanomyces* already present in a wine that has been inoculated with *Saccharomyces cerevisiae* will out compete the *Saccharomyces* strain for nutrients and even inhibit it due to the high levels of acetic acid, decanoic acid and octanoic acid that many strains of *Brettanomyces* can produce.

Once Brett is in a winery, it is very difficult to control even with strict hygiene and the discarding of barrels and equipment that has previously come into contact with "Bretty" wine. This is because many species of *Brettanomyces* can use a wide variety of carbon sources in wine and grape must, including ethanol, for metabolism. Additionally, Brett can produce a wide range of by-products that could influence the wine beyond just the 4-EP and 4-EG compounds previously discussed. Many of these compounds, such as the "footprints" of the 4-EP and 4-EG, will still remain in the wine even after yeast cells die and are removed by racking and sterile filtration.

Cheese

Cheese is a food derived from milk that is produced in a wide range of flavors, textures, and forms by coagulation of the milk protein casein. It comprises proteins and fat from milk, usually the milk of cows, buffalo, goats, or sheep. During production, the milk is usually acidified, and adding the enzyme rennet causes coagulation. The solids are separated and pressed into final form. Some cheeses have molds on the rind or throughout. Most cheeses melt at cooking temperature.

Coulommiers cheese

A platter with cheese and garnishes

A variety of cheeses

Hundreds of types of cheese from various countries are produced. Their styles, textures and flavors depend on the origin of the milk (including the animal's diet), whether they have been pasteurized, the butterfat content, the bacteria and mold, the processing, and aging. Herbs, spices, or wood smoke may be used as flavoring agents. The yellow to red color of many cheeses, such as Red

Leicester, is produced by adding annatto. Other ingredients may be added to some cheeses, such as black pepper, garlic, chives or cranberries.

For a few cheeses, the milk is curdled by adding acids such as vinegar or lemon juice. Most cheeses are acidified to a lesser degree by bacteria, which turn milk sugars into lactic acid, then the addition of rennet completes the curdling. Vegetarian alternatives to rennet are available; most are produced by fermentation of the fungus *Mucor miehei*, but others have been extracted from various species of the *Cynara* thistle family. Cheesemakers near a dairy region may benefit from fresher, lower-priced milk, and lower shipping costs.

Cheese is valued for its portability, long life, and high content of fat, protein, calcium, and phosphorus. Cheese is more compact and has a longer shelf life than milk, although how long a cheese will keep depends on the type of cheese; labels on packets of cheese often claim that a cheese should be consumed within three to five days of opening. Generally speaking, hard cheeses, such as parmesan last longer than soft cheeses, such as Brie or goat's milk cheese. The long storage life of some cheeses, especially when encased in a protective rind, allows selling when markets are favorable.

There is some debate as to the best way to store cheese, but some experts say that wrapping it in cheese paper provides optimal results. Cheese paper is coated in a porous plastic on the inside, and the outside has a layer of wax. This specific combination of plastic on the inside and wax on the outside protects the cheese by allowing condensation on the cheese to be wicked away while preventing moisture from within the cheese escaping.

A specialist seller of cheese is sometimes known as a *cheesemonger*. Becoming an expert in this field requires some formal education and years of tasting and hands-on experience, much like becoming an expert in wine or cuisine. The cheesemonger is responsible for all aspects of the cheese inventory: selecting the cheese menu, purchasing, receiving, storage, and ripening.

Etymology

Cheese on market stand in Basel, Switzerland

The word *cheese* comes from Latin *caseus*, from which the modern word casein is also derived. The earliest source is from the proto-Indo-European root **kwat-*, which means "to ferment, become sour".

More recently, *cheese* comes from *chese* (in Middle English) and *cīese* or *cēse* (in Old English). Similar words are shared by other West Germanic languages—West Frisian *tsiis*, Dutch *kaas*, German *Käse*, Old High German *chāsi*—all from the reconstructed West-Germanic form **kāsī*, which in turn is an early borrowing from Latin.

When the Romans began to make hard cheeses for their legionaries' supplies, a new word started to be used: *formaticum*, from *caseus formatus*, or "molded cheese" (as in "formed", not "moldy"). It is from this word that the French *fromage*, proper Italian *formaggio*, Catalan *formatge*, Breton *fourmaj*, and Provençal *furmo* are derived. Of the Romance languages, Spanish, Portuguese, Romanian, Tuscan and Southern Italian dialects use words derived from *caseus* (*queso*, *queijo*, *caş* and *caso* for example). The word *cheese* itself is occasionally employed in a sense that means "molded" or "formed". *Head cheese* uses the word in this sense.

History

Origins

A piece of soft curd cheese, oven-baked to increase longevity

Cheese is an ancient food whose origins predate recorded history. There is no conclusive evidence indicating where cheesemaking originated, either in Europe, Central Asia or the Middle East, but the practice had spread within Europe prior to Roman times and, according to Pliny the Elder, had become a sophisticated enterprise by the time the Roman Empire came into being.

The earliest evidence of cheese-making in the archaeological record dates back to 5,500 BCE, in what is now Kujawy, Poland, where strainers with milk fats molecules have been found. Earliest proposed dates for the origin of cheesemaking range from around 8000 BCE, when sheep were first domesticated. Since animal skins and inflated internal organs have, since ancient times, provided storage vessels for a range of foodstuffs, it is probable that the process of cheese making was discovered accidentally by storing milk in a container made from the stomach of an animal, resulting in the milk being turned to curd and whey by the rennet from the stomach. There is a legend – with variations – about the discovery of cheese by an Arab trader who used this method of storing milk.

Cheesemaking may have begun independently of this by the pressing and salting of curdled milk to preserve it. Observation that the effect of making cheese in an animal stomach gave more solid and better-textured curds may have led to the deliberate addition of rennet.

Early archeological evidence of Egyptian cheese has been found in Egyptian tomb murals, dating to about 2000 BCE. The earliest cheeses were likely to have been quite sour and salty, similar in texture to rustic cottage cheese or feta, a crumbly, flavorful Greek cheese.

Cheese produced in Europe, where climates are cooler than the Middle East, required less salt for preservation. With less salt and acidity, the cheese became a suitable environment for useful microbes and molds, giving aged cheeses their respective flavors.

The earliest ever discovered preserved cheese was found in the Taklamakan Desert in Xinjiang, China, and it dates back as early as 1615 BCE.

Ancient Greece and Rome

Cheese in a market in Italy

Ancient Greek mythology credited Aristaeus with the discovery of cheese. Homer's *Odyssey* (8th century BCE) describes the Cyclops making and storing sheep's and goats' milk cheese (translation by Samuel Butler):

We soon reached his cave, but he was out shepherding, so we went inside and took stock of all that we could see. His cheese-racks were loaded with cheeses, and he had more lambs and kids than his pens could hold...

When he had so done he sat down and milked his ewes and goats, all in due course, and then let each of them have her own young. He curdled half the milk and set it aside in wicker strainers.

By Roman times, cheese was an everyday food and cheesemaking a mature art. Columella's *De Re Rustica* (circa 65 CE) details a cheesemaking process involving rennet coagulation, pressing of the curd, salting, and aging. Pliny's *Natural History* (77 CE) devotes a chapter (XI, 97) to describing the diversity of cheeses enjoyed by Romans of the early Empire. He stated that the best cheeses came from the villages near Nîmes, but did not keep long and had to be eaten fresh. Cheeses of the Alps and Apennines were as remarkable for their variety then as now. A Ligurian cheese was noted for being made mostly from sheep's milk, and some cheeses produced nearby were stated to weigh as much as a thousand pounds each. Goats' milk cheese was a recent taste in Rome, improved over the "medicinal taste" of Gaul's similar cheeses by smoking. Of cheeses from overseas, Pliny preferred those of Bithynia in Asia Minor.

Cheese, Tacuinum sanitatis Casanatensis (14th century)

Post-Roman Europe

As Romanized populations encountered unfamiliar newly settled neighbors, bringing their own cheese-making traditions, their own flocks and their own unrelated words for *cheese*, cheeses in Europe diversified further, with various locales developing their own distinctive traditions and products. As long-distance trade collapsed, only travelers would encounter unfamiliar cheeses: Charlemagne's first encounter with a white cheese that had an edible rind forms one of the constructed anecdotes of Notker's *Life* of the Emperor.

The British Cheese Board claims that Britain has approximately 700 distinct local cheeses; France and Italy have perhaps 400 each. (A French proverb holds there is a different French cheese for every day of the year, and Charles de Gaulle once asked "how can you govern a country in which there are 246 kinds of cheese?") Still, the advancement of the cheese art in Europe was slow during the centuries after Rome's fall. Many cheeses today were first recorded in the late Middle Ages or after—cheeses like Cheddar around 1500, Parmesan in 1597, Gouda in 1697, and Camembert in 1791.

In 1546 *The Proverbs of John Heywood* claimed "the moon is made of a greene cheese." (*Greene* may refer here not to the color, as many now think, but to being new or unaged.) Variations on this sentiment were long repeated and NASA exploited this myth for an April Fools' Day spoof announcement in 2006.

Modern Era

Until its modern spread along with European culture, cheese was nearly unheard of in east Asian cultures, in the pre-Columbian Americas, and only had limited use in sub-Mediterranean Africa, mainly being widespread and popular only in Europe, the Middle East, the Indian subcontinent, and areas influenced by those cultures. But with the spread, first of European imperialism, and later of Euro-American culture and food, cheese has gradually become known and increasingly popular worldwide.

The first factory for the industrial production of cheese opened in Switzerland in 1815, but large-scale production first found real success in the United States. Credit usually goes to Jesse Williams, a dairy farmer from Rome, New York, who in 1851 started making cheese in an assembly-line fashion using the milk from neighboring farms. Within decades, hundreds of such dairy associations existed.

The 1860s saw the beginnings of mass-produced rennet, and by the turn of the century scientists were producing pure microbial cultures. Before then, bacteria in cheesemaking had come from the environment or from recycling an earlier batch's whey; the pure cultures meant a more standardized cheese could be produced.

Factory-made cheese overtook traditional cheesemaking in the World War II era, and factories have been the source of most cheese in America and Europe ever since.

Cheese production – 2013	
Country	**Production (millions of tonnes)**
United States	5.4
Germany	2.2
France	1.9
Italy	1.2
Netherlands	0.8
World	**21.3**
Source: FAOSTAT of the United Nations	

Production

In 2013, world production of cheese was 21.3 million tonnes, with the United States accounting for 25% (5.4 million tonnes) of the world total followed by Germany, France and Italy (table).

During 2015, Germany, France, Netherlands and Italy exported 10-14% of their produced cheese. The United States, the biggest world producer of cheese, is a marginal exporter (5.3% of total), as most of its production is for the domestic market.

Consumption

France, Iceland, Finland, Denmark and Germany were the highest consumers of cheese in 2014, averaging 25 kg (55 lb) per person.

Processing

Curdling

A required step in cheesemaking is separating the milk into solid curds and liquid whey. Usually this is done by acidifying (souring) the milk and adding rennet. The acidification can be accomplished directly by the addition of an acid, such as vinegar, in a few cases (paneer, queso fresco). More commonly starter bacteria are employed instead which convert milk sugars into lactic acid. The same bacteria (and the enzymes they produce) also play a large role in the eventual flavor of aged cheeses. Most cheeses are made with starter bacteria from the *Lactococcus*, *Lactobacillus*, or

Streptococcus families. Swiss starter cultures also include *Propionibacter shermani*, which produces carbon dioxide gas bubbles during aging, giving Swiss cheese or Emmental its holes (called "eyes").

During industrial production of Emmental cheese, the as-yet-undrained curd is broken by rotating mixers.

Some fresh cheeses are curdled only by acidity, but most cheeses also use rennet. Rennet sets the cheese into a strong and rubbery gel compared to the fragile curds produced by acidic coagulation alone. It also allows curdling at a lower acidity—important because flavor-making bacteria are inhibited in high-acidity environments. In general, softer, smaller, fresher cheeses are curdled with a greater proportion of acid to rennet than harder, larger, longer-aged varieties.

While rennet was traditionally produced via extraction from the inner mucosa of the fourth stomach chamber of slaughtered young, unweaned calves, most rennet used today in cheesemaking is produced recombinantly. The majority of the applied chymosin is retained in the whey and, at most, may be present in cheese in trace quantities. In ripe cheese, the type and provenance of chymosin used in production cannot be determined.

Curd Processing

Cheese factory in the Netherlands

At this point, the cheese has set into a very moist gel. Some soft cheeses are now essentially complete: they are drained, salted, and packaged. For most of the rest, the curd is cut into small cubes. This allows water to drain from the individual pieces of curd.

Some hard cheeses are then heated to temperatures in the range of 35–55 °C (95–131 °F). This forces more whey from the cut curd. It also changes the taste of the finished cheese, affecting both the bacterial culture and the milk chemistry. Cheeses that are heated to the higher temperatures are usually made with thermophilic starter bacteria that survive this step—either *Lactobacilli* or *Streptococci.*

Salt has roles in cheese besides adding a salty flavor. It preserves cheese from spoiling, draws moisture from the curd, and firms cheese's texture in an interaction with its proteins. Some cheeses are salted from the outside with dry salt or brine washes. Most cheeses have the salt mixed directly into the curds.

Other techniques influence a cheese's texture and flavor. Some examples are :

- Stretching: (Mozzarella, Provolone) The curd is stretched and kneaded in hot water, developing a stringy, fibrous body.
- Cheddaring: (Cheddar, other English cheeses) The cut curd is repeatedly piled up, pushing more moisture away. The curd is also mixed (or *milled*) for a long time, taking the sharp edges off the cut curd pieces and influencing the final product's texture.
- Washing: (Edam, Gouda, Colby) The curd is washed in warm water, lowering its acidity and making for a milder-tasting cheese.

Most cheeses achieve their final shape when the curds are pressed into a mold or form. The harder the cheese, the more pressure is applied. The pressure drives out moisture—the molds are designed to allow water to escape—and unifies the curds into a single solid body.

Parmigiano-Reggiano in a modern factory

Ripening

A newborn cheese is usually salty yet bland in flavor and, for harder varieties, rubbery in texture. These qualities are sometimes enjoyed—cheese curds are eaten on their own—but normally cheeses are left to rest under controlled conditions. This aging period (also called ripening, or, from the French, *affinage*) lasts from a few days to several years. As a cheese ages, microbes and enzymes transform texture and intensify flavor. This transformation is largely a result of the breakdown of casein proteins and milkfat into a complex mix of amino acids, amines, and fatty acids.

Some cheeses have additional bacteria or molds intentionally introduced before or during aging. In traditional cheesemaking, these microbes might be already present in the aging room; they are simply allowed to settle and grow on the stored cheeses. More often today, prepared cultures are used, giving more consistent results and putting fewer constraints on the environment where the cheese ages. These cheeses include soft ripened cheeses such as Brie and Camembert, blue cheeses such as Roquefort, Stilton, Gorgonzola, and rind-washed cheeses such as Limburger.

Types

Feta from Greece

There are many types of cheese, with around 500 different varieties recognized by the International Dairy Federation, more than 400 identified by Walter and Hargrove, more than 500 by Burkhalter, and more than 1,000 by Sandine and Elliker. The varieties may be grouped or classified into types according to criteria such as length of ageing, texture, methods of making, fat content, animal milk, country or region of origin, etc.—with these criteria either being used singly or in combination, but with no single method being universally used. The method most commonly and traditionally used is based on moisture content, which is then further discriminated by fat content and curing or ripening methods. Some attempts have been made to rationalise the classification of cheese—a scheme was proposed by Pieter Walstra which uses the primary and secondary starter combined with moisture content, and Walter and Hargrove suggested classifying by production methods which produces 18 types, which are then further grouped by moisture content.

Local cheese at an open-air market in Peru.

Moisture content (soft to hard)

Categorizing cheeses by firmness is a common but inexact practice. The lines between "soft", "semi-soft", "semi-hard", and "hard" are arbitrary, and many types of cheese are made in softer or firmer variations. The main factor that controls cheese hardness is moisture content, which depends largely on the pressure with which it is packed into molds, and on aging time.

Fresh, whey and stretched curd cheeses

The main factor in the categorization of these cheeses is their age. Fresh cheeses without additional preservatives can spoil in a matter of days.

Content (double cream, goat, ewe and water buffalo)

Emmental

Some cheeses are categorized by the source of the milk used to produce them or by the added fat content of the milk from which they are produced. While most of the world's commercially available cheese is made from cows' milk, many parts of the world also produce cheese from goats and sheep. Double cream cheeses are soft cheeses of cows' milk enriched with cream so that their fat content is 60% or, in the case of triple creams, 75%. The use of the terms "double" or "triple" is not meant to give a quantitative reference to the change in fat content, since the fat content of whole cows' milk is 3%-4%.

Soft-ripened and blue-vein

There are at least three main categories of cheese in which the presence of mold is a significant feature: soft ripened cheeses, washed rind cheeses and blue cheeses.

Processed cheeses

Processed cheese is made from traditional cheese and emulsifying salts, often with the addition of milk, more salt, preservatives, and food coloring. It is inexpensive, consistent, and melts smoothly. It is sold packaged and either pre-sliced or unsliced, in a number of varieties. It is also available in aerosol cans in some countries.

Cooking and Eating

At refrigerator temperatures, the fat in a piece of cheese is as hard as unsoftened butter, and its protein structure is stiff as well. Flavor and odor compounds are less easily liberated when cold. For improvements in flavor and texture, it is widely advised that cheeses be allowed to warm up to room temperature before eating. If the cheese is further warmed, to 26–32 °C (79–90 °F), the fats will begin to "sweat out" as they go beyond soft to fully liquid.

Zigerbrüt, cheese grated onto bread through a mill, from the Canton of Glarus in Switzerland.

Above room temperatures, most hard cheeses melt. Rennet-curdled cheeses have a gel-like protein matrix that is broken down by heat. When enough protein bonds are broken, the cheese itself turns from a solid to a viscous liquid. Soft, high-moisture cheeses will melt at around 55 °C (131 °F), while hard, low-moisture cheeses such as Parmesan remain solid until they reach about 82 °C (180 °F). Acid-set cheeses, including halloumi, paneer, some whey cheeses and many varieties of fresh goat cheese, have a protein structure that remains intact at high temperatures. When cooked, these cheeses just get firmer as water evaporates.

Saganaki, lit on fire, served in Chicago.

Some cheeses, like raclette, melt smoothly; many tend to become stringy or suffer from a separation of their fats. Many of these can be coaxed into melting smoothly in the presence of acids or starch. Fondue, with wine providing the acidity, is a good example of a smoothly melted cheese dish. Elastic stringiness is a quality that is sometimes enjoyed, in dishes including pizza and Welsh rarebit. Even a melted cheese eventually turns solid again, after enough moisture is cooked off. The saying "you can't melt cheese twice" (meaning "some things can only be done once") refers to the fact that oils leach out during the first melting and are gone, leaving the non-meltable solids behind.

As its temperature continues to rise, cheese will brown and eventually burn. Browned, partially burned cheese has a particular distinct flavor of its own and is frequently used in cooking (e.g., sprinkling atop items before baking them).

Nutrition and Health

Mozzarella

The nutritional value of cheese varies widely. Cottage cheese may consist of 4% fat and 11% protein while some whey cheeses are 15% fat and 11% protein, and triple-crème cheeses are 36% fat and 7% protein. In general, cheese is a rich source (20% or more of the Daily Value, DV) of calcium, protein, phosphorus, sodium and saturated fat. A 28-gram (0.99 oz) (one ounce) serving of cheddar cheese contains about 7 grams (0.25 oz) of protein and 202 milligrams of calcium. Nutritionally, cheese is essentially concentrated milk: it takes about 200 grams (7.1 oz) of milk to provide that much protein, and 150 grams (5.3 oz) to equal the calcium.

MacroNutrients (grams) of common cheeses per 100gm				
Cheese	**Water**	**Protein**	**Fat**	**Carbs**
Swiss	37.1	26.9	27.8	5.4
Feta	55.2	14.2	21.3	4.1
Cheddar	36.8	24.9	33.1	1.3
Mozarella	50	22.2	22.4	2.2
Cottage	80	11.1	4.3	3.4

Vitamin contents in %DV of common cheeses per 100gm													
Cheese	**A**	**B1**	**B2**	**B3**	**B5**	**B6**	**B9**	**B12**	**Ch.**	**C**	**D**	**E**	**K**
Swiss	17	4	17	0	4	4	1	**56**	2.8	0	11	2	3
Feta	8	10	50	5	10	21	8	28	2.2	0	0	1	2
Cheddar	20	2	22	0	4	4	5	14	3	0	3	1	3
Mozzarella	14	2	17	1	1	2	2	38	2.8	0	0	1	3
Cottage	3	2	10	0	6	2	3	7	3.3	0	0	0	0

Mineral contents in %DV of common cheeses per 100 grams										
Cheese	**Ca**	**Fe**	**Mg**	**P**	**K**	**Na**	**Zn**	**Cu**	**Mn**	**Se**
Swiss	**79**	10	1	**57**	2	8	29	2	0	26

Mineral contents in %DV of common cheeses per 100 grams										
Cheese	**Ca**	**Fe**	**Mg**	**P**	**K**	**Na**	**Zn**	**Cu**	**Mn**	**Se**
Feta	**49**	4	5	**34**	2	46	19	2	1	21
Cheddar	72	4	7	51	3	26	21	2	1	20
Mozzarella	**51**	2	5	**35**	2	26	19	1	1	24
Cottage	8	0	2	16	3	15	3	1	0	14

Ch. = Choline; Ca = Calcium; Fe = Iron; Mg = Magnesium; P = Phosphorus; K = Potassium; Na = Sodium; Zn = Zinc; Cu = Copper; Mn = Manganese; Se = Selenium;

Note : All nutrient values including protein are in %DV per 100 grams of the food item except for Macronutrients. Source : Nutritiondata.self.com

Neonatal Infection and Death

Cheese has the potential for promoting the growth of *Listeria* bacteria. *Listeria monocytogenes* can also cause serious infection in an infant and pregnant woman and can be transmitted to her infant in utero or after birth. The infection has the potential of seriously harming or even causing the death of a preterm infant, an infant of low or very low birth weight, or an infant with an immune system deficiency or a congenital defect of the immune system. The presence of this pathogen can sometimes be determined by the symptoms that appear as a gastrointestinal illness in the mother. The mother can also acquire infection from ingesting food that contains other animal products such as, unpasteurized milk, delicatessen meats, and hot dogs.

Heart Disease

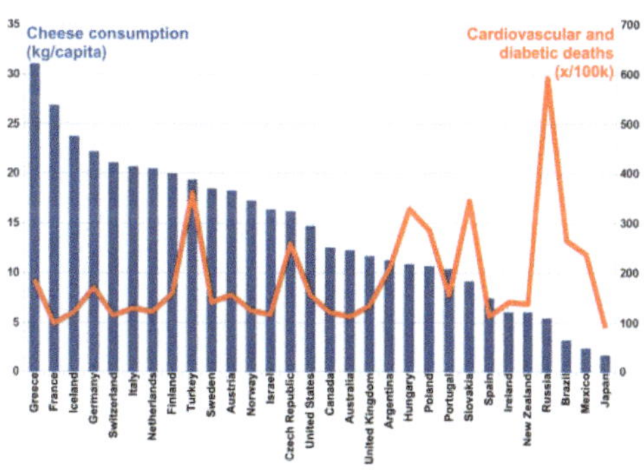

Average cheese consumption and rates of mortality due to cardiovascular disease or diabetes

A review of the medical literature published in 2012 noted that: "Cheese consumption is the leading contributor of SF (saturated fat) in the U.S. diet, and therefore would be predicted to increase LDL-C (LDL cholesterol) and consequently increase the risk of CVD (cardiovascular disease)." It found that: "Based on results from numerous prospective observational studies and meta-analyses, most, but not all, have shown no association and in some cases an inverse relationship between the intake of milk fat containing dairy products and the risk of CVD, CHD (coronary heart disease), and stroke. A limited number of prospective cohort studies found no significant association between the intake of total full-fat dairy products and the risk of CHD or stroke....Most clinical studies showed that full-fat natural cheese, a highly fermented product, significantly lowers LDL-C compared with butter intake of equal total fat and saturated fat content."

Pasteurization

A number of food safety agencies around the world have warned of the risks of raw-milk cheeses. The U.S. Food and Drug Administration states that soft raw-milk cheeses can cause "serious infectious diseases including listeriosis, brucellosis, salmonellosis and tuberculosis". It is U.S. law since 1944 that all raw-milk cheeses (including imports since 1951) must be aged at least 60 days. Australia has a wide ban on raw-milk cheeses as well, though in recent years exceptions have been made for Swiss Gruyère, Emmental and Sbrinz, and for French Roquefort. There is a trend for cheeses to be pasteurized even when not required by law.

Pregnant women may face an additional risk from cheese; the U.S. Centers for Disease Control has warned pregnant women against eating soft-ripened cheeses and blue-veined cheeses, due to the listeria risk, which can cause miscarriage or harm the fetus.

Cultural Attitudes

A cheese merchant in a French market

A traditional Polish sheep's cheese market in Zakopane, Poland

Although cheese is a vital source of nutrition in many regions of the world and is extensively consumed in others, its use is not universal.

Cheese is rarely found in East Asian cuisines, presumably for historical reasons.However, East Asian sentiment against cheese is not universal. In Nepal, the Dairy Development Corporation commercially manufactures cheese made from yak milk and a hard cheese made from either cow or yak milk knows as chhurpi. The national dish of Bhutan, *ema datsi*, is made from homemade yak or mare milk cheese and hot peppers. In Yunnan China, several ethnic minority groups produce Rushan and Rubing from cow's milk. Cheese consumption may be increasing in China, with annual sales doubling from 1996 to 2003 (to a still small 30 million U.S. dollars a year). Certain kinds of Chinese preserved bean curd are sometimes misleadingly referred to in English as "Chinese cheese" because of their texture and strong flavor.

Strict followers of the dietary laws of Islam and Judaism must avoid cheeses made with rennet from animals not slaughtered in a manner adhering to halal or kosher laws. Both faiths allow cheese made with vegetable-based rennet or with rennet made from animals that were processed in a halal or kosher manner. Many less orthodox Jews also believe that rennet undergoes enough processing to change its nature entirely and do not consider it to ever violate kosher law. As cheese is a dairy food, under kosher rules it cannot be eaten in the same meal with any meat.

Rennet derived from animal slaughter, and thus cheese made with animal-derived rennet, is not vegetarian. Most widely available vegetarian cheeses are made using rennet produced by fermentation of the fungus *Mucor miehei*. Vegans and other dairy-avoiding vegetarians do not eat conventional cheese, but some vegetable-based cheese substitutes (soy or almond) are used as substitutes.

Even in cultures with long cheese traditions, consumers may perceive some cheeses that are especially pungent-smelling or mold-bearing varieties such as Limburger or Roquefort, as unpalatable. Such cheeses are an acquired taste because they are processed using molds or microbiological cultures, allowing odor and flavor molecules to resemble those in rotten foods. One author stated: "An aversion to the odor of decay has the obvious biological value of steering us away from possible food poisoning, so it is no wonder that an animal food that gives off whiffs of shoes and soil and the stable takes some getting used to."

Collecting cheese labels is called "tyrosemiophilia".

Yogurt

Yogurt, yoghurt, or yoghourt is a food produced by bacterial fermentation of milk.

The bacteria used to make yogurt are known as "yogurt cultures". Fermentation of lactose by these bacteria produces lactic acid, which acts on milk protein to give yogurt its texture and characteristic tang. Cow's milk is commonly available worldwide, and, as such, is the milk most commonly used to make yogurt. Milk from water buffalo, goats, ewes, mares, camels, and yaks is also used to produce yogurt where available locally. Milk used may be homogenized or not (milk distributed in many parts of the world is homogenized); both types may be used, with substantially different results.

Yogurt is produced using a culture of *Lactobacillus delbrueckii* subsp. *bulgaricus* and *Streptococcus thermophilus* bacteria. In addition, other lactobacilli and bifidobacteria are also sometimes added during or after culturing yogurt. Some countries require yogurt to contain a certain amount of colony-forming units of bacteria; in China, for example, the requirement for the number of lactobacillus bacteria is at least 1×10^6 CFU per gram per milliliter. To produce yogurt, milk is first heated, usually to about 85 °C (185 °F), to denature the milk proteins so that they set together rather than form curds. After heating, the milk is allowed to cool to about 45 °C (113 °F). The bacterial culture is mixed in, and a temperature of 45 °C (113 °F) is maintained for four to seven hours to allow fermentation.

Etymology and Spelling

The word is derived from Turkish: *yoğurt*, and is usually related to the verb *yoğurmak*, "to knead", or "to be curdled or coagulated; to thicken". It may be related to *yoğun*, meaning thick or dense. The letter ğ was traditionally rendered as "gh" in transliterations of Turkish prior to 1928. In older Turkish, the letter denoted a voiced velar fricative, but this sound is elided between back vowels in modern Turkish, in which the word is pronounced.

In English, there are several variations of the spelling of the word, including *yogurt*, *yoghurt* and to a lesser extent *yoghourt*, *yogourt*, *yaghourt*, *yahourth*, *yoghurd*, *joghourt*, and *jogourt*. In the United Kingdom and Australia, *yogurt* and *yoghurt* are both current, *yogurt* being used by the Australian and British dairy councils, and *yoghourt* is an uncommon alternative. In the United States, Canada, and New Zealand, *yogurt* is the usual spelling and *yoghurt* a minor variant.

Historically, there have been cases of yogurt being spelled with a "J" instead of a "Y" (e.g. *jogurt* and *joghurt*) due to alternative transliteration methods. However, there has been a decline in these variations in English speaking countries; in certain European countries, on the other hand, it is still commonly spelled with a "J". Most people tend to spell in the manner shown on the packaging of the major brands in their country.

Whatever the spelling, the word is usually pronounced with a short gland and Wales, and with a long in Scotland, North America, Australia, New Zealand, Ireland and South Africa.

History

Unstirred Turkish *Süzme Yoğurt* (strained yogurt), with a 10% fat content

Analysis of the *L. delbrueckii* subsp. *bulgaricus* genome indicates that the bacterium may have originated on the surface of a plant. Milk may have become spontaneously and unintentionally exposed to it through contact with plants, or bacteria may have been transferred from the udder of domestic milk-producing animals.

The origins of yogurt are unknown, but it is thought to have been invented in Mesopotamia around 5000 BC.

In ancient Indian records, the combination of yogurt and honey is called "the food of the gods". Persian traditions hold that "Abraham owed his fecundity and longevity to the regular ingestion of yogurt".

The cuisine of ancient Greece included a dairy product known as oxygala which is believed to have been a form of yogurt. Galen (AD 129 – c. 200/c. 216) mentioned that oxygala was consumed with honey, similar to the way thickened Greek yogurt is eaten today.

The oldest writings mentioning yogurt are attributed to Pliny the Elder, who remarked that certain "barbarous nations" knew how "to thicken the milk into a substance with an agreeable acidity". The use of yogurt by medieval Turks is recorded in the books *Diwan Lughat al-Turk* by Mahmud Kashgari and *Kutadgu Bilig* by Yusuf Has Hajib written in the 11th century. Both texts mention the word "yogurt" in different sections and describe its use by nomadic Turks. The earliest yogurts were probably spontaneously fermented by wild bacteria in goat skin bags.

Some accounts suggest that Indian emperor Akbar's cooks would flavor yogurt with mustard seeds and cinnamon. Another early account of a European encounter with yogurt occurs in French clinical history: Francis I suffered from a severe diarrhea which no French doctor could cure. His ally Suleiman the Magnificent sent a doctor, who allegedly cured the patient with yogurt. Being grateful, the French king spread around the information about the food which had cured him.

Until the 1900s, yogurt was a staple in diets of people in the Russian Empire (and especially Central Asia and the Caucasus), Western Asia, South Eastern Europe/Balkans, Central Europe, and India. Stamen Grigorov (1878–1945), a Bulgarian student of medicine in Geneva, first examined the microflora of the Bulgarian yogurt. In 1905, he described it as consisting of a spherical and a rod-like lactic acid-producing bacteria. In 1907, the rod-like bacterium was called *Bacillus bulgaricus* (now *Lactobacillus delbrueckii subsp. bulgaricus*). The Russian Nobel laureate and biologist Ilya Ilyich Mechnikov (also known as Élie Metchnikoff), from the Institut Pasteur in Paris, was influenced by Grigorov's work and hypothesized that regular consumption of yogurt was responsible for the unusually long lifespans of Bulgarian peasants. Believing *Lactobacillus* to be essential for good health, Mechnikov worked to popularize yogurt as a foodstuff throughout Europe.

Isaac Carasso industrialized the production of yogurt. In 1919, Carasso, who was from Ottoman Salonika, started a small yogurt business in Barcelona, Spain, and named the business Danone ("little Daniel") after his son. The brand later expanded to the United States under an Americanized version of the name: Dannon.

Yogurt with added fruit jam was patented in 1933 by the Radlická Mlékárna dairy in Prague.

Yogurt was introduced to the United States in the first decade of the twentieth century, influenced by Élie Metchnikoff's *The Prolongation of Life; Optimistic Studies* (1908); it was available in tablet form for those with digestive intolerance and for home culturing. It was popularized by John Harvey Kellogg at the Battle Creek Sanitarium, where it was used both orally and in enemas, and later by Armenian immigrants Sarkis and Rose Colombosian, who started "Colombo and Sons Creamery" in Andover, Massachusetts in 1929. Colombo Yogurt was originally delivered around New England in a horse-drawn wagon inscribed with the Armenian word "madzoon" which was later changed to "yogurt", the Turkish name of the product, as Turkish was the lingua franca between immigrants of the various Near Eastern ethnicities who were the main consumers at that time. Yogurt's popularity in the United States was enhanced in the 1950s and 1960s, when it was presented as a health food. By the late 20th century, yogurt had become a common American food item and Colombo Yogurt was sold in 1993 to General Mills, which discontinued the brand in 2010.

Nutrition and Health

In a 100-gram amount providing 406 kilojoules (97 kcal) of dietary energy, yogurt (plain Greek yogurt from whole milk) is 81% water, 9% protein, 5% fat and 4% carbohydrates, including 4% sugars (table). As a proportion of the Daily Value (DV), a serving of yogurt is a rich source of vitamin B_{12} (31% DV) and riboflavin (23% DV), with moderate content of protein, phosphorus and selenium (14 to 19% DV; table).

Comparison of Whole Dairy Milk and Plain Yogurt from Whole Dairy Milk, One Cup (245 g) Each		
Property	**Milk**	**Yogurt**
kilo calories	146	149
Total Fat	7.9 g	8.5 g
Cholesterol	24.4 mg	11 mg
Sodium	98 mg	113 mg
Phosphorus	222 mg	233 mg
Potassium	349 mg	380 mg
Total Carbohydrates	12.8 g	12 g
Protein	7.9 g	9 g
Vitamin A	249 IU	243 IU
Vitamin C	0.0 mg	1.2 mg
Vitamin D	96.5 IU	~
Vitamin E	0.1 mg	0.1 mg
Vitamin K	0.5 µg	0.5 µg
Thiamine	0.1 mg	0.1 mg
Riboflavin	0.3 mg	0.3 mg
Niacin	0.3 mg	0.2 mg
Vitamin B6	0.1 mg	0.1 mg
Folate	12.2 µg	17.2 µg
Vitamin B12	1.1 µg	0.9 µg
Choline	34.9 mg	1.0 mg
Betaine	1.5 mg	~
Water	215 g	215 g
Ash	1.7 g	1.8 g

Tilde (~) represents missing or incomplete data. – The above shows that there is little difference between whole milk and yogurt made from whole milk with respect to the listed nutritional constituents. The differences may be explained as a result of testing the product after draining liquid whey from the yogurt thereby changing the percentage of that constituent in the final product.

Although yogurt is often associated with probiotics having positive effects on immune, cardiovascular or metabolic health, there is insufficient high-quality clinical evidence to conclude that consuming yogurt lowers risk of diseases or improves health.

Lactose-intolerant individuals may tolerate yogurt better than other dairy products due to the conversion of lactose to the sugars glucose and galactose, and the fermentation of lactose to lactic acid carried out by the bacteria present in the yogurt.

Varieties and Presentation

Tzatziki is a side dish made with yogurt, popular in Greek cuisine, and similar yet thicker than the Turkish Cacik and close to the traditional Bulgarian Milk salad.

Da-hi is a yogurt of the Indian subcontinent, known for its characteristic taste and consistency. The word *da-hi* seems to be derived from the Sanskrit word *dadhi*, one of the five elixirs, or panchamrita, often used in Hindu ritual. *Dahi* also holds cultural symbolism in many homes in the *Mithila* region of Nepal and Bihar. Yogurt balances the palate across regional cuisines throughout India. In the hot and humid south, yogurt and foods made of yogurt are a staple in order to cool down – yogurt rice is always the last dish of the meal. Also, the vegetarian population of India derives some protein from yogurt (other than lentil and beans). Sweet yogurt (*meesti doi* or *meethi dahi*) is common in eastern parts of India, made by fermenting sweetened milk. While cow's milk is considered sacred and is currently the primary ingredient for yogurt, goat and buffalo milk were widely used in the past, and valued for the fat content. Butter and cream were made by churning the yogurt/milk.

Skyr is an Icelandic cultured dairy product, similar to strained yogurt. It has been a part of Icelandic cuisine for over a thousand years. It is traditionally served cold with milk and a topping of sugar.

Cacık, a Turkish cold appetizer made from yogurt

In India and Pakistan, it is often used in cosmetics mixed with turmeric and honey. Sour yogurt, is also used as a hair conditioner by women in many parts of India and Pakistan. *Dahi* is also known as *Mosaru* (Kannada), *Thayir* (Tamil/Malayalam), *doi* (Assamese, Bengali), *dohi* (Odia), *perugu* (Telugu) and *Dhahi* or *Dhaunro.*

Raita is a yogurt-based South Asian/Indian condiment, used as a side dish. The yogurt is seasoned with coriander (cilantro), cumin, mint, cayenne pepper, and other herbs and spices. Vegetables such as cucumber and onions are mixed in, and the mixture is served chilled. Raita has a cooling effect on the palate which makes it a good foil for spicy Indian and Pakistani dishes. Raita is sometimes also referred to as *dahi.*

Dadiah or dadih is a traditional West Sumatran yogurt made from water buffalo milk, fermented in bamboo tubes.

Yogurt is popular in Nepal, where it is served as both an appetizer and dessert. Locally called *dahi,* it is a part of the Nepali culture, used in local festivals, marriage ceremonies, parties, religious occasions, family gatherings, and so on. The most famous type of Nepalese yogurt is called *juju dhau,* originating from the city of Bhaktapur.

In Tibet, yak milk (technically dri milk, as the word yak refers to the male animal) is made into yogurt (and butter and cheese) and consumed.

In Northern Iran, *Mâst Chekide* is a variety of kefir yogurt with a distinct sour taste. It is usually mixed with a pesto-like water and fresh herb purée called delal. Yogurt is a side dish to all Iranian meals. The most popular appetizers are spinach or eggplant borani, *Mâst-o-Khiâr* with cucumber, spring onions and herbs, and *Mâst-Musir* with wild shallots. In the summertime, yogurt and ice cubes are mixed together with cucumbers, raisins, salt, pepper and onions and topped with some croutons made of Persian traditional bread and served as a cold soup. Ashe-Mâst is a warm yogurt soup with fresh herbs, spinach and lentils. Even the leftover water extracted when straining yogurt is cooked to make a sour cream sauce called kashk, which is usually used as a topping on soups and stews.

Matsoni is a Georgian yogurt popular in the Caucasus and Russia. It is used in a wide variety of Georgian dishes and is believed to have contributed to the high life expectancy and longevity in the country. Dannon used this theory in their 1978 TV advertisement called *In Soviet Georgia* where shots of elderly Georgian farmers were interspersed with an off-camera announcer intoning, "In Soviet Georgia, where they eat a lot of yogurt, a lot of people live past 100." Matsoni is also popular in Japan under the name Caspian Sea Yogurt.

Tarator and Cacık are popular cold soups made from yogurt, popular during summertime in Albania, Azerbaijan (known as Dogramac), Bulgaria, Macedonia, Serbia and Turkey. They are made with ayran, cucumbers, dill, salt, olive oil, and optionally garlic and ground walnuts. Tzatziki in Greece and milk salad in Bulgaria are thick yogurt-based salads similar to tarator.

Khyar w Laban (cucumber and yogurt salad) is a popular dish in Lebanon and Syria. Also, a wide variety of local Lebanese and Syrian dishes are cooked with yogurt like "Kibbi bi Laban", etc.

Rahmjoghurt, a creamy yogurt with much higher fat content (10%) than many yogurts offered in English-speaking countries (*Rahm* is German for "cream"), is available in Germany and other countries.

Dovga, a yogurt soup cooked with a variety of herbs and rice is popular in Azerbaijan, often served warm in winter or refreshingly cold in summer.

Yogurt made with unhomogenized milk is sometimes called cream-top yogurt; a layer of cream rises to the top.

Jameed is yogurt which is salted and dried to preserve it. It is popular in Jordan.

Zabadi is the type of yogurt made in Egypt, usually from the milk of the Egyptian water buffalo. It is particularly associated with Ramadan fasting, as it is thought to prevent thirst during all-day fasting.

Sweetened and Flavored Yogurt

To offset its natural sourness, yogurt is also sold sweetened, flavored or in containers with fruit or fruit jam on the bottom. The two styles of yogurt commonly found in the grocery store are set type yogurt and Swiss style yogurt. Set type yogurt is when the yogurt is packaged with the fruit on the bottom of the cup and the yogurt on top. Swiss style yogurt is when the fruit is blended into the yogurt prior to packaging.

Lassi and Moru are common beverages in India. Lassi is stirred liquified curd that is either salted or sweetened with sugar commonly, less commonly honey and often combined with fruit pulp to create flavored lassi. Mango lassi is a western favorite, as is coconut lassi. Consistency can vary widely, with urban and commercial lassis being of uniform texture through being processed, whereas rural and rustic lassi has curds in it, and sometimes has malai (cream) added or removed. Moru is a popular South Indian summer drink, meant to keep drinkers hydrated through the hot and humid summers of the South. It is prepared by considerably thinning down yogurt with water, adding salt (for electrolyte balance) and spices, usually green chili peppers, asafoetida, curry leaves and mustard.

Large amounts of sugar – or other sweeteners for low-energy yogurts – are often used in commercial yogurt. Some yogurts contain added starch, pectin (found naturally in fruit), and/or gelatin to create thickness and creaminess artificially at lower cost. Gelatin is a meat or fish product, therefore vegetarians should avoid products containing it. This type of yogurt is also marketed under the name Swiss-style, although it is unrelated to the way yogurt is eaten in Switzerland. Some yogurts, often called "cream line", are made with whole milk which has not been homogenized so the cream rises to the top.

In the UK, Ireland, France and United States, sweetened, flavored yogurt is the most popular type, typically sold in single-serving plastic cups. Common flavors include vanilla, honey, and toffee, and fruit such as strawberry, cherry, blueberry, blackberry, raspberry, mango and peach. In the early twenty-first century yogurt flavors inspired by desserts, such as chocolate or cheesecake, have been available.

There is concern about the health effects of sweetened yogurt. The United Kingdom and the United States recommend different maximum amounts of daily sugar intake but in both nations many sweetened yogurts have too much.

A 150g (5oz) serving of some 0% fat yogurts can contain as much as 20g (0.7oz) of sugar – the equivalent of five teaspoons, says Action on Sugar – which is about 40% of a woman's daily recommended intake of added sugar (50g or 1.7oz) and about 30% of that for men (70g or 2.5oz).

The American Heart Association recommends that men eat no more than 36 grams of sugar per day, and women no more than 20. Many of the top-selling yogurts have even more than the 19 grams of sugar that is contained in a Twinkie.

Consumers wanting sweetened yogurt are advised to choose yogurt sweetened with sugar substitute and check the contents list to avoid corn syrup, high fructose corn syrup, honey, or sugar.

Strained Yogurt

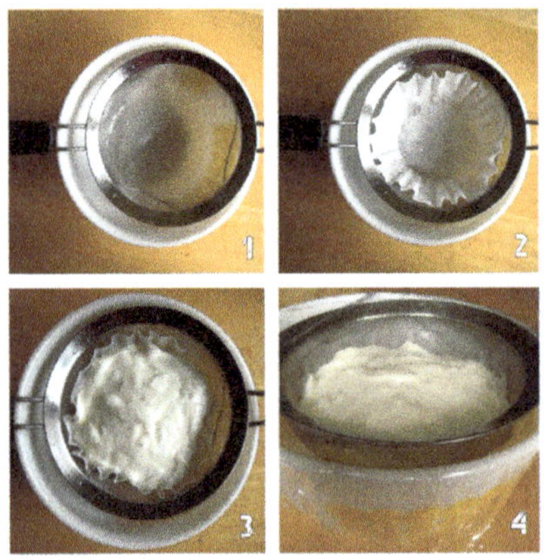

Use coffee filter to strain yogurt in a home refrigerator.

Strained yogurt is yogurt which has been strained through a filter, traditionally made of muslin and more recently of paper or cloth. This removes the whey, giving a much thicker consistency and a distinctive slightly tangy taste. Strained yogurt is becoming more popular with those who make yogurt at home, especially if using skimmed milk which results in a thinner consistency.

Yogurt that has been strained to filter or remove the whey is known as Labneh in Middle Eastern countries. It has a consistency between that of yogurt and cheese. It is popular for sandwiches in Middle Eastern countries. Olive oil, cucumber slices, olives, and various green herbs may be added. It can be thickened further and rolled into balls, preserved in olive oil, and fermented for a few more weeks. It is sometimes used with onions, meat, and nuts as a stuffing for a variety of pies or kibbeh balls.

Some types of strained yogurts are boiled in open vats first, so that the liquid content is reduced. The popular East Indian dessert, a variation of traditional dahi called mishti dahi, offers a thicker, more custard-like consistency, and is usually sweeter than western yogurts.

Strained yogurt is also enjoyed in Greece and is the main component of *tzatziki* (from Turkish "cacık"), a well-known accompaniment to gyros and souvlaki pita sandwiches: it is a yogurt sauce or dip made with the addition of grated cucumber, olive oil, salt and, optionally, mashed garlic.

Srikhand, a popular dessert in India, is made from strained yogurt, saffron, cardamom, nutmeg and sugar and sometimes fruits such as mango or pineapple.

In North America and Britain, strained yogurt is commonly called "Greek yogurt". Strained yogurt is sometimes marketed in North America as "Greek yogurt" and in Britain as "Greek-style yoghurt". In Britain the name "Greek" may only be applied to yogurt made in Greece.

Beverages

PCC Dairy Yogurt Milk, with live cultures, made from water buffalo's cream milk Philippine Carabao Center.

Doogh ("dawghe" in Neo-Aramaic), ayran or dhallë is a yogurt-based, salty drink popular in Iran, Albania, Bulgaria, Turkey, Azerbaijan, Afghanistan, Pakistan, Bangladesh, Macedonia, Uzbekistan, Kazakhstan and Kyrgyzstan. It is made by mixing yogurt with water and (sometimes) salt. The same drink is known *tan* in Armenia; *laban ayran* in Syria and Lebanon; *shenina* in Iraq and Jordan; *laban arbil* in Iraq; *majjiga* (Telugu), *majjige* (Kannada), and *moru* (Tamil and Malayalam) in South India; namkeen *lassi* in Punjab and all over Pakistan.

Ayran

Borhani (or Burhani) is a spicy yogurt drink popular in Bangladesh and parts of Bengal. It is usually served with kacchi biryani at weddings and special feasts. Key ingredients are yogurt blended with mint leaves (mentha), mustard seeds and black rock salt (Kala Namak). Ground roasted cumin, ground white pepper, green chili pepper paste and sugar are often added.

Lassi is a yogurt-based beverage originally from the Indian subcontinent that is usually slightly salty or sweet. Lassi is a staple in Punjab. In some parts of the subcontinent, the sweet version may be commercially flavored with rosewater, mango or other fruit juice to create a very different drink. Salty lassi is usually flavored with ground, roasted cumin and red chilies; this salty variation may also use buttermilk, and in India is interchangeably called *ghol* (Bengal), *mattha* (North India), "majjige" (Karnataka), *majjiga* (Telangana & Andhra Pradesh), *moru* (Tamil Nadu and Kerala), *Dahi paani Chalha* (Odisha), *tak* (Maharashtra), or *chaas* (Gujarat). Lassi is very widely drunk in India, Pakistan, and Bangladesh. Mango Lassi is a popular drink at Indian restaurants in US.

In Bosnia and Herzegovina, Croatia, Macedonia, Montenegro, Serbia, and Slovenia, an unsweetened and unsalted yogurt drink usually called simply *jogurt* is a popular accompaniment to *burek* and other baked goods.

Sweetened yogurt drinks are the usual form in Europe (including the UK) and the US, containing fruit and added sweeteners. These are typically called "drinkable yogurt".

Also available are "yogurt smoothies", which contain a higher proportion of fruit and are more like smoothies. In Ecuador, yogurt smoothies flavored with native fruit are served with pan de yuca as a common type of fast food.

Also in Turkey, yogurt soup or *Yayla Çorbası* is a popular way of consuming yogurt. The soup is a mix of yogurt, rice, flour and dried mint.

Plant-milk Yogurt

A variety of plant-milk yogurts appeared in the 2000s, using soy milk, rice milk, and nut milks such as almond milk and coconut milk. So far the most widely sold variety of plant milk yogurts is soy yogurt. These yogurts are suitable for vegans, people with intolerance to dairy milk, and those who prefer plant milks.

Plant-milk yogurt

Making Yogurt at Home

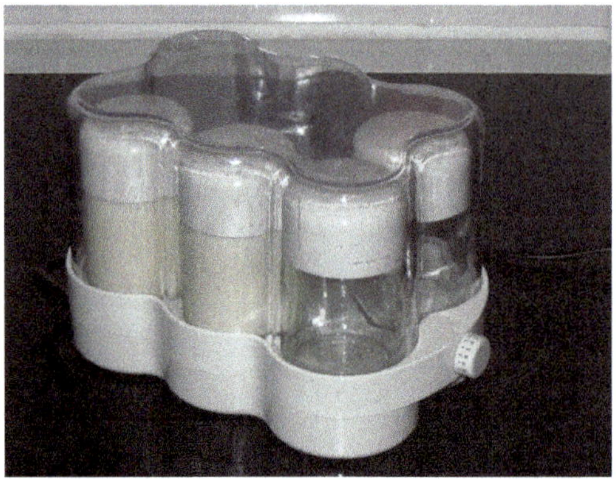

Commercially available yogurt maker

Yogurt is made by heating milk to a temperature that denaturates its proteins (scalding), essential for making yogurt, cooling it to a temperature that will not kill the live microorganisms that turn the milk into yogurt, inoculating certain bacteria (starter culture), usually *Streptococcus thermophilus* and *Lactobacillus bulgaricus*, into the milk, and finally keeping it warm for several hours. The milk may be held at 85 °C (185 °F) for a few minutes, or boiled (giving a somewhat different result). It must be cooled to 50 °C (122 °F) or somewhat less, typically 40–46 °C (104–115 °F). Starter culture must then be mixed in well, and the mixture must be kept undisturbed and warm for several hours, ranging from 5 to 12, with longer fermentation producing a more acid yogurt. The starter culture may be a small amount of live yogurt; dried starter culture is available commercially.

Home yogurt maker

Milk with a higher concentration of solids than normal milk may be used; the higher solids content produces a firmer yogurt. Solids can be increased by adding dried milk.

The yogurt-making process provides two significant barriers to pathogen growth, heat and acidity (low pH). Both are necessary to ensure a safe product. Acidity alone has been questioned by recent outbreaks of food poisoning by *E. coli O157:H7* that is acid-tolerant. *E. coli O157:H7* is easily destroyed by pasteurization (heating); the initial heating of the milk kills pathogens as well as denaturing proteins. The microorganisms that turn milk into yogurt can tolerate higher temperatures than most pathogens, so that a suitable temperature not only encourages the formation of yogurt, but inhibits pathogenic microorganisms.

Once the yogurt has formed it can, if desired, be strained to reduce the whey content and thicken it.

Bread

Bread is a staple food prepared from a dough of flour and water, usually by baking. Throughout recorded history it has been popular around the world and is one of the oldest artificial foods, having been of importance since the dawn of agriculture.

Proportions of types of flour and other ingredients vary widely, as do modes of preparation. As a result, types, shapes, sizes, and textures of breads differ around the world. Bread may be leavened by processes such as reliance on naturally occurring sourdough microbes, chemicals, industrially produced yeast, or high-pressure aeration. Some bread is cooked before it can leaven, including for traditional or religious reasons. Non-cereal ingredients such as fruits, nuts and fats may be included. Commercial bread commonly contains additives to improve flavor, texture, color, shelf life, and ease of manufacturing.

Bread is served in various forms with any meal of the day. It is eaten as a snack, and used as an ingredient in other culinary preparations, such as sandwiches, and fried items coated in bread crumbs to prevent sticking. It forms the bland main component of bread pudding, as well as of stuffings designed to fill cavities or retain juices that otherwise might drip out.

Bread has a social and emotional significance beyond its importance as nourishment. It plays essential roles in religious rituals and secular culture. Its prominence in daily life is reflected in language, where it appears in proverbs, colloquial expressions ("He stole the bread from my mouth"), in prayer ("Give us this day our daily bread") and in the etymology of words, such as "companion" (from Latin *com* "with" + *panis* "bread").

Etymology

The Old English word for bread was *hlaf* (*hlaifs* in Gothic: modern English *loaf*), which appears to be the oldest Teutonic name. Old High German *hleib* and modern German *Laib* derive from this Proto-Germanic word, which was borrowed into Slavic (Polish *chleb*, Russian *khleb*) and Finnic (Finnish *leipä*, Estonian *leib*) languages as well. The Middle and Modern English word bread appears in Germanic languages, such as Frisian *brea*, Dutch *brood*, German *Brot*, Swedish *bröd*, and Norwegian and Danish *brød*; it may be related to *brew* or perhaps to *break*, originally meaning "broken piece", "morsel".

History

Bread shop, Tacuinum Sanitatis from Northern Italy, beginning of the 15th century

Bread is one of the oldest prepared foods. Evidence from 30,000 years ago in Europe revealed starch residue on rocks used for pounding plants. It is possible that during this time, starch extract from the roots of plants, such as cattails and ferns, was spread on a flat rock, placed over a fire and cooked into a primitive form of flatbread. Around 10,000 BC, with the dawn of the Neolithic age and the spread of agriculture, grains became the mainstay of making bread. Yeast spores are ubiquitous, including on the surface of cereal grains, so any dough left to rest leavens naturally.

There were multiple sources of leavening available for early bread. Airborne yeasts could be harnessed by leaving uncooked dough exposed to air for some time before cooking. Pliny the Elder reported that the Gauls and Iberians used the foam skimmed from beer to produce "a lighter kind of bread than other peoples." Parts of the ancient world that drank wine instead of beer used a paste composed of grape juice and flour that was allowed to begin fermenting, or wheat bran steeped in wine, as a source for yeast. The most common source of leavening was to retain a piece of dough from the previous day to use as a form of sourdough starter, as Pliny also reported.

In 1961 the Chorleywood bread process was developed, which used the intense mechanical working of dough to dramatically reduce the fermentation period and the time taken to produce a loaf. The process, whose high-energy mixing allows for the use of lower protein grain, is now widely used around the world in large factories. As a result, bread can be produced very quickly and at low costs to the manufacturer and the consumer. However, there has been some criticism of the effect on nutritional value.

Types

Brown bread (left) and whole grain bread

Bread is the staple food of the Middle East, North Africa, Europe, and in European-derived cultures such as those in the Americas, Australia, and Southern Africa, in contrast to East Asia where rice is the staple. Bread is usually made from a wheat-flour dough that is cultured with yeast, allowed to rise, and finally baked in an oven. The addition of yeast to the bread explains the air pockets commonly found in bread. Owing to its high levels of gluten (which give the dough sponginess and elasticity), common or bread wheat is the most common grain used for the preparation of bread, which makes the largest single contribution to the world's food supply of any food.

Bread is also made from the flour of other wheat species (including spelt, emmer, einkorn and kamut). Non-wheat cereals including rye, barley, maize (corn), oats, sorghum, millet and rice have been used to make bread, but, with the exception of rye, usually in combination with wheat flour as they have less gluten.

Gluten-free breads have been created for people affected by gluten-related disorders such as coeliac disease and non-coeliac gluten sensitivity, who may benefit from a gluten-free diet. Gluten-free bread is made with ground flours from a variety of materials such as almonds, rice, sorghum, corn, or legumes such as beans, but since these flours lack gluten they may not hold their shape as they rise and their crumb may be dense with little aeration. Additives such as xanthan gum, guar gum, hydroxypropyl methylcellulose (HPMC), corn starch, or eggs are used to compensate for the lack of gluten.

Properties

Chemical Composition

In wheat, phenolic compounds are mainly found in hulls in the form of insoluble bound ferulic acid, where it is relevant to wheat resistance to fungal diseases.

Rye bread contains phenolic acids and ferulic acid dehydrodimers.

Three natural phenolic glucosides, secoisolariciresinol diglucoside, p-coumaric acid glucoside and ferulic acid glucoside, can be found in commercial breads containing flaxseed.

Culinary Uses

Bread pudding

Bread can be served at many temperatures; once baked, it can subsequently be toasted. It is most commonly eaten with the hands, either by itself or as a carrier for other foods. Bread can be dipped into liquids such as gravy, olive oil, or soup; it can be topped with various sweet and savory spreads, or used to make sandwiches containing meats, cheeses, vegetables, and condiments.

Bread is used as an ingredient in other culinary preparations, such as the use of breadcrumbs to provide crunchy crusts or thicken sauces, sweet or savoury bread puddings, or as a binding agent in sausages and other ground meat products.

Nutritional Significance

Nutritionally, bread is known as an ample source for the grains category of nutrition. Serving size of bread is standard through ounces, counting one slice of bread (white processed bread) as 1 oz. Also, bread is considered a good source of carbohydrates through the whole grains, nutrients such as magnesium, iron, selenium, B vitamins, and dietary fiber.

Crust

The bread crust is formed from surface dough during the cooking process. It is hardened and browned through the Maillard reaction using the sugars and amino acids and the intense heat at the bread surface. The crust of most breads is harder, and more complexly and intensely flavored, than the rest. Old wives tales suggest that eating the bread crust makes a person's hair curlier. Additionally, the crust is rumored to be healthier than the rest. Some studies have shown that this is true as the crust has more dietary fiber and antioxidants such as pronyl-lysine, which is being researched for its potential colorectal cancer inhibitory properties.

Preparation

Steps in bread making, here for an unleavened Chilean tortilla

Doughs are usually baked, but in some cuisines breads are steamed (e.g., mantou), fried (e.g., puri), or baked on an unoiled frying pan (e.g., tortillas). It may be leavened or unleavened (e.g. matzo). Salt, fat and leavening agents such as yeast and baking soda are common ingredients, though bread may contain other ingredients, such as milk, egg, sugar, spice, fruit such as raisins, vegetables such as onion, nuts such as walnut or seeds such as poppy.

Baking bread in East Timor

Formulation

Professional bread recipes are stated using the baker's percentage notation. The amount of flour is denoted to be 100%, and the other ingredients are expressed as a percentage of that amount by weight. Measurement by weight is more accurate and consistent than measurement by volume, particularly for dry ingredients. The proportion of water to flour is the most important measurement in a bread recipe, as it affects texture and crumb the most. Hard wheat flours absorb about 62% water, while softer wheat flours absorb about 56%. Common table breads made from these doughs result in a finely textured, light bread. Most artisan bread formulas contain anywhere from 60 to 75% water. In yeast breads, the higher water percentages result in more CO_2 bubbles and a coarser bread crumb. One pound (450 g) of flour yields a standard loaf of bread or two French loaves.

Calcium propionate is commonly added by commercial bakeries to retard the growth of molds.

Flour

Flour is grain ground to a powdery consistency. Flour provides the primary structure, starch and protein to the final baked bread. The protein content of the flour is the best indicator of the quality of the bread dough and the finished bread. While bread can be made from all-purpose wheat flour, a specialty bread flour, containing more protein (12–14%), is recommended for high-quality bread. If one uses a flour with a lower protein content (9–11%) to produce bread, a shorter mixing time is required to develop gluten strength properly. An extended mixing time leads to oxidization of the dough, which gives the finished product a whiter crumb, instead of the cream color preferred by most artisan bakers.

Wheat flour, in addition to its starch, contains three water-soluble protein groups (albumin, globulin, and proteoses) and two water-insoluble protein groups (glutenin and gliadin). When flour is mixed with water, the water-soluble proteins dissolve, leaving the glutenin and gliadin to form the structure of the resulting bread. When relatively dry dough is worked by kneading, or wet dough is allowed to rise for a long time, the glutenin forms strands of long, thin, chainlike molecules, while the shorter gliadin forms bridges between the strands of glutenin. The resulting networks of strands produced by these two proteins are known as gluten. Gluten development improves if the dough is allowed to autolyse.

Liquids

Water, or some other liquid, is used to form the flour into a paste or dough. The weight of liquid required varies between recipes, but a ratio of 3 parts liquid to 5 parts flour is common for yeast breads. Recipes that use steam as the primary leavening method may have a liquid content in excess of 1 part liquid to 1 part flour. Instead of water, recipes may use liquids such as milk or other dairy products (including buttermilk or yoghurt), fruit juice, or eggs. These contribute additional sweeteners, fats, or leavening components, as well as water.

Fats or Shortenings

Fats, such as butter, vegetable oils, lard, or that contained in eggs, affect the development of gluten in breads by coating and lubricating the individual strands of protein. They also help to hold the structure together. If too much fat is included in a bread dough, the lubrication effect causes the protein structures to divide. A fat content of approximately 3% by weight is the concentration that produces the greatest leavening action. In addition to their effects on leavening, fats also serve to tenderize breads and preserve freshness.

Bread Improvers

Bread improvers and dough conditioners are often used in producing commercial breads to reduce the time needed for rising and to improve texture and volume. The substances used may be oxidising agents to strengthen the dough or reducing agents to develop gluten and reduce mixing time, emulsifiers to strengthen the dough or to provide other properties such as making slicing easier, or enzymes to increase gas production.

Salt is often added to enhance flavor and restrict yeast activity. It also affects the crumb and the overall texture by stabilizing and strengthening the gluten. Some artisan bakers forego early addition of salt to the dough, whether wholemeal or refined, and wait until after a 20-minute rest to allow the dough to autolyse.

Leavening

A dough trough once used for leavening bread from Aberdour Castle, Fife, Scotland.

Leavening is the process of adding gas to a dough before or during baking to produce a lighter, more easily chewed bread. Most bread eaten in the West is leavened.

Chemicals

A simple technique for leavening bread is the use of gas-producing chemicals. There are two common methods. The first is to use baking powder or a self-raising flour that includes baking powder. The second is to include an acidic ingredient such as buttermilk and add baking soda; the reaction of the acid with the soda produces gas. Chemically leavened breads are called *quick breads* and *soda breads*. This method is commonly used to make muffins, pancakes, American-style biscuits, and quick breads such as banana bread.

Yeast

Compressed fresh yeast

Many breads are leavened by yeast. The yeast most commonly used for leavening bread is *Saccharomyces cerevisiae,* the same species used for brewing alcoholic beverages. This yeast ferments some of the carbohydrates in the flour, including any sugar, producing carbon dioxide. Commercial bakers often leaven their dough with commercially produced baker's yeast. Baker's yeast has the advantage of producing uniform, quick, and reliable results, because it is obtained from a pure culture. Many artisan bakers produce their own yeast with a growth culture. If kept in the right conditions, it provides leavening for many years.

The baker's yeast and sourdough methods follow the same pattern. Water is mixed with flour, salt and the leavening agent. Other additions (spices, herbs, fats, seeds, fruit, etc.) are not needed to bake bread, but are often used. The mixed dough is then allowed to rise one or more times (a longer rising time results in more flavor, so bakers often "punch down" the dough and let it rise again), then loaves are formed, and (after an optional final rising time) the bread is baked in an oven.

Many breads are made from a "straight dough", which means that all of the ingredients are combined in one step, and the dough is baked after the rising time; others are made from a "pre-ferment" in which the leavening agent is combined with some of the flour and water a day or so ahead of baking and allowed to ferment overnight. On the day of baking, the rest of the ingredients are added, and the process continues as with straight dough. This produces a more flavorful bread with better texture. Many bakers see the starter method as a compromise between the reliable results of baker's yeast and the flavor and complexity of a longer fermentation. It also allows the baker to use only a minimal amount of baker's yeast, which was scarce and expensive when it first became available. Most yeasted pre-ferments fall into one of three categories: "poolish" or "pouliche", a loose-textured mixture composed of roughly equal amounts of flour and water (by weight); "biga", a stiff mixture with a higher proportion of flour; and "pâte fermentée", which is simply a portion of dough reserved from a previous batch.

Before first rising

After first rising

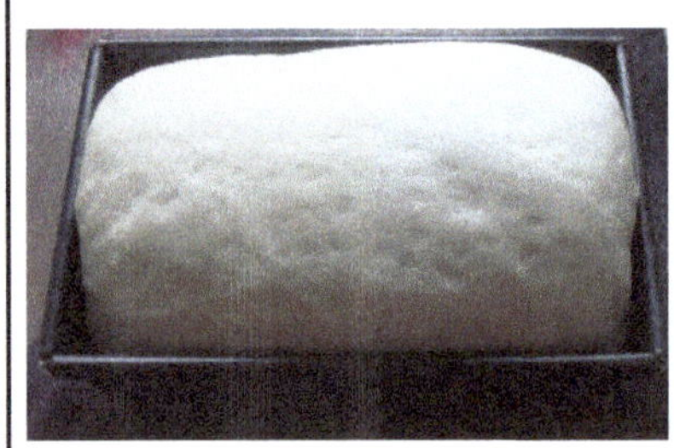
After proofing, ready to bake

Sourdough

Sourdough is a type of bread produced by a long fermentation of dough using naturally occurring yeasts and lactobacilli. It usually has a mildly sour taste because of the lactic acid produced during anaerobic fermentation by the lactobacilli.

Sourdough breads are made with a sourdough starter. The starter cultivates yeast and lactobacilli in a mixture of flour and water, making use of the microorganisms already present on flour; it does not need any added yeast. A starter may be maintained indefinitely by regular additions of flour and water. Some bakers have starters many generations old, which are said to have a special taste or texture. At one time, all yeast-leavened breads were sourdoughs. Recently there has been a revival of sourdough bread in artisan bakeries.

Traditionally, peasant families throughout Europe baked on a fixed schedule, perhaps once a week. The starter was saved from the previous week's dough. The starter was mixed with the new ingredients, the dough was left to rise, and then a piece of it was saved (to be the starter for next week's bread).

Sourdough loaves

Steam

The rapid expansion of steam produced during baking leavens the bread, which is as simple as it is unpredictable. Steam-leavening is unpredictable since the steam is not produced until the bread is baked. Steam leavening happens regardless of the raising agents (baking soda, yeast, baking powder, sour dough, beaten egg white) included in the mix. The leavening agent either contains air bubbles or generates carbon dioxide. The heat vaporises the water from the inner surface of the bubbles within the dough. The steam expands and makes the bread rise. This is the main factor in the rising of bread once it has been put in the oven. CO_2 generation, on its own, is too small to account for the rise. Heat kills bacteria or yeast at an early stage, so the CO_2 generation is stopped.

Bacteria

Salt-rising bread employs a form of bacterial leavening that does not require yeast. Although the leavening action is inconsistent, and requires close attention to the incubating conditions, this bread is making a comeback for its cheese-like flavor and fine texture.

Aeration

Aerated bread was leavened by carbon dioxide being forced into dough under pressure. From the mid 19th to mid 20th centuries bread made this way was somewhat popular in the United Kingdom, made by the Aerated Bread Company and sold in its high-street tearooms. The company was founded in 1862, and ceased independent operations in 1955.

The Pressure-Vacuum mixer was later developed by the Flour Milling and Baking Research Association for the Chorleywood bread process. It manipulates the gas bubble size and optionally the composition of gases in the dough via the gas applied to the headspace. The organic baker Andrew Whitely, writing in *The Independent*, called the process "the covert corruption of our daily food".

Cultural Significance

Bread has a significance beyond mere nutrition in many cultures because of its history and contemporary importance. Bread is also significant in Christianity as one of the elements (alongside wine) of the Eucharist, and in other religions including Paganism.

In many cultures, bread is a metaphor for basic necessities and living conditions in general. For example, a "bread-winner" is a household's main economic contributor and has little to do with actual bread-provision. This is also seen in the phrase "putting bread on the table". The Roman poet Juvenal satirized superficial politicians and the public as caring only for "*panem et circenses*" (bread and circuses). In Russia in 1917, the Bolsheviks promised "peace, land, and bread." The term "breadbasket" denotes an agriculturally productive region. In Slavic cultures bread and salt is offered as a welcome to guests. In India, life's basic necessities are often referred to as "roti, kapra aur makan" (bread, cloth, and house).

Dark sprouted bread

Words for bread, including "dough" and "bread" itself, are used in English-speaking countries as synonyms for money. A remarkable or revolutionary innovation may be called the best thing since "sliced bread".

References

- Klein, Donald W.; Lansing M.; Harley, John (2006). Microbiology (6th ed.). New York: McGraw-Hill. ISBN 978-0-07-255678-0.
- AP Biology. Anestis, Mark. 2nd Edition. McGraw-Hill Professional. 2006. ISBN 978-0-07-147630-0. p. 61
- An Introduction (10 ed.). San Francisco, CA 94111, USA: Pearson Benjamin Cummings. p. 135. ISBN 978-0-321-58202-7.
- Tortora, Gerard J.; Funke, Berdell R.; Case, Christine L. (2010). "5". Microbiology
- Voet, Donald; Voet, Judith G. (1995). Biochemistry (2nd ed.). New York, NY: John Wiley & Sons. ISBN 978-0-471-58651-7.
- Stryer, Lubert (1995). Biochemistry (fourth ed.). New York - Basingstoke: W. H. Freeman and Company. ISBN 978-0716720096.
- Life, the science of biology. Purves, William Kirkwood. Sadava, David. Orians, Gordon H. 7th Edition. Macmillan Publishers. 2004. ISBN 978-0-7167-9856-9. pp. 139–140
- Introductory Botany: plants, people, and the Environment. Berg, Linda R. Cengage Learning, 2007. ISBN 978-0-534-46669-5. p. 86
- Madigan, Michael T.; Martinko, John M.; Parker, Jack (1996). Brock biology of microorganisms (8th ed.).

Prentice Hall. ISBN 978-0-13-520875-5.

- New beer in an old bottle: Eduard Buchner and the Growth of Biochemical Knowledge. Cornish-Bowden, Athel. Universitat de Valencia. 1997. ISBN 978-84-370-3328-0. p. 25.
- The enigma of ferment: from the philosopher's stone to the first biochemical Nobel prize. Lagerkvist, Ulf. World Scientific Publishers. 2005. ISBN 978-981-256-421-4. p. 7.
- Hui YH, Meunier-Goddik L, Josephsen J, Nip WK, Stanfield PS (2004). Handbook of Food and Beverage Fermentation Technology. CRC Press. pp. 27 and passim. ISBN 978-0-8247-5122-7.
- J. Robinson (ed) "The Oxford Companion to Wine" Third Edition pg 267-269 Oxford University Press 2006 ISBN 0-19-860990-6
- J. Robinson Jancis Robinson's Wine Course Third Edition pg 74-84 Abbeville Press 2003 ISBN 0-7892-0883-0.

5

Food Preservation and Safety

Food preservation prevents the growth of microorganisms on food. The traditional techniques mentioned are drying, cooling, freezing, boiling, sugaring and pickling. Other topics discussed in the section are vacuum packing, preservatives, camping food, food safety, ISO 22000, food grading and beef carcass classifications. The aspects elucidates in this chapter are of vital importance, and provides a better understanding of food preservation.

Food Preservation

Food preservation involves preventing the growth of bacteria, fungi (such as yeasts), or other micro-organisms (although some methods work by introducing benign bacteria or fungi to the food), as well as slowing the oxidation of fats that cause rancidity. Food preservation may also include processes that inhibit visual deterioration, such as the enzymatic browning reaction in apples after they are cut during food preparation.

Many processes designed to preserve food will involve a number of food preservation methods. Preserving fruit by turning it into jam, for example, involves boiling (to reduce the fruit's moisture content and to kill bacteria, etc.), sugaring (to prevent their re-growth) and sealing within an airtight jar (to prevent recontamination). Some traditional methods of preserving food have been shown to have a lower energy input and carbon footprint, when compared to modern methods.

Some methods of food preservation are known to create carcinogens. In 2015, the International Agency for Research on Cancer of the World Health Organization classified processed meat, i.e. meat that has undergone salting, curing, fermenting, and smoking, as "carcinogenic to humans".

Maintaining or creating nutritional value, texture and flavor is an important aspect of food preservation, although, historically, some methods drastically altered the character of the food being preserved. In many cases these changes have come to be seen as desirable qualities – cheese, yogurt and pickled onions being common examples.

Traditional Techniques

New techniques of food preservation became available to the home chef from the dawn of agriculture until the Industrial Revolution.

Drying

Drying is one of the oldest techniques used to hamper the decomposition of food products. As early as 12,000 B.C., Middle Eastern and Oriental cultures were drying foods using the power of the sun.

Vegetables and fruits are naturally dried by the sun and wind, but "still houses" were built in areas that did not have enough sunlight to dry things. A fire would be built inside the building to provide the heat to dry the various fruits, vegetables, and herbs.

Cooling

Cooling preserves food by slowing down the growth and reproduction of micro-organisms and the action of enzymes that cause food to rot. The introduction of commercial and domestic refrigerators drastically improved the diets of many in the Western world by allowing foods such as fresh fruit, salads and dairy products to be stored safely for longer periods, particularly during warm weather.

Freezing

Freezing is also one of the most commonly used processes, both commercially and domestically, for preserving a very wide range of foods, including prepared foods that would not have required freezing in their unprepared state. For example, potato waffles are stored in the freezer, but potatoes themselves require only a cool dark place to ensure many months' storage. Cold stores provide large-volume, long-term storage for strategic food stocks held in case of national emergency in many countries.

Boiling

Boiling liquid food items can kill any existing microbes. Milk and water are often boiled to kill any harmful microbes that may be present in them.

Heating

Heating to temperatures which are sufficient to kill microorganisms inside the food is a method used with perpetual stews. Milk is also boiled before storing to kill many microorganisms.

Salting

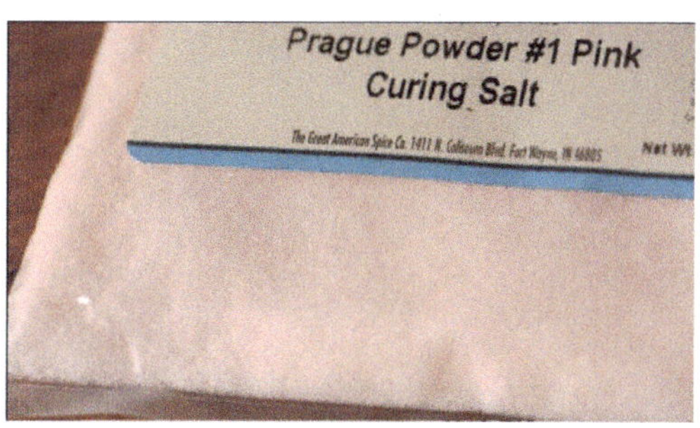

Bag of Prague powder #1, also known as "curing salt" or "pink salt". It is typically a combination of salt and sodium nitrite, with the pink color added to distinguish it from ordinary salt.

Salting or curing draws moisture from a substance through a process of osmosis. substances are cured with salt or sugar, or a combination of the two. Nitrates and nitrites are also often used to cure meat and contribute the characteristic pink colour. It was a main method of preservation in medieval times and around the 1700s.

Sugaring

The earliest cultures have used sugar as a preservative, and it was commonplace to store fruit in honey. Similar to pickled foods, sugar cane was brought to Europe through the trade routes. In northern climates without sufficient sun to dry foods, preserves are made by heating the fruit with sugar. "Sugar tends to draw water from the microbes (plasmolysis). This process leaves the microbial cells dehydrated, thus killing them. In this way, the food will remain safe from microbial spoilage." Sugar is used to preserve fruits, either in an anti-microbial syrup with fruit such as apples, pears, peaches, apricots and plums, or in crystallized form where the preserved material is cooked in sugar to the point of crystallization and the resultant product is then stored dry. This method is used for the skins of citrus fruit (candied peel), angelica and ginger. Also sugaring can be used in jam jellies.

Smoking

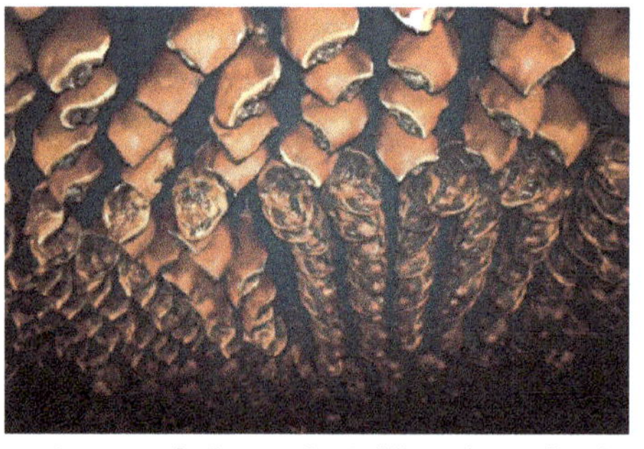

Smoked ham in a smokehouse in Schleswig-Holstein, Germany

Smoking is used to lengthen the shelf life of perishable food items. This effect is achieved by exposing the food to smoke from burning plant materials such as wood. Smoke deposits a number of pyrolysis products onto the food, including the phenolssyringol, guaiacol and catechol. These compounds aid in the drying and preservation of meats and other foods. Most commonly subjected to this method of food preservation are meats and fish that have undergone curing. Fruits and vegetables like paprika, cheeses, spices, and ingredients for making drinks such as malt and tea leaves are also smoked, but mainly for cooking or flavoring them. It is one of the oldest food preservation methods, which probably arose after the development of cooking with fire.

Pickling

Pickling is a method of preserving food in an edible anti-microbial liquid. Pickling can be broadly classified into two categories: chemical pickling and fermentation pickling.

In chemical pickling, the food is placed in an edible liquid that inhibits or kills bacteria and other micro-organisms. Typical pickling agents include brine (high in salt), vinegar, alcohol, and vegetable oil, especially olive oil but also many other oils. Many chemical pickling processes also involve heating or boiling so that the food being preserved becomes saturated with the pickling agent. Common chemically pickled foods include cucumbers, peppers, corned beef, herring, and eggs, as well as mixed vegetables such as piccalilli.

In fermentation pickling, the food itself produces the preservation agent, typically by a process that produces lactic acid. Fermented pickles include sauerkraut, nukazuke, kimchi, surströmming,

Lye

Sodium hydroxide (lye) makes food too alkaline for bacterial growth. Lye will saponify fats in the food, which will change its flavor and texture. Lutefisk uses lye in its preparation, as do some olive recipes. Modern recipes for century eggs also call for lye.

Canning

Preserved food

Canning involves cooking food, sealing it in sterile cans or jars, and boiling the containers to kill or weaken any remaining bacteria as a form of sterilization. It was invented by the French confectioner Nicolas Appert. By 1806, this process was used by the French Navy to preserve meat, fruit, vegetables, and even milk. Although Appert had discovered a new way of preservation, it wasn't understood until 1864 when Louis Pasteur found the relationship between microorganisms, food spoilage, and illness.

Foods have varying degrees of natural protection against spoilage and may require that the final step occur in a pressure cooker. High-acid fruits like strawberries require no preservatives to can and only a short boiling cycle, whereas marginal vegetables such as carrots require longer boiling and addition of other acidic elements. Low-acid foods, such as vegetables and meats, require pressure canning. Food preserved by canning or bottling is at immediate risk of spoilage once the can or bottle has been opened.

Lack of quality control in the canning process may allow ingress of water or micro-organisms. Most such failures are rapidly detected as decomposition within the can causes gas production and the can will swell or burst. However, there have been examples of poor manufacture (underprocessing) and poor hygiene allowing contamination of canned food by the obligate anaerobe-*Clostridium botulinum*, which produces an acute toxin within the food, leading to severe illness or death. This organism produces no gas or obvious taste and remains undetected by taste or smell. Its toxin is denatured by cooking, however. Cooked mushrooms, handled poorly and then canned, can support the growth of Staphylococcus aureus, which produces a toxin that is not destroyed by canning or subsequent reheating.

Jellying

Food may be preserved by cooking in a material that solidifies to form a gel. Such materials include gelatin, agar, maize flour, and arrowroot flour. Some foods naturally form a protein gel when cooked, such as eels and elvers, and sipunculid worms, which are a delicacy in Xiamen, in the Fujian province of the People's Republic of China. Jellied eels are a delicacy in the East End of London, where they are eaten with mashed potatoes. Potted meats in aspic (a gel made from gelatine and clarified meat broth) were a common way of serving meat off-cuts in the UK until the 1950s. Many jugged meats are also jellied.

A traditional British way of preserving meat (particularly shrimp) is by setting it in a pot and sealing it with a layer of fat. Also common is potted chicken liver; compare pâté.

Jugging

Meat can be preserved by jugging. Jugging is the process of stewing the meat (commonly game or fish) in a covered earthenware jug or casserole. The animal to be jugged is usually cut into pieces, placed into a tightly-sealed jug with brine or gravy, and stewed. Red wine and/or the animal's own blood is sometimes added to the cooking liquid. Jugging was a popular method of preserving meat up until the middle of the 20th century.

Burial

Burial of food can preserve it due to a variety of factors: lack of light, lack of oxygen, cool temperatures, pH level, or desiccants in the soil. Burial may be combined with other methods such as salting or fermentation. Most foods can be preserved in soil that is very dry and salty (thus a desiccant) such as sand, or soil that is frozen.

Many root vegetables are very resistant to spoilage and require no other preservation than storage in cool dark conditions, for example by burial in the ground, such as in a storage clamp. Century eggs are created by placing eggs in alkaline mud (or other alkaline substance), resulting in their "inorganic" fermentation through raised pH instead of spoiling. The fermentation preserves them and breaks down some of the complex, less flavorful proteins and fats into simpler, more flavorful ones. Cabbage was traditionally buried in the fall in northern farms in the U.S. for preservation. Some methods keep it crispy while other methods produce sauerkraut. A similar process is used in the traditional production of kimchi. Sometimes meat is buried under conditions that cause preservation. If buried on hot coals or ashes, the heat can kill pathogens, the dry ash can desiccate, and the earth can block oxygen and further contamination. If buried where the earth is very cold, the earth acts like a refrigerator.

In Orissa, India, it is practical to store rice by burying it underground. This method helps to store for three to six months during the dry season.

Curing

The earliest form of curing was dehydration. To accelerate this process, salt is usually added. In the culinary world, it was common to choose raw salts from various sources (rock salt, sea salt, etc.). More modern "examples of salts that are used as preservatives include sodium chloride (NaCl), sodium nitrate ($NaNO_3$) and sodium nitrite ($NaNO_2$). Even at mild concentrations (up to 2%), sodium chloride, found in many food products, is capable of neutralizing the antimicrobial character of natural compounds."

Fermentation

Some foods, such as many cheeses, wines, and beers, use specific micro-organisms that combat spoilage from other less-benign organisms. These micro-organisms keep pathogens in check by creating an environment toxic for themselves and other micro-organisms by producing acid or alcohol. Methods of fermentation include, but are not limited to, starter micro-organisms, salt, hops, controlled (usually cool) temperatures and controlled (usually low) levels of oxygen. These methods are used to create the specific controlled conditions that will support the desirable organisms that produce food fit for human consumption.

Fermentation is the microbial conversion of starch and sugars into alcohol. Not only can fermentation produce alcohol, but it can also be a valuable preservation technique. Fermentation can also make foods more nutritious and palatable. For example, drinking water in the Middle Ages was dangerous because it often contained pathogens that could spread disease. When the water is made into beer, the boiling during the brewing process kills any bacteria in the water that could make people sick. Additionally, the water now has the nutrients from the barley and other ingredients, and the microorganisms can also produce vitamins as they ferment.

Industrial/Modern Techniques

Techniques of food preservation were developed in research laboratories for commercial applications.

Pasteurization

Pasteurization is a process for preservation of liquid food. It was originally applied to combat the souring of young local wines. Today, the process is mainly applied to dairy products. In this method, milk is heated at about 70 °C for 15 to 30 seconds to kill the bacteria present in it and cooling it quickly to 10 °C to prevent the remaining bacteria from growing. The milk is then stored in sterilized bottles or pouches in cold places. This method was invented by Louis Pasteur, a French chemist, in 1862.

Vacuum Packing

Vacuum-packing stores food in a vacuum environment, usually in an air-tight bag or bottle. The vacuum environment strips bacteria of oxygen needed for survival. Vacuum-packing is commonly used for storing nuts to reduce loss of flavor from oxidization. A major drawback to vacuum packaging, at the consumer level, is that vacuum sealing can deform contents and rob certain foods, such as cheese, of its flavor.

Artificial Food Additives

Preservative food additives can be *antimicrobial*, which inhibit the growth of bacteria or fungi, including mold, or *antioxidant*, such as oxygen absorbers, which inhibit the oxidation of food constituents. Common antimicrobial preservatives include calcium propionate, sodium nitrate, sodium nitrite, sulfites (sulfur dioxide, sodium bisulfite, potassium hydrogen sulfite, etc.) and disodium EDTA. Antioxidants include BHA and BHT. Other preservatives include formaldehyde (usually in solution), glutaraldehyde (kills insects), ethanol, and methylchloroisothiazolinone.

Radiation

Irradiation of food is the exposure of food to ionizing radiation. The two types of ionizing radiation used are beta particles (high-energy electrons) and gamma rays (emitted from radioactive sources as cobalt-60 or cesium-137). Treatment effects include killing bacteria, molds, and insect pests, reducing the ripening and spoiling of fruits, and at higher doses inducing sterility. The technology may be compared to pasteurization; it is sometimes called "cold pasteurization", as the product is not heated.

The irradiation process is not directly related to nuclear energy, but does use radioactive isotopes produced in nuclear reactors. Cobalt-60, for example does not occur naturally and can only be produced through neutron bombardment of cobalt-59. Ionizing radiation at high energy levels is hazardous to life (hence its usefulness in sterilisation); for this reason, irradiation facilities have a heavily shielded irradiation room where the process takes place. Radiation safety procedures are used to ensure that neither the workers in such facilities nor the environment receives any radiation dose above administrative limits. Irradiated food does not and cannot become radioactive, and national and international expert bodies have declared food irradiation as wholesome. However, the wholesomeness of consuming such food is disputed by opponents and consumer organizations. National and international expert bodies have declared food irradiation as "wholesome"; organizations of the United Nations, such as the World Health Organization and Food and Agriculture Organization, endorse food irradiation. International legislation on whether food may be irradiated or not varies worldwide from no regulation to full banning. Irradiation may allow lower-quality or contaminated foods to be rendered marketable.

Approximately 500,000 tons of food items are irradiated per year worldwide in over 40 countries. These are mainly spices and condiments with an increasing segment of fresh fruit irradiated for fruit fly quarantine.

Pulsed electric Field Electroporation

Pulsed electric field (PEF) electroporation is a method for processing cells by means of brief pulses of a strong electric field. PEF holds potential as a type of low-temperature alternative pasteurization process for sterilizing food products. In PEF processing, a substance is placed between two electrodes, then the pulsed electric field is applied. The electric field enlarges the pores of the cell membranes, which kills the cells and releases their contents. PEF for food processing is a developing technology still being researched. There have been limited industrial applications of PEF processing for the pasteurization of fruit juices. To date, several PEF treated juices are available on the market in Europe. Furthermore, for several years a juice pasteurization application in the US has used PEF. For cell disintegration purposes especially potato processors show great interest in PEF technology as an efficient alternative for their preheaters. Potato applications are already operational in the US and Canada. There are also commercial PEF potato applications in various countries in Europe, as well as in Australia, India and China.

Modified Atmosphere

Modifying atmosphere is a way to preserve food by operating on the atmosphere around it. Salad crops that are notoriously difficult to preserve are now being packaged in sealed bags with an atmosphere modified to reduce the oxygen (O_2) concentration and increase the carbon dioxide (CO_2) concentration. There is concern that, although salad vegetables retain their appearance and texture in such conditions, this method of preservation may not retain nutrients, especially vitamins. There are two methods for preserving grains with carbon dioxide. One method is placing a block of dry ice in the bottom and filling the can with the grain. Another method is purging the container from the bottom by gaseous carbon dioxide from a cylinder or bulk supply vessel.

Carbon dioxide prevents insects and, depending on concentration, mold and oxidation from damaging the grain. Grain stored in this way can remain edible for approximately five years.

Nitrogen gas (N_2) at concentrations of 98% or higher is also used effectively to kill insects in the grain through hypoxia. However, carbon dioxide has an advantage in this respect, as it kills organisms through hypercarbia and hypoxia (depending on concentration), but it requires concentrations of above 35%, or so. This makes carbon dioxide preferable for fumigation in situations where a hermetic seal cannot be maintained.

Controlled Atmospheric Storage (CA): "CA storage is a non-chemical process. Oxygen levels in the sealed rooms are reduced, usually by the infusion of nitrogen gas, from the approximate 21 percent in the air we breathe to 1 percent or 2 percent. Temperatures are kept at a constant 0 to 2 °C (32 to 36 °F). Humidity is maintained at 95 percent and carbon dioxide levels are also controlled. Exact conditions in the rooms are set according to the apple variety. Researchers develop specific regimens for each variety to achieve the best quality. Computers help keep conditions constant." "Eastern Washington, where most of Washington's apples are grown, has enough warehouse storage for 181 million boxes of fruit, according to a report done in 1997 by managers for the Washington State Department of Agriculture Plant Services Division. The storage capacity study shows that 67 percent of that space —enough for 121,008,000 boxes of apples — is CA storage."

Air-tight storage of grains (sometimes called hermetic storage) relies on the respiration of grain, insects, and fungi that can modify the enclosed atmosphere sufficiently to control insect pests. This is a method of great antiquity, as well as having modern equivalents. The success of the method relies on having the correct mix of sealing, grain moisture, and temperature.

A patented process uses fuel cells to exhaust and automatically maintain the exhaustion of oxygen in a shipping container, containing, for example, fresh fish.

Nonthermal Plasma

This process subjects the surface of food to a "flame" of ionized gas molecules, such as helium or nitrogen. This causes micro-organisms to die off on the surface.

High-pressure Food Preservation

High-pressure food preservation or pascalization refers to the use of a food preservation technique that makes use of high pressure. "Pressed inside a vessel exerting 70,000 pounds per square inch

(480 MPa) or more, food can be processed so that it retains its fresh appearance, flavor, texture and nutrients while disabling harmful microorganisms and slowing spoilage." By 2005, the process was being used for products ranging from orange juice to guacamole to deli meats and widely sold.

Biopreservation

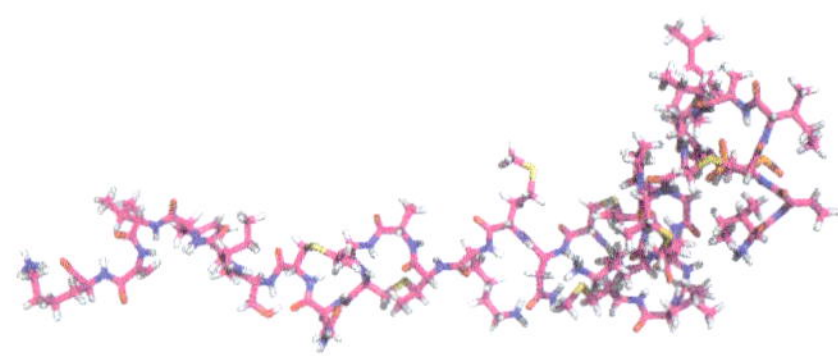

3D stick model of nisin. Some lactic acid bacteria manufacture nisin. It is a particularly effective preservative.

Biopreservation is the use of natural or controlled microbiota or antimicrobials as a way of preserving food and extending its shelf life. Beneficial bacteria or the fermentation products produced by these bacteria are used in biopreservation to control spoilage and render pathogens inactive in food. It is a benign ecological approach which is gaining increasing attention.

Of special interest are lactic acid bacteria (LAB). Lactic acid bacteria have antagonistic properties that make them particularly useful as biopreservatives. When LABs compete for nutrients, their metabolites often include active antimicrobials such as lactic acid, acetic acid, hydrogen peroxide, and peptidebacteriocins. Some LABs produce the antimicrobial nisin, which is a particularly effective preservative.

These days, LAB bacteriocins are used as an integral part of hurdle technology. Using them in combination with other preservative techniques can effectively control spoilage bacteria and other pathogens, and can inhibit the activities of a wide spectrum of organisms, including inherently resistant Gram-negative bacteria.

Hurdle Technology

Hurdle technology is a method of ensuring that pathogens in food products can be eliminated or controlled by combining more than one approach. These approaches can be thought of as "hurdles" the pathogen has to overcome if it is to remain active in the food. The right combination of hurdles can ensure all pathogens are eliminated or rendered harmless in the final product.

Hurdle technology has been defined by Leistner (2000) as an intelligent combination of hurdles that secures the microbial safety and stability as well as the organoleptic and nutritional quality and the economic viability of food products. The organoleptic quality of the food refers to its sensory properties, that is its look, taste, smell, and texture.

Examples of hurdles in a food system are high temperature during processing, low temperature during storage, increasing the acidity, lowering the water activity or redox potential, and the presence of preservatives or biopreservatives. According to the type of pathogens and how risky they are, the intensity of the hurdles can be adjusted individually to meet consumer preferences in an economical way, without sacrificing the safety of the product.

Principal hurdles used for food preservation (after Leistner, 1995)		
Parameter	**Symbol**	**Application**
High temperature	F	Heating
Low temperature	T	Chilling, freezing
Reduced water activity	a_w	Drying, curing, conserving
Increased acidity	pH	Acid addition or formation
Reduced redox potential	E_h	Removal of oxygen or addition of ascorbate
Biopreservatives		Competitive flora such as microbialfermentation
Other preservatives		Sorbates, sulfites, nitrites

Pickling

Pickling is the process of preserving or expanding the lifespan of food by either anaerobicfermentation in brine or immersion in vinegar. The resulting food is called a *pickle*, or, to prevent ambiguity, prefaced with the adjective *pickled*. The pickling procedure will typically affect the food's texture and flavor. In East Asia, vinaigrette (vegetable oil and vinegar) is also used as a pickling medium. Foods that are pickled include meats, fruits, eggs, and vegetables.

Another distinguishing characteristic is a pH 4.6 or lower, which is sufficient to kill most bacteria. Pickling can preserve perishable foods for months. Antimicrobial herbs and spices, such as mustard seed, garlic, cinnamon or cloves, are often added. If the food contains sufficient moisture, a pickling brine may be produced simply by adding dry salt. For example, Germansauerkraut and Koreankimchi are produced by salting the vegetables to draw out excess water. Natural fermentation at room temperature, by lactic acid bacteria, produces the required acidity. Other pickles are made by placing vegetables in vinegar. Unlike the canning process, pickling (which includes fermentation) does not require that the food be completely sterile before it is sealed. The acidity or salinity of the solution, the temperature of fermentation, and the exclusion of oxygen determine which microorganisms dominate, and determine the flavor of the end product.

When both salt concentration and temperature are low, *Leuconostoc mesenteroides* dominates, producing a mix of acids, alcohol, and aroma compounds. At higher temperatures *Lactobacillus plantarum* dominates, which produces primarily lactic acid. Many pickles start with *Leuconostoc*, and change to *Lactobacillus* with higher acidity.

History

Pickling began 4000 years ago using cucumbers native to India. This was used as a way to preserve food for out-of-season use and for long journeys, especially by sea. Salt pork and salt beef were common staples for sailors before the days of steam engines. Although the process was invented to preserve foods, pickles are also made and eaten because people enjoy the resulting flavors. Pickling may also improve the nutritional value of food by introducing B vitamins produced by bacteria.

Etymology

The term *pickle* is derived from the Dutch word *pekel*, meaning *brine*. In the U.S. and Canada, and sometimes Australia and New Zealand, the word *pickle* alone almost always refers to a pickled cucumber, except when it is used figuratively. It may also refer to other types of pickles such as "pickled onion", "pickled cauliflower", etc. In the UK, pickle, as in a "cheese and pickle sandwich", may also refer to Ploughman's pickle, a kind of chutney.

Popularity of Pickles Around the World

South Asia

South Asia has a large variety of pickles (known as *achar* in Assamese, Punjabi, Hindi, Bengali, *uppinakaayi* in Kannada, *lonacha* in Marathi, *uppilittathu* or *achar* in Malayalam*oorukai* in Tamil, *ooragaya* in Telugu), which are mainly made from varieties of mango, lemon, lime, goongura(a sour leafy shrub), tamarind and Indian gooseberry (amla), chilli. Vegetables such as brinjal, carrots, cauliflower, tomato, bitter gourd, green tamarind, ginger, garlic, onion, and citron are also occasionally used. These fruits and vegetables are generally mixed with ingredients like salt, spices, and vegetable oils and are set to mature in a moistureless medium.

In Pakistan, pickles are known locally as *achaar* (in Urdu) and come in a variety of flavors. A popular item is the traditional mixed Hyderabadi pickle, a common delicacy prepared from an assortment of fruits (most notably mangoes) and vegetables blended with selected spices.

In Sri Lanka, *achcharu* is traditionally prepared from carrots, onions, and ground dates that are mixed with mustard powder, ground pepper, crushed ginger, garlic, and vinegar, and left to sit in a clay pot.

Southeast Asia

Singapore, Indonesian and Malaysian pickles, called *acar*, are typically made out of cucumber, carrot, bird's eye chilies, and shallots, these items being seasoned with vinegar, sugar and salt. Fruits, such as papaya and pineapple, are also sometimes pickled.

Kimchi is a very common side dish in Korea.

In the Philippines, *achara* is primarily made out of green papaya, carrots, and shallots, with cloves of garlic and vinegar. Other versions could include ginger, bell peppers, daikon, cucumbers or bamboo shoots. Separately, in some provinces, unripe mangoes or *burong mangga*, unripe tomatoes, green apple guavas, jicama turnips, bitter gourd and other fruit and vegetables are also pickled. Siling labuyo, sometimes with garlic and red onions, are also pickled in bottled vinegar. The spiced vinegar itself is a staple condiment in Filipino cuisine.

In Vietnam, vegetable pickles are called *dưa muối* ("salted vegetables") or *dưa chua* ("sour vegetables"). In Burma, tea leaves are pickled to produce lahpet, which has strong social and cultural importance.

East Asia

China is home to a huge variety of pickled vegetables, including radish, *baicai* (Chinese cabbage, notably *suan cai, la bai cai, pao cai,* and Tianjin preserved vegetable), *zha cai,* chili pepper, and cucumbers, among many others.

Japanese*tsukemono* (pickled foods) include *takuan* (daikon), *umeboshi* (ume plum), *gari&beni shoga* (ginger), turnip, cucumber, and Chinese cabbage.

The Korean staple kimchi is usually made from pickled napa cabbage and radish, but is also made from green onions, garlic stems, chives and a host of other vegetables. Kimchi is popular throughout East Asia. Jangajji is another example of pickled vegetables.

Western Asia

Torshi, traditional pickles in Southeast Europe, Western Asia and the Caucasus

In Iran, Turkey, Arab countries, the Balkans, and the Caucasus, pickles (called torshi in Persian, *turşu* in Turkish language and *mekhallel* in Arabic) are commonly made from turnips, peppers, carrots, green olives, cucumbers, cabbage, green tomatoes, lemons, and cauliflower.

Europe

Central and Eastern Europe

Coriander seeds are one of the spices popularly added to pickled vegetables in Europe.

In Hungary the main meal *(lunch)* usually goes with some kind of pickles *(savanyúság)* but they are commonly consumed at other times of the day too. The most commonly consumed pickles are sauerkraut*(savanyú káposzta)*, the different kinds of pickled cucumbers and peppers and *csalamádé* but tomatoes, carrots, beetroot, baby corn, onions, garlic, certain squashes and melons and a few fruits like plums and apples are used to make pickles too. Stuffed pickles are specialties usually made of peppers or melons pickled after being stuffed with a cabbage filling. Pickled plum stuffed with garlic is a unique Hungarian type of pickle just like *csalamádé* and leavened cucumber *(kovászos uborka)*. *Csalamádé* a type of mixed pickle made of cabbage, cucumber, paprika, onion, carrot, tomatoes and bay leaf mixed up with vinegar as the fermenting agent. Leavened cucumber, unlike other types of pickled cucumbers that are around all year long, is rather a seasonal pickle produced in the summer. Cucumbers, spices, herbs and slices of bread are put in a glass jar with salt water and kept in direct sunlight for a few days. The yeast from the bread, along with other pickling agents and spices fermented under the hot sun, give the cucumbers a unique flavor, texture and slight carbonation. Its juice can be used to make a special type of spritzer*('Újházy fröccs')* instead of carbonated water. It is common for Hungarian households to produce their own pickles. Different regions or towns have their special recipes unique to them. Among them all the Vecsési Sauerkraut *(Vecsési savanyú káposzta)* is the most famous.

Pickled tomatoes are common in CIS

Romanian pickles (murături) are made out of beetroot, cucumbers, green tomatoes (*gogonele*), carrots, cabbage, garlic , sauerkraut (bell peppers stuffed with cabbage), bell peppers, melons, mushrooms, turnips, celery and cauliflower. Meat, like pork, can also be preserved in salt and lard.

Polish and Czech traditional pickles are cucumbers and sauerkraut, but other pickled fruits and vegetables, including plums, pumpkins and mushrooms are also common.

Russian, Ukrainian and Belarusian pickled items include beets, mushrooms, tomatoes, sauerkraut, cucumbers, ramsons, garlic, eggplant (which is typically stuffed with julienned carrots), custard squash, and watermelon. In these countries countries garden produce is commonly pickled using salt, dill, blackcurrant leaves, bay leaves and garlic and is stored in a cool, dark place.

Southern Europe

An Italian pickled vegetable dish is giardiniera, which includes onions, carrots, celery and cauliflower. Many places in southern Italy, particularly in Sicily, pickle eggplants and hot peppers.

In Albania, Bulgaria, Serbia, Macedonia and Turkey, mixed pickles, known as *turshi* or *turshu* form popular appetizers, which are typically eaten with *rakia*. Pickled green tomatoes, cucumbers, carrots, bell peppers, peppers, eggplants, and sauerkraut are also popular.

Turkish pickles, called *turşu*, are made out of vegetables, roots, and fruits such as peppers, cucumber, Armenian cucumber, cabbage, tomato, eggplant (aubergine), carrot, turnip, beetroot, green almond, baby watermelon, baby cantaloupe, garlic, cauliflower, bean and green plum. A mixture of spices flavor the pickles.

In Greece, pickles, are made out of carrots, celery, eggplants stuffed with diced carrots, cauliflower, tomatoes, and peppers.

Northern Europe

In Britain, pickled onions and pickled eggs are often sold in pubs and fish and chip shops. Pickled beetroot, walnuts, and gherkins, and condiments such as Branston Pickle and piccalilli are typically eaten as an accompaniment to pork pies and cold meats, sandwiches or a ploughman's lunch. Other popular pickles in the UK are pickled mussels, cockles, red cabbage, mango chutney, sauerkraut, and olives. Rollmops are also quite widely available under a range of names from various producers both within and out of the UK.

Pickled herring, rollmops, and salmon are popular in Scandinavia. Pickled cucumbers and red garden beets are important as condiments for several traditional dishes. Pickled capers are also common in Scandinavian cuisine.

United States and Canada

A dish of giardiniera

In the United States and Canada, pickled cucumbers (most often referred to simply as "pickles" in Canada and the United States), olives, and sauerkraut are most popular, although pickles popular in other nations are also available.

Canadian pickling is similar to that of Britain. Through the winter, pickling is an important method of food preservation. Pickled cucumbers, onions, and eggs are very popular. Chow Chow is a tart vegetable mix popular in the Maritime Provinces and the Southern United States, similar to picalily. Pickled fish is very popular, as in Scotland. Meat is often also pickled or preserved in different brines throughout the winter, most prominently in the harsh climate of Newfoundland.

Giardiniera, a mixture of pickled peppers, celery and olives, is a popular condiment in Chicago and other cities with large Italian-American populations, and is often consumed with Italian beef sandwiches. Pickled eggs are common in the Upper Peninsula of Michigan. Pickled herring is available in the Upper Midwest. Pennsylvania Dutch Country has a strong tradition of pickled foods, including chow-chow and red beet eggs. In the Southern United States, pickled okra and watermelon rind are popular, as are deep-fried pickles and pickled pig's feet, pickled chicken eggs, pickled quail eggs, pickled Garden vegetables and pickled sausage. In Mexico, chili peppers, particularly of the Jalapeño and serrano varieties, pickled with onions, carrots and herbs form common condiments. Various pickled vegetables, fish, or eggs may make a side dish to a Canadian lunch or dinner.

In the United States, National Pickle Day is a popular food holiday every year on November 14.

Mexico, Central America, and South America

In the Mesoamerican region pickling is known as "encurtido" or "curtido" for short. The pickles or "curtidos" as known in Latin America are served cold, as an appetizer, as a side dish or as a tapas dish in Spain. In several Central American countries it is prepared with cabbage, onions, carrots, lemon, vinegar, oregano, and salt. In Mexico, "curtido" consists of carrots, onions, and jalapeño peppers and used to accompany meals still common in taquerias and restaurants. In order to prepare a carrot "curtido" simply add carrots to vinegar and other ingredients that are common to the region such as chilli, tomato, and onions. Varies depending on the food, in the case of sour. Another example of a type of pickling which involves the pickling of meats or seafood is the "escabeche" or "ceviches" popular in Peru, Ecuador, and throughout Latin America and the Caribbean. These dishes include the pickling of pig's feet, pig's ears, and gizzards prepared as an "escabeche" with spices and seasonings to flavor it. The ceviches consists of shrimp, octopus, and various fishes seasoned and served cold.

Process

Vase by Bát Tràng porcelain for pickling

In traditional pickling fruit or vegetables are submerged in a brine (20-40 grams/L of salt (3.2–6.4 oz/imp gal or 2.7–5.3 oz/US gal)) or shredded and salted as in sauerkraut preparation and held underwater by flat stones layered on top. Alternatively a lid with an airtrap or a tight lid could be used if the lid is able to release pressure which might result from carbon dioxide build up. When using an open container it should be covered with cloth to keep insects away. Mold or (white) kahm yeast may form on the surface and should be removed. Kahm yeast is mostly harmless but can impart an off-taste.

In chemical pickling, the jar and lid are first boiled in order to sterilize them. The fruits or vegetables to be pickled are then added to the jar along with brine, vinegar, or both, as well as spices, and are then allowed to mature until the desired taste is obtained.

The food can be pre-soaked in brine before transferring to vinegar. This reduces the water content of the food which would otherwise dilute the vinegar. This method is particularly useful for fruit and vegetables with a high natural water content.

In commercial pickling, a preservative like sodium benzoate or EDTA may also be added to enhance shelf life. In fermentation pickling, the food itself produces the preservation agent, typically by a process involving Lactobacillus bacteria that produce lactic acid as the preservative agent.

Alum is used in pickling to promote crisp texture and approved as a food additive by the United States Food and Drug Administration.

"Refrigerator pickles" are unfermented pickles made by marinating fruit or vegetables in a seasoned vinegar solution. They must be stored under refrigeration or undergo canning to achieve long-term storage.

Possible Health Hazards of Pickled Vegetables

The *British Journal of Cancer* released an online 2009 meta-analysis of research on pickles as increasing the risks of esophageal cancer. The report, citing limited data in a statistical meta analysis, indicates a potential two-fold increased risk of oesophageal cancer associated with Asian pickled vegetable consumption. Results from the research are described as having "high heterogeneity" and the study said that further well-designed prospective studies were warranted.

Some common fungi can facilitate the formation of N-nitroso compounds, which are strong oesophageal carcinogens in several animal models. Roussin red methyl ester, a non-alkylating nitroso compound with tumour-promoting effect in vitro, was identified in pickles from Linxian in much higher concentrations than in samples from low-incidence areas. Fumonisinmycotoxins have been shown to cause liver and kidney tumours in rodents.

Frozen Food

A frozen processed foods aisle at a supermarket in Canada

Cutting frozen tuna using a bandsaw in the Tsukiji fish market in Tokyo, Japan (2002)

Freezingfood preserves it from the time it is prepared to the time it is eaten. Since early times, farmers, fishermen, and trappers have preserved their grains and produce in unheated buildings during the winter season. Freezing food slows down decomposition by turning residual moisture into ice, inhibiting the growth of most bacterial species. In the food commodity industry, there are two processes: mechanical and cryogenic (or flash freezing). The freezing kinetics is important to preserve the food quality and texture. Quicker freezing generates smaller ice crystals and maintains cellular structure. Cryogenic freezing is the quickest freezing technology available due to the ultra low liquid nitrogen temperature −196 °C (−320 °F).

Preserving food in domestic kitchens during the 20th and 21st centuries is achieved using household freezers. Accepted advice to householders was to freeze food on the day of purchase. An initiative by a supermarket group in 2012 (backed by the UK's Waste & Resources Action Programme) promotes advising the freezing of food "as soon as possible up to the product's 'use by' date". The Food Standards Agency was reported as supporting the change, providing the food had been stored correctly up to that time.

Preservatives

Frozen products do not require any added preservatives because microorganisms do not grow when the temperature of the food is below −9.5 °C (15 °F), which is sufficient on its own in preventing food spoilage. Long-term preservation of food may call for food storage at even lower temperatures. Carboxymethylcellulose (CMC), a tasteless and odorless stabilizer, is typically added to frozen food because it does not adulterate the quality of the product.

History

Natural food freezing (using winter frosts) had been in use by tribes in cold climates for centuries. By 1885 a small number of chicken and geese were being shipped from Russia to London in insulated cases using this technique. By March 1899, the "British Refrigeration and Allied Interests" reported that a food importing business, "Baerselman Bros", was shipping some 200,000 frozen geese and chickens per week from three Russian depots to New Star Wharf, Lower Shadwell, London over three or four winter months. This trade in frozen food was enabled by the introduction of Linde cold air freezing plants in three Russian depots and the London warehouse. The Shadwell warehouse stored the frozen goods until they were shipped to markets in London, Birmingham, Liverpool and Manchester. The techniques were later expanded into the meat packing industry.

From 1929, Clarence Birdseye introduced "flash freezing" to the American public. Birdseye first became interested in food freezing during fur-trapping expeditions to Labrador in 1912 and 1916, where he saw the natives use natural freezing to preserve foods. More advanced attempts include food frozen for Eleanor Roosevelt on her trip to Russia. Other experiments, involving orange juice, ice cream and vegetables were conducted by the military near the end of World War II.

Technology

The freezing technique itself, just like the frozen food market, is developing to become faster, more efficient and more cost-effective.

Mechanical freezers were the first to be used in the food industry and are used in the vast majority of freezing / refrigerating lines. They function by circulating a refrigerant, normally ammonia, around the system, which withdraws heat from the food product. This heat is then transferred to a condenser and dissipated into air or water. The refrigerant itself, now a high pressure, hot liquid, is directed into an evaporator. As it passes through an expansion valve, it is cooled and then vaporises into a gaseous state. Now a low pressure, low temperature gas again, it can be reintroduced into the system.

Cryogenic or (flash freezing) of food is a more recent development, but is used by many leading food manufacturers all over the world. Cryogenic equipment uses very low temperature gases – usually liquid nitrogen or solid carbon dioxide – which are applied directly to the food product.

Packaging

Frozen food packaging must maintain its integrity throughout machine filling, sealing, freezing, storage, transportation, thawing, and often cooking. As many frozen foods are cooked in a microwave oven, manufacturers have developed packaging that can go straight from freezer to the microwave.

In 1974, the first differential heating container (DHC) was sold to the public. A DHC is a sleeve of metal designed to allow frozen foods to receive the correct amount of heat. Various sized apertures were positioned around the sleeve. The consumer would put the frozen dinner into the sleeve according to what needed the most heat. This ensured proper cooking.

Today there are multiple options for packaging frozen foods. Boxes, cartons, bags, pouches, Boil-in-Bags, lidded trays and pans, crystallized PET trays, and composite and plastic cans.

Scientists are continually researching new aspects of frozen food packaging. Active packaging offers a host of new technologies that can actively sense and then neutralize the presence of bacteria or other harmful species. Active packaging can extend shelf-life, maintain product safety, and help preserve the food over a longer period of time. Several functions of active packaging are being researched:

- Oxygen scavengers
- Time Temperature Indicators and digital temperature data loggers
- Antimicrobials
- Carbon Dioxide controllers
- Microwave susceptors
- Moisture control: Water activity, Moisture vapor transmission rate, etc.
- Flavor enhancers
- Odor generators
- Oxygen-permeable films
- Oxygen generators
- Validation of cold chain

Effects on Nutrients

Vitamin Content of Frozen Foods

- Vitamin C: Usually lost in a higher concentration than any other vitamin. A study was performed on peas to determine the cause of vitamin C loss. A vitamin loss of ten percent occurred during the blanching phase with the rest of the loss occurring during the cooling and washing stages. The vitamin loss was not actually accredited to the freezing process. Another experiment was performed involving peas and lima beans. Frozen and canned vegetables were both used in the experiment. The frozen vegetables were stored at −23 °C (−10 °F) and the canned vegetables were stored at room temperature 24 °C (75 °F). After 0, 3, 6, and 12 months of storage, the vegetables were analyzed with and without cooking. O'Hara, the scientist performing the experiment said, "From the view point of the vitamin content of the two vegetables when they were ready for the plate of the consumer, there did not appear to be any marked advantages attributable to method of preservation, frozen storage, processed in a tin, or processed in glass."
- Vitamin B1 (Thiamin): A vitamin loss of 25 percent is normal. Thiamin is easily soluble in water and is destroyed by heat.
- Vitamin B2 (Riboflavin): Not much research has been done to see how much freezing affects Riboflavin levels. Studies that have been performed are inconclusive; one study found an 18 percent vitamin loss in green vegetables, while another determined a 4 percent loss. It is commonly accepted that the loss of Riboflavin has to do with the preparation for freezing rather than the actual freezing process itself.
- Vitamin A (Carotene): There is little loss of carotene during preparation for freezing and freezing of most vegetables. Much of the vitamin loss is incurred during the extended storage period.

Effectiveness

A frozen food warehouse at McMurdo Station, Antarctica

Freezing is an effective form of food preservation because the pathogens that cause food spoilage are killed or do not grow very rapidly at reduced temperatures. The process is less effective in food preservation than are thermal techniques, such as boiling, because pathogens are more likely to be able to survive cold temperatures rather than hot temperatures. One of the problems surrounding the use of freezing as a method of food preservation is the danger that pathogens deactivated (but not killed) by the process will once again become active when the frozen food thaws.

Foods may be preserved for several months by freezing. Long-term frozen storage requires a constant temperature of –18 °C (0 °F) or less.

Quality

The speed of the freezing has a direct impact on the size and the number of ice crystals formed within a food product's cells and extracellular space. Slow freezing leads to fewer but larger ice crystals while fast freezing leads to smaller but more numerous ice crystals. Large ice crystals can puncture the walls of the cells of the food product which will cause a degradation of the texture of the product as well as the loss of its natural juices during thawing. That is why there will be a qualitative difference observed between food products frozen by ventilated mechanical freezing, non-ventilated mechanical freezing or cryogenic freezing with liquid nitrogen.

Reaction

According to a study, an American consumes on average 71 frozen foods a year, most of which are pre-cooked frozen meals. Many food critics host shows are dedicated to the tasting and reviewing of frozen foods, the web show *Freezer burns* being one of the more notable ones.

Gelatin

Gelatin or gelatine (from Latin: *gelatus* meaning "stiff", "frozen") is a translucent, colorless, brittle (when dry), flavorless food derived from collagen obtained from various animal raw materials. It is commonly used as a gelling agent in food, pharmaceutical drugs, photography, and cosmetic manufacturing. Substances containing gelatin or functioning in a similar way are called "gelatinous". Gelatin is an irreversibly hydrolyzed form of collagen, wherein the hydrolysis results in the reduction of protein fibrils into smaller peptides, which will have broad molecular weight ranges associated with physical and chemical methods of denaturation, based on the process of hydrolysis. It is found in most gummy candy, as well as other products such as marshmallows, gelatin desserts, and some ice creams, dips, and yogurts. Gelatin for recipe use comes in the form of sheets, granules, or powder. Instant types can be added to the food as they are; others need to be soaked in water beforehand.

Sheet (or leaf) gelatin for cooking

Composition and Properties

Gelatin is a mixture of peptides and proteins produced by partial hydrolysis of collagen extracted from the skin, bones, and connective tissues of animals such as domesticated cattle, chicken, pigs, and fish. During hydrolysis, the natural molecular bonds between individual collagen strands are broken down into a form that rearranges more easily. Its chemical composition is, in many respects, closely similar to that of its parent collagen. Photographic and pharmaceutical grades of gelatin generally are sourced from beef bones and pig skin.

Gelatin readily dissolves in hot water and sets to a gel on cooling. When added directly to cold water, it does not dissolve well, however. Gelatin also is soluble in most polar solvents. Gelatin solutions show viscoelastic flow and streaming birefringence. Solubility is determined by the method of manufacture. Typically, gelatin can be dispersed in a relatively concentrated acid. Such dispersions are stable for 10–15 days with little or no chemical changes and are suitable for coating purposes or for extrusion into a precipitating bath.

The mechanical properties of gelatin gels are very sensitive to temperature variations, the previous thermal history of the gels, and the amount of time elapsing. These gels exist over only a small temperature range, the upper limit being the melting point of the gel, which depends on gelatin grade and concentration, but typically, is less than 35 °C (95 °F) and the lower limit the freezing point at which ice crystallizes. The upper melting point is below human body temperature, a factor that is important for mouthfeel of foods produced with gelatin. The viscosity of the gelatin-water mixture is greatest when the gelatin concentration is high and the mixture is kept cool at about 4 °C (39 °F). The gel strength is quantified using the Bloom test.

Production

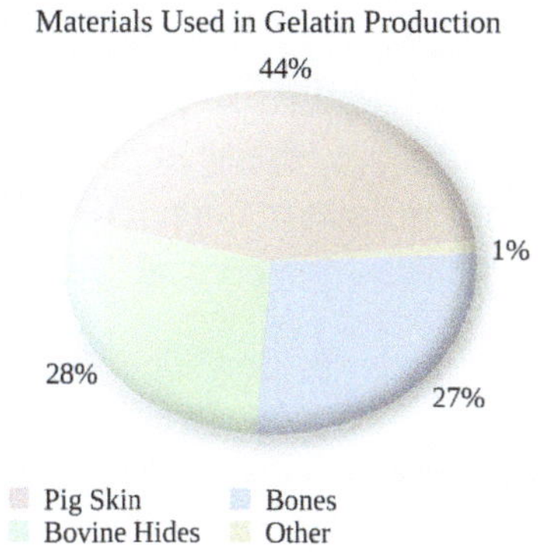

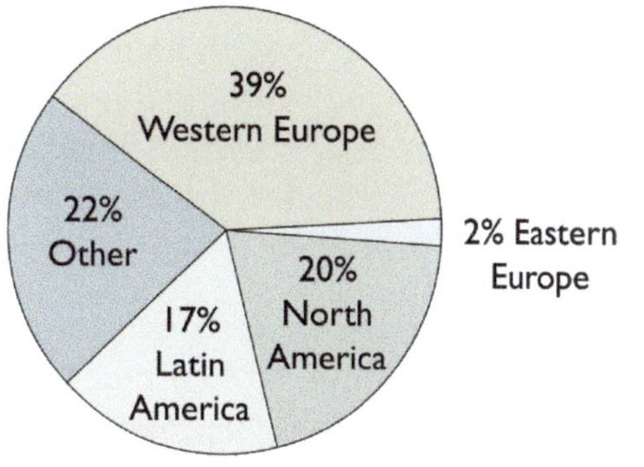

Gelatin Production by Geography

The worldwide production amount of gelatin is about 375,000–400,000 tonnes per year ($830\times10^{\Delta 6}$–$880\times10^{\Delta 6}$ lb/a). On a commercial scale, gelatin is made from by-products of the meat and leather industries. Recently, fish by-products also have been considered suitable because they eliminate some of the religious obstacles surrounding gelatin consumption. Gelatin is derived from pork skins, pork, and cattle bones, or split cattle hides. The raw materials are prepared by different curing, acid, and alkali processes that are employed to extract the dried collagen hydrolysate. These processes may take several weeks, and differences in such processes have great effects on the properties of the final gelatin products.

Gelatin also can be prepared in the home. Boiling certain cartilaginous cuts of meat or bones results in gelatin being dissolved into the water. Depending on the concentration, the resulting stock (when cooled) will form a jelly or gel naturally. This process is used for aspic.

While many processes exist whereby collagen may be converted to gelatin, they all have several factors in common. The intermolecular and intramolecular bonds that stabilize insoluble collagen must be broken, and also, the hydrogen bonds that stabilize the collagen helix must be broken. The manufacturing processes of gelatin consists of three main stages:

1. Pretreatments to make the raw materials ready for the main extraction step and to remove impurities that may have negative effects on physiochemical properties of the final gelatin product
2. The main extraction step, which usually is done with hot water or dilute acid solutions as a multistage extraction to hydrolyze collagen into gelatin
3. The refining and recovering treatments including filtration, clarification, evaporation, sterilization, drying, rutting, grinding, and sifting to remove the water from the gelatin solution, to blend the gelatin extracted, and to obtain dried, blended and ground final product

Pretreatments

If the raw material used in the production of the gelatin is derived from bones, dilute acid solutions are used to remove calcium and other salts. Hot water or several solvents may be used to reduce the fat content, which should not exceed 1% before the main extraction step. If the raw material consists of hides and skin; size reduction, washing, removal of hair from hides, and degreasing are necessary to prepare the hides and skins for the main extraction step.

Collagen hydrolysis is performed by one of three different methods: acid-, alkali-, and enzymatic hydrolysis. Acid treatment is especially suitable for less fully crosslinked materials such as pig skin collagen and normally requires 10 to 48 hours. Alkali treatment is suitable for more complex collagen such as that found in bovine hides and requires more time, normally several weeks. The purpose of the alkali treatment is to destroy certain chemical crosslinks still present in collagen. Within the gelatin industry, the gelatin obtained from acid-treated raw material has been called type-A gelatin and the gelatin obtained from alkali-treated raw material is referred to as type-B gelatin.

Enzymatic hydrolysis of collagen for gelatin extraction is relatively new.The treatment time is shorter than that required for alkali treatment, however, and results in almost complete conversion to the pure product. The physical properties of the final gelatin product are considered better.

Extraction

After preparation of the raw material, i.e., reducing cross-links between collagen components and removing some of the impurities such as fat and salts, partially purified collagen is converted into gelatin by extraction with either water or acid solutions at appropriate temperatures. All industrial processes are based on neutral or acid pH values because although alkali treatments speed up conversion, they also promote degradation processes. Acidic extraction conditions are extensively used in the industry, but the degree of acid varies with different processes. This extraction step is a multistage process, and the extraction temperature usually is increased in later extraction steps, which ensures minimum thermal degradation of the extracted gelatin.

Recovery

This process includes several steps such as filtration, evaporation, drying, grinding, and sifting. These operations are concentration-dependent and also dependent on the particular gelatin used. Gelatin degradation should be avoided and minimized, so the lowest temperature possible is used for the recovery process. Most recoveries are rapid, with all of the processes being done in several stages to avoid extensive deterioration of the peptide structure. A deteriorated peptide structure would result in a low gel strength, which is not generally desired.

Uses

Early History of Food Applications

The first use of gelatin in foods is attributed to Medieval Britain (1400s) when cattle hooves were boiled to produce a gel. Further commercial development occurred in 1754 when a British manufacturing patent was issued. Food applications in the USA and France during 1800-1900 appear to have established the versatility of gelatin, including the origin of its popularity in the USA as Jell-O. Over middle-late 1800s, Charles and Rose Knox of New York manufactured and marketed gelatin powder, diversifying the appeal and applications of gelatin.

Culinary Uses

Probably best known as a gelling agent in cooking, different types and grades of gelatin are used in a wide range of food and nonfood products: common examples of foods that contain gelatin are gelatin desserts, trifles, aspic, marshmallows, candy corn, and confections such as Peeps, gummy bears, fruit snacks, and jelly babies. Gelatin may be used as a stabilizer, thickener, or texturizer in foods such as yogurt, cream cheese, and margarine; it is used, as well, in fat-reduced foods to simulate the mouthfeel of fat and to create volume. It also is used in the production of several types of Chinese soup dumplings, specifically Shanghainese soup dumplings, or *xiaolongbao*, as well as *Shengjian mantou*, a type of fried and steamed dumpling. The fillings of both are made by combining ground pork with gelatin cubes, and in the process of cooking, the gelatin melts, creating a soupy interior with a characteristic gelatinous stickiness.

Eggs in aspic

Gelatin is used for the clarification of juices, such as apple juice, and of vinegar.

Isinglass, from the swim bladders of fish, is still used as a fining agent for wine and beer. Besides hartshorn jelly, from deer antlers (hence the name "hartshorn"), isinglass was one of the oldest sources of gelatin.

Technical Uses

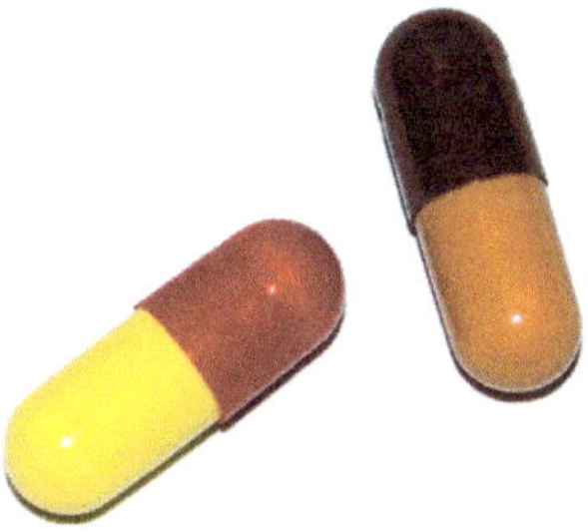

Capsules made of gelatin

- Certain professional and theatrical lighting equipment use color gels to change the beam color. Historically, these were made with gelatin, hence the term, color gel.
- Gelatin typically constitutes the shells of pharmaceutical capsules to make them easier to swallow. Hypromellose is a vegetarian-acceptable alternative to gelatin, but is more expensive to produce.
- Some animal glues such as hide glue may be unrefined gelatin.
- It is used to hold silver halide crystals in an emulsion in virtually all photographic films and photographic papers. Despite some efforts, no suitable substitutes with the stability and low cost of gelatin have been found.

- Used as a carrier, coating, or separating agent for other substances, for example, it makes β-carotene water-soluble, thus imparting a yellow color to any soft drinks containing beta-carotene.
- Gelatin is closely related to bone glue and is used as a binder in match heads and sandpaper.
- Cosmetics may contain a nongelling variant of gelatin under the name hydrolyzed collagen.
- Gelatin was first used as an external surface sizing for paper in 1337 and continued as a dominant sizing agent of all European papers through the mid-nineteenth century. In modern times, it is mostly found in watercolor paper, and occasionally in glossy printing papers, artistic papers, and playing cards, and it maintains the wrinkles in crêpe paper.

Dietary Restrictions and Gelatin Substitutes

The consumption of gelatin from particular animals may be forbidden by religious rules or cultural taboos. For example, Jewishkosher and Islamichalal customs require gelatin from sources other than pigs, such as cattle or fish, and from animals slaughtered ritually. Roma people are cautious of gelatin products that may have been made from horses, as their culture forbids the consumption of horses. Some companies specify the source of the gelatin used. Vegans and many vegetarians choose not to eat foods containing gelatin made from animals. Likewise, Hindu and Jain customs may require gelatin alternatives from sources other than animals, as many Hindus are vegetarian. Hindus who are not vegetarians often consume gelatin from all sources except cattle, which are considered sacred. Since many people are not aware about gelatin's sources, however, they unknowingly consume it. Other people simply consider gelatin unpalatable due to the ingredients used in its production.

Partial alternatives to gelatins derived from animals include the seaweed extracts agar and carrageenan, as well as pectin and konjac. Research into synthetic collagen is ongoing; as of 2011, partial success has been achieved in replicating collagen's structure using self-assembling peptides.

Protein Content

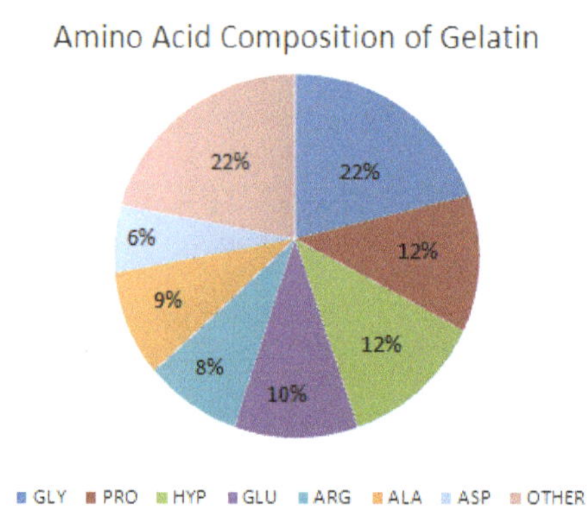

Amino acid composition

Although gelatin is 98–99% protein by dry weight, it has little additional nutritional value, varying according to the source of the raw material and processing technique.

Safety Concerns

Since 1997, the U.S. Food and Drug Administration (FDA), with support from the TSE (transmissible spongiform encephalopathy) Advisory Committee, has been monitoring the potential risk of transmitting animal diseases, especially bovine spongiform encephalopathy (BSE), commonly known as mad cow disease. The FDA study concluded: "...steps such as heat, alkaline treatment, and filtration could be effective in reducing the level of contaminating TSE agents; however, scientific evidence is insufficient at this time to demonstrate that these treatments would effectively remove the BSE infectious agent if present in the source material."

The Scientific Steering Committee (SSC) of the European Union in 2003 stated that the risk associated with bovine bone gelatin is very low or zero. In 2006, the European Food Safety Authority (EFSA) stated that the SSC opinion was confirmed, that the BSE risk of bone-derived gelatin was very small, and removed support for the 2003 request to exclude the skull and vertebrae of bovine origin older than 12 months from the material used in gelatin manufacturing.

Today, all reputable gelatin manufacturers follow the Quality Management System according to ISO 9001. In this way, all process steps are documented. For pharmaceutical-grade gelatins, strict regulations from the FDA, the European CPMP, and European Pharmacopoeia must be met. A detailed overview of the regulatory requirements for gelatin production may be found in the *Gelatine Handbook*, pp. 99–101.

Health Effects

In 2011, reviewing the clinical data that had been published to date, the European Food Safety Authority Panel on Dietetic Products, Nutrition, and Allergies concluded that "a cause and effect relationship has not been established between the consumption of collagen hydrolysate and maintenance of joints". A 2012 review also found insufficient evidence to support its use for osteoarthritis.

Nonetheless, in 2013, Health Canada approved a label for "hydrolyzed collagen" specifying that the label may say that it is useful to supplement dietary amino acid intake and, that it "helps to reduce joint pain associated with osteoarthritis."

Vacuum Packing

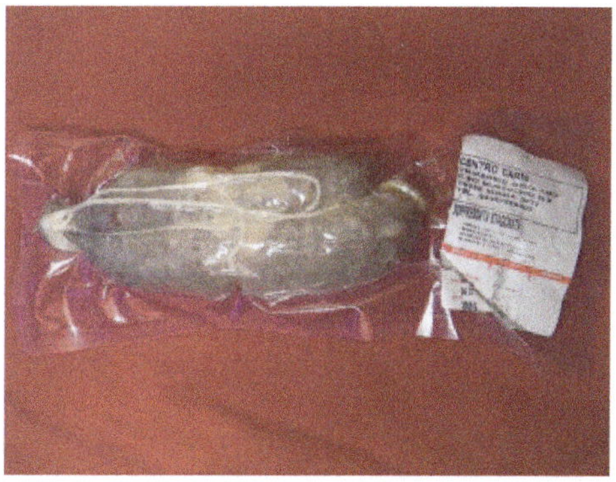

Vacuum packing is a method of packaging that removes air from the package prior to sealing. This method involves (manually or automatically) placing items in a plastic film package, removing air from inside, and sealing the package.Shrink film is sometimes used to have a tight fit to the contents. The intent of vacuum packing is usually to remove oxygen from the container to extend the shelf life of foods and, with flexible package forms, to reduce the volume of the contents and package.

Vacuum packing reduces atmospheric oxygen, limiting the growth of aerobic bacteria or fungi, and preventing the evaporation of volatile components. It is also commonly used to store dry foods over a long period of time, such as cereals, nuts, cured meats, cheese, smoked fish, coffee, and potato chips (crisps). On a more short term basis, vacuum packing can also be used to store fresh foods, such as vegetables, meats, and liquids, because it inhibits bacterial growth.

Vacuum packing greatly reduces the bulk of non-food items. For example, clothing and bedding can be stored in bags evacuated with a domestic vacuum cleaner or a dedicated vacuum sealer. This technique is sometimes used to compact household waste, for example where a charge is made for each full bag collected.

Vacuum packaging products, using plastic bags, canisters, bottles, or mason jars, are available for home use.

For delicate food items which might be crushed by the vacuum packing process (such as potato chips), an alternative is to replace the interior gas with nitrogen. This has the same effect of inhibiting deterioration due to the removal of oxygen.

External Sealers

External vacuum sealers involve a bag being attached to the vacuum-sealing machine externally. The machine will remove the air and seal the bag, which is all done outside the machine. A heat sealer is often used to seal the pack.

Single Vacuum Chamber Machines

This video shows vacuum packaging of organic rice.

Single chamber sealers require the entire product to be placed within the machine. Like external sealers, a plastic bag is typically used for packaging. Once the product is placed in the machine, the lid is closed and air is removed. Then, there is a heat seal inside the chamber that will seal the bag, after sealing the bag the chamber is refilled with air by the automatic opening of a vent to the outside. This oncoming pressure squeezes all remaining air in the bag. The lid is then opened and the product removed. Chamber sealers are typically used for low-to-medium-volume packaging, and also have the capability to vacuum seal liquids.

Tabletop Vacuum Packaging Machine

Double Vacuum Chamber Machines

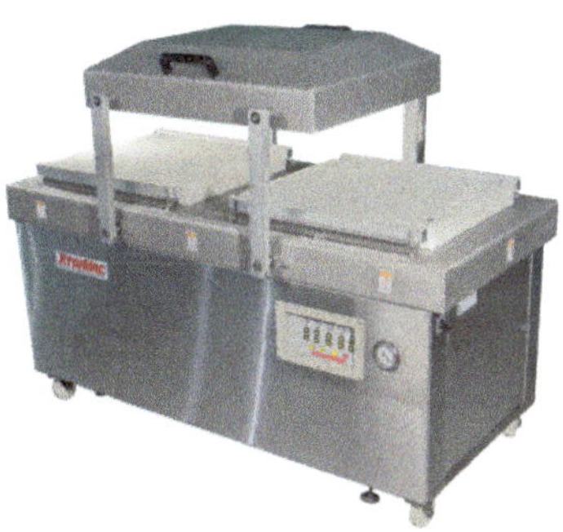

Double Chamber Vacuum Packaging Machine

Double chamber sealers require the entire product to be placed in a plastic bag within the machine. Once the product is placed in the machine on the seal bar, the lid is closed and air is removed. Then a seal bar inside the chamber seals the product in the bag, after sealing the bag the chamber is refilled with air by the automatic opening of a vent to the outside. This oncoming pressure squeezes all remaining air in the bag. The lid is then opened and the product removed. Double chamber sealers are typically used for medium-volume packaging, and also have the capability to vacuum seal liquids. The lid generally swings from one side to another, increasing production speed over a single chamber model. Double chamber vacuum packaging machines generally have either spring-weighted lids or fully automatic lids.

Double chamber vacuum packaging machines are commonly used for:

- Fresh Meat
- Processed Meat
- Cheese (hard and soft)
- Candy & Chocolate
- Empty Cans (it's a example of atmospheric pressure)

Automatic Belt Vacuum Chamber Machines

Automatic belt chamber sealers require the entire product to be placed in a plastic bag or flow wrapped pouch within the machine. The product travels on the conveyor belt, it is automatically positioned in the machine on the seal bar, the lid is closed and air is removed. Then a seal bar inside the chamber seals the product in the bag. After sealing the bag, the chamber is refilled with air by the automatic opening of a vent to the outside. This oncoming pressure squeezes all remaining air in the bag. The lid is then opened and the product removed. Automatic belt vacuum chamber machines are typically used for high-speed packaging of large items, and also have the capability to vacuum seal liquids. The lid generally travels straight up and down.

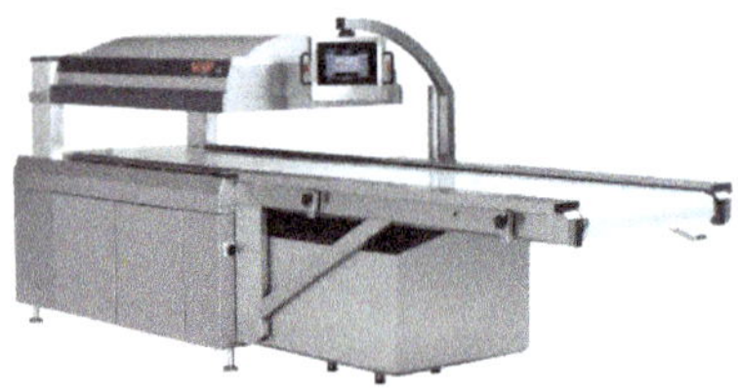

Automatic Belt Vacuum Chamber Machine. Automatic belt vacuum chamber machines offer vastly increased speed and automation and accommodate large products.

Automatic belt vacuum chamber packaging machines are commonly used for:

- Fresh Meat (large portions)
- Processed Meat
- Large Sausage logs
- Cheese (hard and soft)

Thermoforming (Rollstock) Vacuum Packaging Machines

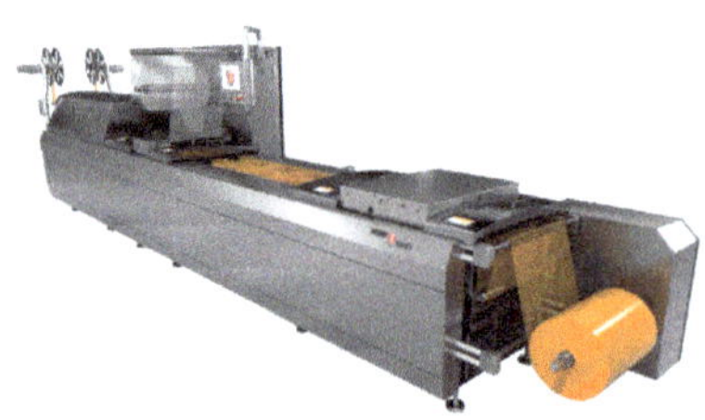

Thermoform packaging machines are used in larger production facilities for vacuum packaging products.

Vacuum Packaging in large production facilities can be done with thermoforming machines. These are Form-Fill-Seal style machines that form the package from rolls of packaging film (webbing). Products are loaded into the thermoformed pockets, the top web is laid and sealed under a vacuum, producing vacuum packaged products. Thermoforming can greatly increase packaging production speed. Thermoformed plastics can be customized for size, color, clarity, and shape to fit products perfectly, creating a consistent appearance. Some common uses for Thermoforming in vacuum packaging include:

- Fresh & Marinated Meat
- Sausage
- Cheese
- Candy / Chocolate
- Grain

- Grab-and-Go Snacks (beef jerky, snack sticks)
- Pharmaceutical and Medical Products
- Coins / Collectables

Shelf Life

Depending on the product, the shelf life of vacuum packaged products can exceed normal bagged or wrapped packages. Beef can last up to six weeks refrigerated, and much longer when frozen.

High Barrier Shrink Vacuum Bags

The amount of shelf life enhanced by a vacuum bag is dependent on the structure in the material. A standard vacuum bag is composed of a PA/PE structure where PA is for puncture resistance and PE is for sealing. The high barrier category includes the usage of more layers focused on the prevention of oxygen permeability, and therefore shelf life protection. There are two materials used in high barrier structures, polyvinylidene chloride (PVDC) and ethylene vinyl alcohol (EVOH). Shelf life indication can be effectively measured by how many cubic centimeters of oxygen can permeate through 1 square meter of material over a 24-hour period. A standard PA/PE bag allows on average 100 cubic centimeters, PVDC allows on average over 10, and EVOH on average 1 cubic centimeter. Multi-layer structures allow the ability to use strong oxygen-barrier materials for enhanced shelf life protection. The PremiumPack structure is a good example of EVOH based high barrier shrink material.

Preventing Freezer Burn

When foods are frozen without preparation, freezer burn can occur. It happens when the surface of the food is dehydrated, and this leads to a dried and leathery appearance. Freezer burn also ruins the flavor and texture of foods. Vacuum packing reduces freezer burn by preventing the food from exposure to the cold, dry air.

Sous-vide Cooking

Vacuum packaging also allows for a special cooking method, sous-vide. Sous-vide, French for *under vacuum*, involves poaching food that is vacuum sealed in a plastic bag.

Food Safety

In an oxygen-depleted environment, anaerobic bacteria can proliferate, potentially causing food-safety issues. Vacuum packing is often used in combination with other packaging and food processing techniques.

Preservative

A preservative is a substance or a chemical that is added to products such as food, beverages, pharmaceutical drugs, paints, biological samples, cosmetics, wood, and many other products to prevent decomposition by microbial growth or by undesirable chemical changes. In general, preservation is implemented in two modes, chemical and physical. Chemical preservation entails adding chemical compounds to the product. Physical preservation entails processes such as refrigeration or drying. Preservative food additives reduce the risk of foodborne infections, decrease microbial spoilage, and preserve fresh attributes and nutritional quality. Some physical techniques for food preservation include dehydration, UV-C radiation, freeze-drying, and refrigeration. Chemical preservation and physical preservation techniques are sometimes combined.

Antimicrobial Additives

Antimicrobial preservatives prevent degradation by bacteria. This method is the most traditional and ancient type of preserving—ancient methods such as pickling and adding honey prevent microorganism growth by modifying the pH level. The most commonly used antimicrobial preservative is lactic acid. Common antimicrobial preservatives are presented in the table.Nitrates and nitrites are also antimicrobial. The detailed mechanism of these chemical compounds range from inhibiting growth of the bacteria to the inhibition of specific enzymes.

E number	chemical compound	comment
E201 – E203	benzoic acid, sodium benzoate	used in acidic foods such as jams, salad dressing, juices, pickles, carbonated drinks, soy sauce
E214 – E219	hydroxybenzoate and derivatives	stable at a broad pH range
E270	lactic acid	-
E249 – E250	nitrite	used in meats to prevent botulism toxin
E251 – E252	nitrate	used in meats
E280 – E283	propionic acid and sodium propionate	baked goods
E220 – E227	sulfur dioxide and sulfites	common for fruits
E200 – E203	sorbic acid and sodium sorbate	common for cheese, wine, baked goods

Antioxidants

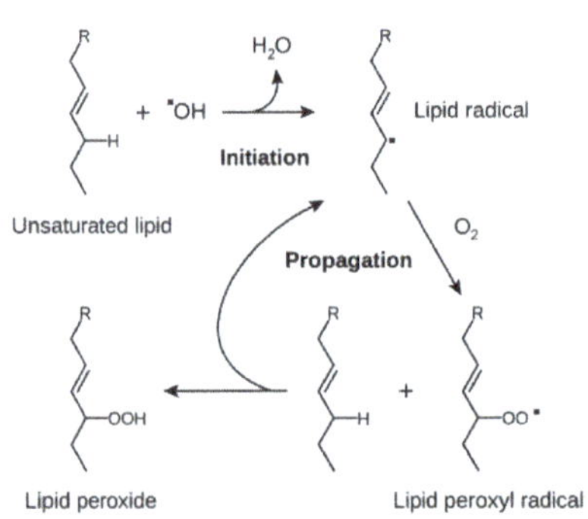

The free radical pathway for the first phase of the oxidative rancidification of fats. This process is slowed by antioxidants.

The oxidation process spoils most food, especially those with a high fat content. Fats quickly turn rancid when exposed to oxygen. Antioxidants prevent or inhibit the oxidation process. The most common antioxidant additives are ascorbic acid (vitamin C) and ascorbates. Thus, antioxidants are commonly added to oils, cheese, and chips. Other antioxidants include the phenol derivatives BHA, BHT, TBHQ and propyl gallate. These agents suppress the formation of hydroperoxides. Other preservatives include ethanol and methylchloroisothiazolinone.

E number	chemical compound	comment
E300-304	ascorbic acid, sodium ascorbate	cheese, chips
E321	butylated hydroxytoluene, butylated hydroxyanisole	also used in food packaging
E310-312	gallic acid and sodium gallate	oxygen scavenger
E220 – E227	sulfur dioxide and sulfites	beverages, wine
E306 – E309	tocopherols	vitamin E activity

A variety of agents are added to sequester (deactivate) metal ions that otherwise catalyze the oxidation of fats. Common sequestering agents are disodium EDTA, citric acid (and citrates), tartaric acid, and lecithin.

Nonsynthetic Compounds for Food Preservation

Citric and ascorbic acids target enzymes that degrade fruits and vegetables, e.g., phenolase which turns surfaces of cut apples and potatoes brown. Ascorbic acid and tocopherol, which are vitamins, are common preservatives. Smoking entails exposing food to a variety of phenols, which are antioxidants. Natural preservatives include rosemary extract, hops, salt, sugar, vinegar, alcohol, diatomaceous earth and castor oil.

Traditional preservatives, such as sodium benzoate have raised health concerns in the past. Benzoate was shown in a study to cause hypersensitivity in some asthma sufferers. This has caused reexamination of natural preservatives which occur in vegetables.

History

Preservatives have been used since prehistoric times. Smoked meat for example has phenols and other chemicals that retard spoilage. The preservation of foods has evolved greatly over the centuries and has been instrumental in increasing food security. The use of preservatives other than traditional oils, salts, paints, etc. in food began in the late 19th century, but was not widespread until the 20th century.

The use of food preservatives varies greatly depending on the country. Many developing countries that do not have strong governments to regulate food additives face either harmful levels of preservatives in foods or a complete avoidance of foods that are considered unnatural or foreign. These countries have also proven useful in case studies surrounding chemical preservatives, as they have been only recently introduced. In urban slums of highly populated countries, the knowledge about contents of food tends to be extremely low, despite consumption of these imported foods.

Public Awareness of Food Preservation

Public awareness of food preservatives is uneven. Americans have a perception that food-borne illnesses happen more often in other countries. This may be true, but the occurrence of illnesses, hospitalizations, and deaths are still high. It is estimated by the Center for Disease Control (CDC) that each year there are 76 million illnesses, 325,000 hospitalizations, and 5,000 deaths linked to food-borne illness.

The increasing demand for ready-to-eat fresh food products has led to challenges for food distributors regarding the safety and quality of their foods. Artificial preservatives meet some of these challenges by preserving freshness for longer periods of time, but these preservatives can cause negative side-effects as well. Sodium nitrite is a preservative used in lunch meats, hams, sausages, hot dogs, and bacon to prevent botulism. It serves the important function of controlling the bacteria that cause botulism, but sodium nitrite can react with proteins, or during cooking at high heats, to form carcinogenic N-nitrosamines.It has also been linked to cancer in lab animals. The commonly used sodium benzoate has been found to extend the shelf life of bottled tomato paste to 40 weeks without loss of quality. However, it can form the carcinogenbenzene when combined with vitamin C. Many food manufacturers have reformed their products to eliminate this combination, but a risk still exists. Consumption of sodium benzoate may also cause hyperactivity. For over 30 years, there has been a debate about whether or not preservatives and other food additives can cause hyperactivity. Studies have found that there may be increases in hyperactivity amongst children who consume artificial colorings and benzoate preservatives and who are already genetically predisposed to hyperactivity, but these studies were not entirely conclusive. Hyperactivity only increased moderately, and it was not determined if the preservatives, colorings, or a combination of the two were responsible for the increase.

Active Packaging

The terms active packaging, intelligent packaging, and smart packaging refer to packaging systems used with foods, pharmaceuticals, and several other types of products. They help extend shelf life, monitor freshness, display information on quality, improve safety, and improve convenience.

The terms are closely related. *Active packaging* usually means having active functions beyond the inert passive containment and protection of the product.*Intelligent* and *smart* packaging usually involve the ability to sense or measure an attribute of the product, the inner atmosphere of the package, or the shipping environment. This information can be communicated to users or can trigger active packaging functions. Programmable matter, smart materials, etc can be employed in packages.

Depending on the working definitions, some traditional types of packaging might be considered as "active" or "intelligent". More often, the terms are used with new technologically advanced systems: microelectronics, computer applications, nanotechnology, etc.

Moisture Control

For many years, desiccants have been used to actively control the water vapor in a closed package.

A desiccant is a hygroscopic substance usually in a porous pouch or sachet which is placed inside a sealed package. They have been used to reduce corrosion of machinery and electronics and to extend the shelf life of moisture sensitive foods and drugs.

Corrosion

Corrosion inhibitors can be applied to items to help prevent rust and corrosion. Volatile corrosion inhibitors (VCI) or vapor phase corrosion inhibitors can be provided inside a package in a pouch or can be incorporated in a saturated overwrap of special paper. Many of these are organic salts that condense on the metal to resist corrosion. Some films also have VCI emitting capability.

Films are available with copper ions in the polymer structure, These neutralize the corrosive gas in a package and deter rust.

VCI create a neutral environment in the packaging. It works on the principle of difference in vapour pressure and causes reaction with Metals and Non metals and also with Moisture to render it active to corrosion. There are different forms of VCI's available like Papers, Plastics, HDPE Papers, Oils, Foams, Chips, Aluminum Barrier Foils, Bubble, Emitters etc. that can prevent corrosion at many stages.

Oxygen Control

Oxygen scavengers or oxygen absorbers help remove oxygen from a closed package. Some are small packets or sachets containing powdered iron: as the iron rusts, oxygen is removed from the surrounding atmosphere. Newer systems are on cards or can be built into package films or molded structures. In addition, the physical characteristics of the packaging itself (oxygen transmission rate - OTR) can dictate how effective an oxygen absorber can be, and how long it will stay effective. Packaging with a low OTR will let less oxygen in the closed package through the polymer barrier itself.

Atmosphere

With some products, such as cheese, it has long been common to flush the package with nitrogen prior to sealing: the inert nitrogen is absorbed into the cheese, allowing a tight shrink film package. The nitrogen removes oxygen and "actively" interacts with the cheese to make the package functional.

More recently, other mixtures of gas have been used inside the package to extend the shelf life. The gas mixture depends on the specific product and its degradation mechanisms. Some package components have been developed that incorporate active chemistry to help maintain certain atmospheres in packages.

Oxygen scavengers, carbon dioxide generators, ethanol generators, etc. are available to help keep the atmosphere in a package at specified conditions.

Temperature Monitor

Some temperature indicators give a visual signal that a specified temperature has been exceeded. Others, Time temperature indicators, signal when a critical accumulation of temperature deviation over time has been exceeded. When the mechanism of the indicator is tuned to the mechanism of product degradation, these can provide valuable signals for consumers.

Digital temperature data loggers record the temperatures encountered throughout the shipment. This data can be used to predict product degradation and help determine if the product is suited for normal sale or if expedited sale is required. They also determine the time of the temperature excess: this can be used to direct corrective action.

Thermochromic inks are sometimes used to signal temperature excess or change. Some are reversible while others have a permanent change of color. These can be used alone or with other packaging functions such as barcodes.

The inks can also signal a desired temperature for consumers. For example, one type of beer can has ink that graphically shows when an ideal drinking temperature is achieved.

Controlling Package Temperatures

Self-heating field rations for up to 18 soldiers

For critical vaccines, insulated shipping containers are passive packaging to help control the temperatures fluctuations seen even with a controlled cold chain. In addition, gel packs are often used to actively keep the temperature of the contents within specified acceptable temperature ranges.

Some newer packages have the ability to heat or cool the product for the consumer. These have segregated compartments where exothermic or endothermic reactions provide the desired effect. Self-heating cans are available for several products.

Dispensing

Some packages have closures or other dispensing systems that actively change the contents from a liquid to an aerosol. These are used for products ranging from precision inhalers for medications to spray bottles of household cleaners.

Some dispensing packages for two-part epoxy adhesives do more than passively contain the two components. When dispensed, some packages meter and mix the two components so the adhesive is fully functioning at the point of application.

The ability of a package to fully empty or dispense a viscous liquid is somewhat dependent on the surface energy of the inner walls of the container. The use of superhydrophobic surfaces is useful but can be further improved by using new lubricant-impregnated surfaces.

RFID

Radio-frequency identification chips are becoming more common as smart labels that are used to track and trace packages and unit loads throughout distribution. Newer developments include recording the temperature history of shipments and other intelligent packaging functions.

Security

A variety of security printing methods, security holograms, and specialized labels are available to help confirm that the product in the package is not counterfeit. RFID chips are being used in this application also.

Electronic article surveillance (on the product or on the package) is used to help counter shoplifting.

Microwave Packaging

Metallised films are used as a susceptor for cooking in microwave ovens. These increase the heating capacity and help make foods crisp and brown. Plastic microwavable containers are also used for microwave cooking.

Shock and Vibration

Shock detectors have been available for many years. These are attached to the package or to the product in the package to determine if an excessive shock has been encountered. The mechanisms of these *shock overload* devices have been spring-mass systems, magnets, drops of red dye, and several others.

Recently, digital shock and vibration data loggers have been available to more accurately record the shocks and vibrations of shipment. These are used to monitor critical shipments to determine if extra inspection and calibration is required. They are also used to monitor the types of shocks and vibrations encountered in transit for use in package testing in a laboratory.

Other Developments

Some engineered packaging films contain enzymes, anti-bacterial agents, scavengers, and other active components to help control food degradation and extend shelf life.

Edible films have been developed to allow consumers to eat the package along with the product.

Special packaging has been developed for shipping organs which keeps them alive during extended shipments. The organs are alive and fresh for transplanting.

Regulations

Active packaging is often designed to actively interact with the contents of the package. Thus extra care is often needed for active or smart packagings that are food contact materials.

Food packagers take extra care with some types of active packaging. For example, when the oxygen atmosphere in a package is reduced for extending shelf life, controls for anaerobic bacteria need to be considered. Also when a controlled atmosphere reduces the appearance of food degradation, consumers need to retain a means of determining whether actual degradation is present.

Camping Food

Freeze-dried bacon bars that can be used as camping food

Backcountry camping food includes ingredients used to prepare food suitable for backcountry camping and backpacking. The foods differ substantially from the ingredients found in a typical home kitchen. The primary differences relate to campers' and backpackers' special needs for foods that have appropriate cooking time, perishability, weight, and nutritional content.

A freeze-dried ice cream sandwich

To address these needs, camping food is often made up of either freeze-dried, precooked or dehydrated ingredients. Many campers use a combination of these foods.

Meal and Ingredient Requirements

Limited Cooking Time

Due to the difficulty of carrying large amounts of cooking fuel, campers often require their meals to cook in a short amount of time (5–20 minutes). Many campers prefer a 'just add boiling water' method of cooking, while others enjoy a more involved, and therefore often higher quality meal. The amount of cooking time can be disregarded if campers are able to cook over a campfire, however, due to the possibility of a burn-ban being in place, campers do not often rely on this option.

Shelf-stability

Camping foods are often shelf-stable, that is, they require no refrigeration. Campers may be outdoors for days or weeks at a time, and will often pack food for the entire trip. Campers will sometimes take fresh food that can be consumed in the first day or two of a hike but will usually not risk carrying perishable food beyond that timeframe. Campers hiking in the snow or other cold conditions or campers with access to a cold water source may be able to store perishable food in the snow or secured in a bag and kept in the cold water to act as a refrigeration source.

Lightweight

Backpackers must carry everything with them so they require all of their gear and food to be as lightweight as possible. Campers often turn to freeze-dried and dehydrated meals and ingredients for this reason, but they will also sometimes take a pouch of tuna or some other ingredient with a high water content with them as a treat, providing that the item has nutritional value.

Nutrition Content

Backpackers, canoeists, climbers and other outdoor enthusiasts often cover many miles everyday, consuming thousands of calories to keep their energy level high. Backpackers require an average of 480 calories per hour as well as higher sodium levels. To ensure their bodies are properly nourished, campers must pay close attention to their meal plans.

Ingredients

To prepare meals that work well outdoors, campers have begun to employ a variety of techniques. All campers are advised to prepare meals that are made of easy to prepare ingredients.

Freeze-dried Ingredients

Freeze dried eggs can be shelf stable for up to 25 years

Freeze-drying requires the use of heavy machinery and is not something that most campers are able to do on their own. Freeze-dried ingredients are often considered superior to dehydrated ingredients however, because they rehydrate at camp faster and retain more flavor than their dehydrated counterparts. Freeze-dried ingredients take so little time to rehydrate that they can often be eaten without cooking them first and have a texture similar to a crunchy chip.

Small amounts of freeze-dried ingredients are sometimes available for sale from emergency supply outlets or from stores specific to camping. Freeze-dried ingredients that have not been combined into a meal are often hard to find, however, and are often sought out by campers.

Dehydrated Meals & Ingredients

Dehydration can reduce the weight of the food by sixty to ninety percent by removing water through evaporation. Some foods dehydrate well, such as onions, peppers, and tomatoes. Dehydration often produces a more compact, albeit slightly heavier, end result than freeze-drying.

Full meals or individual ingredients may be dehydrated. Dehydration of individual ingredients allows the flexibility to cook different meals based on available ingredients, while precooked and dehydrated meals offer greater convenience. Several cookbooks and online grocery stores specialize in dehydrated foods. A Fork in the Trail and Another Fork in the Trail are both backcountry cookbooks that focus on dehydrating full meals. For campers wishing to save time, Packit Gourmet is one source that provides dehydrated vegetables for use in camping recipes.

Precooked Foods

MRE entree

Surplus military Meals, Meals, Ready-to-Eat (MREs) are sometimes used by campers. These meals contain precooked foods in retort pouches. A retort pouch is a plastic and metal foil laminate pouch that is used as an alternative to traditional industrial canning methods.

Contents of a MRE package

Common Ingredients

Uncooked instant ramen

The final type of ingredients available to campers are those that are typically found in the grocery store. Some examples of these types of food are polenta, grits, quick-cooking pasta (such as angel hair pasta), ramen, instant potatoes, dried soups, jerky and pouch meats such as tuna, SPAM or salmon.

When using these common ingredients, campers often repackage them to reduce packaging or combine them into a meal-ready package, therefore reducing prep-time at camp. The main requirement that campers look for in these types of ingredients is the cook-time with 20 minutes being the longest amount of cook-time that most campers will tolerate.

Backcountry Cooking Methods

A cast iron *potjie* on a fire, very similar to a Dutch oven

Camping Stoves and Cookware

There is a large variety of camping stoves on the market ranging in specialty from being extremely lightweight to focusing on using very little fuel. The majority of campers rely on a stove for their cooking needs as they boast several advantages over cooking over a campfire. Since most camping stoves have an adjustable heat source, they can be much easier to use than a campfire. The ability to quickly adjust the flame to reduce your pot from a boil to a simmer, for example, is considered invaluable to many campers. Campfires can take a long time to start and get to a point where they are suitable for cooking over. Since a cook-stove can be ready in minutes, this is an advantage for many campers. Many types of cookware exist for outdoor cooking.

Campfire

Wilderness areas can often have a burn-ban, prohibiting people from starting a fire. If a camper were to rely on the campfire method as their only source for cooking heat, they could find themselves in an unlucky situation. Cooking over a campfire can lead to pots and pans darkened with soot. Soot can be extremely difficult to remove and, if left on the pan, can easily rub off onto clothing or the inside of the backpack. Campers have discovered methods of preventing this problem, such as coating the pans with cooking oil, to make the soot easier to remove.

Campers relying on the use of a campfire do not have to carry the extra weight of a cook stove and may rely on a campfire to reduce their pack weight. Campfires provide a great amount of warmth while cook stoves provide none. On cold days, a campfire is often welcome. Leave No Trace discourages the use of a campfire as a source of heat. Campers making a campfire in the same location time after time can deplete the available wood in the area, which impacts the natural habitat of the animals. Campers are also more likely to inadvertently leave food scraps around the fire pit, which could attract animals.

Chemical Heaters

Some camping food is ready to eat and may be warmed using chemical heaters, such as the flameless heaters used in MREs, or a self-contained chemical heater built into the food packaging itself.

Solar Cooking

Solar cooking provides clean and safe alternative to campfire. Using solar cookers is easy and inexpensive since they do not require fuel to work. Most solar cookers also provide minimum required temperature during cloudy days to prepare the food. Despite many advantages that solar cooking provides it is unusable during the nighttime and it will not provide heat and protection against wild animals like campfire does.

Home Canning

A 1914 advertisement for a combination steam canner and roaster. The described method will not protect against botulism.

Home canning or bottling, also known colloquially as putting up or processing, is the process of preserving foods, in particular, fruits, vegetables, and meats, by packing them into glass jars and then heating the jars to kill the organisms that would create spoilage.

North America

Green beans in a pressure canner ready to be processed

In North America, home canning is usually done in Mason jars, which have thicker walls than single-use commercial glass jars. Unless the food being preserved has a high acid content (pH <4.6) or salt or sugar content resulting in water availability<0.85, such as pickles or jellies, the filled jars are also processed under pressure in a canner, a specialized type of pressure cooker. Ordinary pressure cookers are not recommended for canning as their smaller size and the reduced thickness of the cooker wall will not allow for the correct building up and reducing time of pressure, which is factored into the overall processing time and therefore will not destroy all the harmful microorganisms. The goal in using a pressure canner is to achieve a "botulinum cook" of 121 °C for 3 minutes, throughout the entire volume of canned product. Canners often incorporate racks to hold Mason jars, and pressure canners are capable of achieving the elevated temperatures needed to prevent spoilage.

The most common configuration is a Mason jar with a flat lid and screw ring. The lid is generally made of plated or painted steel, with an elastomeric washer or gasket bonded to the underside of the rim. The lid also incorporates a slightly dimpled shape, which acts as an indicator of the vacuum (or lack thereof) inside a sealed jar. A newer reusable lid is now available that uses a flat plastic disk with a reusable rubber gasket. The ring threads onto the top of the jar over the lid to hold it in place while the jar cools after processing; the ring can be removed once a vacuum has been established in the jar. Jars are commonly in either pint or quart capacities, with two opening diameters, known as "standard" and "wide mouth".

When a jar has cooled and is properly sealed, pressing the dimple on the lid will not make any sound. An improperly sealed jar will allow the dimple to move up and down, sometimes making a popping noise. Lack of this noise does not necessarily indicate that the food in the jar is properly preserved. Typically, during the cooling process, a properly sealed lid will pop once as the pressure inside the jar is reduced enough that atmospheric pressure pushes the lid inward.

Older variations had a ceramic seal inside a one-piece zinc lid. Another method that is no longer recommended was the use of layer of hot paraffin wax poured directly over the top of the food (especially jams and jellies) to seal it from air, thus reducing growth of aerobic microorganisms like mold.

While it is possible to safely preserve many kinds of foodstuffs, home canning can expose consumers to botulism and other kinds of food poisoning if done incorrectly. The most common source of food-borne botulism is home-canned foods prepared in an unsafe manner. Safety measures must be taken when performing home canning, since ingestion of toxin in food produced by *Clostridium botulinum* can cause death. Because of the high risk of illness or death associated with improper canning techniques, the United States Department of Agriculture (USDA) considers it critical that consumers who intend to can at home obtain proper and current information from a reliable source. At the basis of these recommendations is the balance between bringing the food to a high enough temperature for a long enough time that spoilage and disease-producing microorganisms are killed, while not heating the food so much that it loses nutritive value or palatability.

United Kingdom

A Dixie Can Sealer for home use. Now in Thinktank, Birmingham Science Museum.

In the United Kingdom home bottling is done with Kilner jars in a similar way to the Mason jars in the USA, and although old-style Kilner jars have a glass lid without a "dimple" more recent varieties do. Most home bottling is done using the "open kettle method", with hot food ladled into hot jars and lids placed on jars, with no water bath sterilization processing of the product afterward.

Australia

Weck glass with rubber seal and steel tension clips. The red tongue showing downwards indicates an intact sealing

In Australia the most popular home canning system is Fowler's Vacola. This system uses glass jars, single use seals, metal lids, and a water bath canning sterilization process. During the canning process the lids are secured by metal tension clips which are removed once a vacuum seal has formed. Fowler's Vacola products are still produced and are available from some hardware stores. Used equipment is frequently sold on online auction sites and in opportunity shops.

Germany

In Germany the most popular home canning system is Weck jar. This system uses glass jars (80ml-2.5l) and glass bottles (200ml-1l), reusable rubber seals, glass lids, and a water bath canning sterilization process. During the canning process the lids are secured by steel tension clips which can be removed once a vacuum seal has formed. During storage the vertical position of the rubber seal's external tongue indicates the status of the sealing. Weck products are still produced and are available from German hardware stores. Used equipment is frequently sold on online auction sites and in opportunity shops. They can be used for water bath canning and pressure canning. For water bath canning use two steel tension clips and for pressure canning use three steel tension clips. Remove them after vacuum has formed.

Standard Jar Sizes

United States:

- 4 ounce (jelly)
- 8 ounce (jelly)
- 8 ounce (half US pint)
- 16 ounce (US pint)
- 24 ounce (US pint and a half)
- 32 ounce (US quart)
- 64 ounce (half US gallon)

Metric:

- 250 ml

Food Safety

Food safety is a scientific discipline describing handling, preparation, and storage of food in ways that prevent foodborne illness. This includes a number of routines that should be followed to avoid potentially severe health hazards. In this way food safety often overlaps with food defense to prevent harm to consumers. The tracks within this line of thought are safety between industry and the market and then between the market and the consumer. In considering industry to market practices, food safety considerations include the origins of food including the practices relating to food labeling, food hygiene, food additives and pesticide residues, as well as policies on biotechnology and food and guidelines for the management of governmental import and export inspection and certification systems for foods. In considering market to consumer practices, the usual thought is that food ought to be safe in the market and the concern is safe delivery and preparation of the food for the consumer.

Food can transmit disease from person to person as well as serve as a growth medium for bacteria that can cause food poisoning. In developed countries there are intricate standards for food preparation, whereas in lesser developed countries the main issue is simply the availability of adequate safe water, which is usually a critical item. In theory, food poisoning is 100% preventable. The five key principles of food hygiene, according to WHO, are:

1. Prevent contaminating food with pathogens spreading from people, pets, and pests.
2. Separate raw and cooked foods to prevent contaminating the cooked foods.
3. Cook foods for the appropriate length of time and at the appropriate temperature to kill pathogens.
4. Store food at the proper temperature.
5. Do use safe water and safe raw materials.

Issues

Food safety issues and regulations concern:

- Agriculture and animal husbandry practices
- Food manufacturing practices
- Food additives
- Novel foods
- Genetically modified foods
- Food labels

ISO 22000

ISO 22000 is a standard developed by the International Organization for Standardization dealing with food safety. This is a general derivative of ISO 9000. ISO 22000 standard: The ISO 22000 international standard specifies the requirements for a food safety management system that involves interactive communication, system management, prerequisite programs, HACCP principles.

Incidence

A 2003 World Health Organization (WHO) report concluded that about 30% of reported food poisoning outbreaks in the WHO European Region occur in private homes. According to the WHO and CDC, in the USA alone, annually, there are 76 million cases of foodborne illness leading to 325,000 hospitalizations and 5,000 deaths.

Regulations by Jurisdiction and Agency

WHO and FAO

In 2003, the WHO and FAO published the Codex Alimentarius which serves as an guideline to food safety.

However, according to Unit 04 - Communication of Health & Consumers Directorate-General of the European Commission (SANCO): "The Codex, while being recommendations for voluntary application by members, Codex standards serve in many cases as a basis for national legislation. The reference made to Codex food safety standards in the World Trade Organizations' Agreement on Sanitary and Phytosanitary measures (SPS Agreement) means that Codex has far reaching implications for resolving trade disputes. WTO members that wish to apply stricter food safety measures than those set by Codex may be required to justify these measures scientifically." So, an agreement made in 2003, signed by all member states, inclusive all EU, in the codex Stan Codex 240 – 2003 for coconut milk, sulphite containing additives like E223 and E 224 are allowed till 30 mg/kg, does NOT mean, they are allowed into the EU. Only for the latter the EU amended its regulations with (EU) No 583/2012 per 2 July 2012 to allow this additive, already used for decades and absolutely necessary.

Australia

Food Standards Australia New Zealand requires all food businesses to implement food safety systems. These systems are designed to ensure food is safe to consume and halt the increasing incidence of food poisoning, and they include basic food safety training for at least one person in each business. Food safety training is delivered in various forms by, among other organisations, Registered Training Organizations (RTOs), after which staff are issued a nationally recognised unit of competency code on their certificate. Basic food safety training includes:

- Understanding the hazards associated with the main types of food and the conditions to prevent the growth of bacteria which can cause food poisoning and to prevent illness.

- Potential problems associated with product packaging such as leaks in vacuum packs, damage to packaging or pest infestation, as well as problems and diseases spread by pests.

- Safe food handling. This includes safe procedures for each process such as receiving, re-packing, food storage, preparation and cooking, cooling and re-heating, displaying products, handling products when serving customers, packaging, cleaning and sanitizing, pest control, transport and delivery. Also covers potential causes of cross contamination.

- Catering for customers who are particularly at risk of food-borne illness, as well as those with allergies or intolerance.
- Correct cleaning and sanitizing procedures, cleaning products and their correct use, and the storage of cleaning items such as brushes, mops and cloths.
- Personal hygiene, hand washing, illness, and protective clothing.

Food safety standards and requirements are set out at the national level in the Food Standards Code, and brought Into force in each state by state-based Acts and Regulations. Legislation means that people responsible for selling or serving unsafe food may be liable for heavy fines.

China

Food safety is a growing concern in Chinese agriculture. The Chinese government oversees agricultural production as well as the manufacture of food packaging, containers, chemical additives, drug production, and business regulation. In recent years, the Chinese government attempted to consolidate food regulation with the creation of the State Food and Drug Administration in 2003, and officials have also been under increasing public and international pressure to solve food safety problems. However, it appears that regulations are not well known by the trade. Labels used for "green" food, "organic" food and "pollution-free" food are not well recognized by traders and many are unclear about their meaning. A survey by the World Bank found that supermarket managers had difficulty in obtaining produce that met safety requirements and found that a high percentage of produce did not comply with established standards.

Traditional marketing systems, whether in China or the rest of Asia, presently provide little motivation or incentive for individual farmers to make improvements to either quality or safety as their produce tends to get grouped together with standard products as it progresses through the marketing channel. Direct linkages between farmer groups and traders or ultimate buyers, such as supermarkets, can help avoid this problem. Governments need to improve the condition of many markets through upgrading management and reinvesting market fees in physical infrastructure. Wholesale markets need to investigate the feasibility of developing separate sections to handle fruits and vegetables that meet defined safety and quality standards.

European Union

The parliament of the European Union (EU) makes legislation in the form of directives and regulations, many of which are mandatory for member states and which therefore must be incorporated into individual countries' national legislation. As a very large organisation that exists to remove barriers to trade between member states, and into which individual member states have only a proportional influence, the outcome is often seen as an excessively bureaucratic 'one size fits all' approach. However, in relation to food safety the tendency to err on the side of maximum protection for the consumer may be seen as a positive benefit. The EU parliament is informed on food safety matters by the European Food Safety Authority.

Individual member states may also have other legislation and controls in respect of food safety, provided that they do not prevent trade with other states, and can differ considerably in their internal structures and approaches to the regulatory control of food safety.

From 13 December 2014, new legislation - the EU Food Information for Consumers Regulation 1169/2011 - require food businesses to provide allergy information on food sold unpackaged, in for example catering outlets, deli counters, bakeries and sandwich bars.

France

Agence nationale de sécurité sanitaire de l'alimentation, de l'environnement et du travail (anses) is a French governmental agency dealing with food safety.

Germany

The Federal Ministry of Food, Agriculture and Consumer Protection (BMEL) is a Federal Ministry of the Federal Republic of Germany. History: Founded as Federal Ministry of Food, Agriculture and Foresting in 1949, this name did not change until 2001. Then the name changed to Federal Ministry of Consumer Protection, Food and Agriculture. At the 22nd of November 2005, the name got changed again to its current state: Federal Ministry of Food, Agriculture and Consumer Protection. The reason for this last change was that all the resorts should get equal ranking which was achieved by sorting the resorts alphabetically. Vision: A balanced and healthy diet with safe food, distinct consumer rights and consumer information for various areas of life, and a strong and sustainable agriculture as well as perspectives for our rural areas are important goals of the Federal Ministry of Food, Agriculture and Consumer Protection (BMELV). The Federal Office of Consumer Protection and Food Safety is under the control of the Federal Ministry of Food, Agriculture and Consumer Protection. It exercises several duties, with which it contributes to safer food and thereby intensifies health-based consumer protection in Germany. Food can be manufactured and sold within Germany without a special permission, as long as it does not cause any damage on consumers' health and meets the general standards set by the legislation. However, manufacturers, carriers, importers and retailers are responsible for the food they pass into circulation. They are obliged to ensure and document the safety and quality of their food with the use of in-house control mechanisms.

Hong Kong

In Hong Kong SAR, the Centre for Food Safety is in charge of ensuring food sold is safe and fit for consumption.

India

Food Safety and Standards Authority of India, established under the Food Safety and Standards Act, 2006, is the regulating body related to food safety and laying down of standards of food in India.

New Zealand

The New Zealand Food Safety Authority (NZFSA), or Te Pou Oranga Kai O Aotearoa is the New Zealand government body responsible for food safety. NZFSA is also the controlling authority for imports and exports of food and food-related products. The NZFSA as of 2012 is now a division of the Ministry for Primary Industries (MPI) and is no longer its own organization.

Pakistan

Pakistan does not have an integrated legal framework but has a set of laws, which deals with various aspects of food safety. These laws, despite the fact that they were enacted long time ago, have tremendous capacity to achieve at least minimum level of food safety. However, like many other laws, these laws remain very poorly enforced. There are four laws that specifically deal with food safety. Three of these laws directly focus issues related to food safety, while the fourth, the Pakistan Standards and Quality Control Authority Act, is indirectly relevant to food safety.

The Pure Food Ordinance 1960 consolidates and amends the law in relation to the preparation and the sale of foods. All provinces and some northern areas have adopted this law with certain amendments. Its aim is to ensure purity of food being supplied to people in the market and, therefore, provides for preventing adulteration. The Pure Food Ordinance 1960 does not apply to cantonment areas. There is a separate law for cantonments called "The Cantonment Pure Food Act, 1966". There is no substantial difference between the Pure Food Ordinance 1960 and The Cantonment Pure Food Act. Even the rules of operation are very much similar.

Pakistan Hotels and Restaurant Act, 1976 applies to all hotels and restaurants in Pakistan and seeks to control and regulate the rates and standard of service(s) by hotels and restaurants. In addition to other provisions, under section 22(2), the sale of food or beverages that are contaminated, not prepared hygienically or served in utensils that are not hygienic or clean is an offense. There are no express provisions for consumer complaints in the Pakistan Restaurants Act, 1976, Pakistan Penal Code, 1860 and Pakistan Standards and Quality Control Authority Act, 1996. The laws do not prevent citizens from lodging complaints with the concerned government officials; however, the consideration and handling of complaints is a matter of discretion of the officials.

South Korea

Korea Food & Drug Administration

Korea Food & Drug Administration (KFDA) is working for food safety since 1945. It is part of the Government of South Korea.

Food Import Report: According to Food Import Report, it is supposed to report or register what you import. Competent authority is as follows:

Product	Authority
Imported Agricultural Products, Processed Foods, Food Additives, Utensils, Containers & Packages or Health Functional Foods	KFDA (Korea Food and Drug Administration)
Imported Livestock, Livestock products (including Dairy products)	NVRQS (National Veterinary Research and Quarantine Service)
Packaged meat, milk & dairy products (butter, cheese), hamburger patties, meat ball and other processed products which are stipulated by Livestock Sanitation Management Act	NVRQS (National Veterinary Research and Quarantine Service)
Imported Marine products; fresh, chilled, frozen, salted, dehydrated, eviscerated marine produce which can be recognized its characteristics	NFIS (National Fisheries Products Quality Inspection Service)

IOAS-Organic Certification Bodies Registered in KFDA: "Organic" or related claims can be labelled on food products when organic certificates are considered as valid by KFDA. KFDA admits organic certificates which can be issued by 1) IFOAM (International Federation of Organic Agriculture Movement) accredited certification bodies 2) Government accredited certification bodies – 328 bodies in 29 countries have been registered in KFDA.

National Institute of Food and Drug Safety Evaluation

National Institute of Food and Drug Safety Evaluation (NIFDS) is functioning as well. The National Institute of Food and Drug Safety Evaluation is a national organization for toxicological tests and research. Under the Korea Food & Drug Administration, the Institute performs research on toxicology, pharmacology, and risk analysis of foods, drugs, and their additives. The Institute strives primarily to understand important biological triggering mechanisms and improve assessment methods of human exposure, sensitivities, and risk by (1) conducting basic, applied, and policy research that closely examines biologically triggering harmful effects on the regulated products such as foods, food additives, and drugs, and operating the national toxicology program for the toxicological test development and inspection of hazardous chemical substances assessments. The Institute ensures safety by investigation and research on safety by its own researchers, contract research by external academicians and research centers.

Taiwan

In Taiwan, the Ministry of Health and Welfare in charge of Food and Drug Safety, also evaluate the catering industry to maintenance the food product quality. Currently, US$29.01 million of budget is allocated each year for food safety-related efforts.

United Kingdom

In the UK the Food Standards Agency is an independent government department responsible for food safety and hygiene across the UK. They work with businesses to help them produce safe food, and with local authorities to enforce food safety regulations. In 2006 food hygiene legislation changed and new requirements came into force. The main requirement resulting from this change is that if you own or run a food business in the UK, you must have a documented Food Safety Management System, which is based on the principles of Hazard Analysis Critical Control Point HACCP.

United States

The US food system is regulated by numerous federal, state and local officials. It has been criticized as lacking in "organization, regulatory tools, and not addressing food borne illness."

Federal Level Regulation

The Food and Drug Administration publishes the Food Code, a model set of guidelines and procedures that assists food control jurisdictions by providing a scientifically sound technical and legal basis for regulating the retail and food service industries, including restaurants, grocery stores and institutional foodservice providers such as nursing homes. Regulatory agencies at all levels of government in the United States use the FDA Food Code to develop or update food safety rules in their jurisdictions that are consistent with national food regulatory policy. According to the FDA,

48 of 56 states and territories, representing 79% of the U.S. population, have adopted food codes patterned after one of the five versions of the Food Code, beginning with the 1993 edition.

In the United States, federal regulations governing food safety are fragmented and complicated, according to a February 2007 report from the Government Accountability Office. There are 15 agencies sharing oversight responsibilities in the food safety system, although the two primary agencies are the U.S. Department of Agriculture (USDA) Food Safety and Inspection Service (FSIS), which is responsible for the safety of meat, poultry, and processed egg products, and the Food and Drug Administration (FDA), which is responsible for virtually all other foods.

The Food Safety and Inspection Service has approximately 7,800 inspection program personnel working in nearly 6,200 federally inspected meat, poultry and processed egg establishments. FSIS is charged with administering and enforcing the Federal Meat Inspection Act, the Poultry Products Inspection Act, the Egg Products Inspection Act, portions of the Agricultural Marketing Act, the Humane Slaughter Act, and the regulations that implement these laws. FSIS inspection program personnel inspect every animal before slaughter, and each carcass after slaughter to ensure public health requirements are met. In fiscal year (FY) 2008, this included about 50 billion pounds of livestock carcasses, about 59 billion pounds of poultry carcasses, and about 4.3 billion pounds of processed egg products. At U.S. borders, they also inspected 3.3 billion pounds of imported meat and poultry products.

Industry Pressure

There have been concerns over the efficacy of safety practices and food industry pressure on U.S. regulators. A study reported by Reuters found that "the food industry is jeopardizing U.S. public health by withholding information from food safety investigators or pressuring regulators to withdraw or alter policy designed to protect consumers". A survey found that 25% of U.S. government inspectors and scientists surveyed have experienced during the past year corporate interests forcing their food safety agency to withdraw or to modify agency policy or action that protects consumers. Scientists have observed that management undercuts field inspectors who stand up for food safety against industry pressure. According to Dr. Dean Wyatt, a USDA veterinarian who oversees federal slaughter house inspectors, "Upper level management does not adequately support field inspectors and the actions they take to protect the food supply. Not only is there lack of support, but there's outright obstruction, retaliation and abuse of power." A growing number of food and beverage manufacturers are improving food safety standards by incorporating a food safety management system which automates all steps in the food quality management process.

State and Local Regulation

A number of U.S. states have their own meat inspection programs that substitute for USDA inspection for meats that are sold only in-state. Certain state programs have been criticized for undue leniency to bad practices.

However, other state food safety programs supplement, rather than replace, Federal inspections, generally with the goal of increasing consumer confidence in the state's produce. For example, state health departments have a role in investigating outbreaks of food-borne disease bacteria, as in the case of the 2006 outbreak of *Escherichia coli* O157:H7 (a pathogenic strain of the ordinarily harmless bacteria, *E. coli*) from processed spinach. Health departments also promote better food processing practices to eliminate these threats.

In addition to the US Food and Drug Administration, several states that are major producers of fresh fruits and vegetables (including California, Arizona and Florida) have their own state programs to test produce for pesticide residues.

Restaurants and other retail food establishments fall under state law and are regulated by state or local health departments. Typically these regulations require official inspections of specific design features, best food-handling practices, and certification of food handlers. In some places a letter grade or numerical score must be prominently posted following each inspection. In some localities, inspection deficiencies and remedial action are posted on the Internet.

Vietnam

The Vietnam Food Administration is responsible for managing food hygiene, safety, and quality and has made significant progress since its establishment in 1999. Food safety remains a high priority in Vietnam with the growth of export markets and increasing food imports raising the need to rapidly build capacity of the Food Administration in order to reduce threats of foodborne disease. The Food Administration has demonstrated commitment to the food safety challenges it faces, and has embarked on an innovative capacity building activity with technical assistance from the World Health Organization.

Consumer Labeling

United Kingdom

Foodstuffs in the UK have one of two labels to indicate the nature of the deterioration of the product and any subsequent health issues. EHO Food Hygiene certification is required to prepare and distribute food. While there is no specified expiry date of such a qualification the changes in legislation it is suggested to update every five years.

Best before indicates a future date beyond which the food product may lose quality in terms of taste or texture amongst others, but does not imply any serious health problems if food is consumed beyond this date (within reasonable limits).

Use by indicates a legal date beyond which it is not permissible to sell a food product (usually one that deteriorates fairly rapidly after production) due to the potential serious nature of consumption of pathogens. Leeway is sometimes provided by producers in stating display until dates so that products are not at their limit of safe consumption on the actual date stated (this latter is voluntary and not subject to regulatory control). This allows for the variability in production, storage and display methods.

United States

With the exception of infant formula and baby foods which must be withdrawn by their expiration date, Federal law does not require expiration dates. For all other foods, except dairy products in some states, freshness dating is strictly voluntary on the part of manufacturers. In response to consumer demand, perishable foods are typically labelled with a **Sell by** date. It is up to the consumer to decide how long after the Sell by date a package is usable. Other common dating statements are Best if used by, Use-by date, Expiration date, Guaranteed fresh <date>, and Pack date.

Australia and New Zealand

Guide to Food Labelling and Other Information Requirements: This guide provides background information on the general labelling requirements in the Code. The information in this guide applies both to food for retail sale and to food for catering purposes. Foods for catering purposes means those foods for use in restaurants, canteens, schools, caterers or self-catering institutions, where food is offered for immediate consumption. Labelling and information requirements in the new Code apply both to food sold or prepared for sale in Australia and New Zealand and food imported into Australia and New Zealand. Warning and Advisory Declarations, Ingredient Labelling, Date Marking, Nutrition Information Requirements, Legibility Requirements for Food Labels, Percentage Labelling, Information Requirements for Foods Exempt from Bearing a Label.

Food Sampling

a food scientist considers canned peas

Food sampling is a process used to check that a food is safe and that it does not contain harmful contaminants, or that it contains only permitted additives at acceptable levels, or that it contains the right levels of key ingredients and its label declarations are correct, or to know the levels of nutrients present.

A food sample is carried out by subjecting the product to physical analysis. Analysis may be undertaken by or on behalf of a manufacturer regarding their own product, or for official food law enforcement or control purposes, or for research or public information.

To undertake any analysis, unless the whole amount of food to be considered is very small so that the food can be used for testing in its entirety, it is usually necessary for a portion of it to be taken (e.g. a small quantity from a full production batch, or a portion of what is on sale in a shop) – this process is known as food sampling.

In most cases with food to be analysed there are two levels of sampling – the first being selection of a portion from the whole, which is then submitted to a laboratory for testing, and the second being the laboratory's taking of the individual amounts necessary for individual tests that may be applied. It is the former that is 'food sampling': the latter is analytical laboratory 'sub-sampling', often relying upon initial homogenisation of the entire submitted sample.

Where it is intended that the results of any analysis to relate to the food as a whole it is crucially important that the sample is representative of that whole – and the results of any analysis can only be meaningful if the sampling is undertaken effectively. This is true whether the 'whole' is a manufacturer's entire production batch, or where it is a single item but too large to all be used for the test.

Factors relevant in considering the representativeness of a sample include the homogeneity of the food, the relative sizes of the sample to be taken and the whole, the potential degree of variation of the parameter(s) in question through the whole, and the significance and intended use of the analytical result.

Sampling by Manufacturers

Food manufacturers and producers would need to satisfy themselves that any sample taken for analysis is sufficiently representative of the food for the analytical result to be meaningful. This is true whether the data are to be used as the basis of labelling declarations, assurance of compliance with legislative or other standards, monitoring of production as part of HACCP (Hazard Analysis and Critical Control Points), or for routine quality control.

In the United Kingdom although various guidance is available, either from manufacturers' associations or from sources of standards such as British Standards Institution (such as British Standard BS6001), some of which may be relevant to certain food types. It is largely down to manufacturers to make their own evaluations of need and suitability. This must be translated into an assessment both of sample portion size and number, and the frequency of taking samples.

Testing and Methods of Analysis

To ensure food safety and quality, some food samples which are perishable require certain tests and analyses. The following tests and analyses can be conducted:

- Food allergen testing
- Food chemical analysis
- Food contact tests
- Food contaminant testing
- Nutritional analysis and testing
- GMO testing
- Melamine contamination testing
- Microbiological tests

- Spiral plating for bacterial count
- Pesticide residue testing
- Veterinary drug residue testing
- PCR food testing

Food Law Enforcement (UK)

In the United Kingdom, enforcement is under the Food Safety Act 1990. Food sampling is undertaken primarily by local authorities and port health authorities for submission to public analysts for analysis. Much of the legislation relates to food as supplied to a consumer, meaning that every portion of a size of perishable food and foods at risk as may be supplied to a consumer has to comply, so that in such cases the sample submitted for analysis could simply be an entire consumer-sized portion. There are exceptions, however, such as the sampling of nut products for the presence of aflatoxins, which stipulate a primary sample size related to the size of the consignment – with associated requirements for initial homogenisation to produce a smaller sample to be sent for analysis.

The Food Safety Act 1990 affords a right for defence analysis, and for referee analysis in case of disputed analytical results, by stipulating that except where to do so would prevent effective analysis the sample must be divided into three parts. The UK Food Standards Agency provides supplementary guidance to the enforcement authorities to assist with the sampling process and associated decisions by sampling officers.

There is no set frequency or rate for the sampling of food for law enforcement in the UK. Between the 1930s and 1990s there had been a guideline minimum rate for sampling for chemical analysis (not including samples for microbiological examination) of 2.5 samples per annum per 1000 head of population, however that was an arbitrary figure and more recent thinking suggested that the selection of a frequency for sampling should be based on risk. In this context risk includes all 'consumer protection' issues such as pecuniary disadvantage from substandard or counterfeit products, as well as risk to health. The Association of Public Analysts was commissioned by the Food Standards Agency to look into this, culminating in a scheme for *Risk Based Sampling*, though it has not yet been adopted by the enforcement authorities.

Hazard Analysis and Critical Control Points

Hazard analysis and critical control points or HACCP is a systematic preventive approach to food safety from biological, chemical, and physical hazards in production processes that can cause the finished product to be unsafe, and designs measurements to reduce these risks to a safe level. In this manner, HACCP is referred as the prevention of hazards rather than finished product inspection. The HACCP system can be used at all stages of a food chain, from food production and preparation processes including packaging, distribution, etc. The Food and Drug Administration (FDA) and the United States Department of Agriculture (USDA) require mandatory HACCP programs for juice and meat as an effective approach to food safety and protecting public health. Meat HACCP systems are regulated by the USDA, while seafood and juice are regulated by the FDA. All other food companies in the United States that are required to register with the FDA under the Public Health Security and Bioterrorism Preparedness and Response Act of 2002, as well as firms outside the US that export food to the US, are transitioning to mandatory Hazard Analysis and Risk-based Preventive Controls (HARPC) plans.

HACCP is believed to stem from a production process monitoring used during World War II because traditional "end of the pipe" testing on artillery shell's firing mechanisms could not be performed, and a large percentage of the artillery shells made at the time were either duds or misfiring. HACCP itself was conceived in the 1960s when the US National Aeronautics and Space Administration (NASA) asked Pillsbury to design and manufacture the first foods for space flights. Since then, HACCP has been recognized internationally as a logical tool for adapting traditional inspection methods to a modern, science-based, food safety system. Based on risk-assessment, HACCP plans allow both industry and government to allocate their resources efficiently in establishing and auditing safe food production practices. In 1994, the organization of *International HACCP Alliance* was established initially for the US meat and poultry industries to assist them with implementing HACCP and now its membership has been spread over other professional/ industrial areas.

Hence, HACCP has been increasingly applied to industries other than food, such as cosmetics and pharmaceuticals. This method, which in effect seeks to plan out unsafe practices based on science, differs from traditional "produce and sort" quality control methods that do nothing to prevent hazards from occurring and must identify them at the end of the process. HACCP is focused only on the health safety issues of a product and not the quality of the product, yet HACCP principles are the basis of most food quality and safety assurance systems, and the United States, HACCP compliance is regulated by 21 CFR part 120 and 123. Similarly, FAO/WHO published a guideline for all governments to handle the issue in small and less developed food businesses.

History

In the early 1960s, a collaborated effort between the Pillsbury Company, NASA, and the U.S. Army Laboratories began with the objective to provide safe food for space expeditions. People involved in this collaboration included Herbert Hollander, Mary Klicka, and Hamed El-Bisi of the United States Army Laboratories in Natick, Massachusetts, Dr. Paul A. Lachance of the Manned Spacecraft Center in Houston, Texas, and Howard E. Baumann representing Pillsbury as its lead scientist.

In order to ensure that the food that would be sent to space was safe, Lachance imposed strict microbial requirements, including pathogen limits (including *E. coli*, *Salmonella*, and *Clostridium botulinum*). Using the traditional end product testing method, it was soon realized that almost all of the food manufactured was being used on testing and very little was left for actual use. It was realized that a new approach was needed.

NASA's own requirements for Critical Control Points (CCP) in engineering management would be used as a guide for food safety. CCP derived from Failure mode and effects analysis (FMEA) from NASA via the munitions industry to test weapon and engineering system reliability. Using that information, NASA and Pillsbury required contractors to identify "critical failure areas" and eliminate them from the system, a first in the food industry then. Baumann, a microbiologist by training, was so pleased with Pillsbury's experience in the space program that he advocated for his company to adopt what would become HACCP at Pillsbury.

Soon thereafter, Pillsbury was confronted with a food safety issue of its own when glass contamination was found in farina, a cereal commonly used in infant food. Baumann's leadership promoted HACCP in Pillsbury for producing commercial foods, and applied to its own food production. This led to a panel discussion at the 1971 National Conference on Food Protection that included examining CCPs and Good Manufacturing Practices in producing safe foods. Several botulism cases were attributed to under-processed low-acid canned foods in 1970-71. The United States Food and Drug Administration (FDA) asked Pillsbury to organize and conduct a training program on the inspection of canned foods for FDA inspectors. This 21-day program was first held in September 1972 with 11 days of classroom lecture and 10 days of canning plant evaluations. Canned food regulations (21CFR 108, 21 CFR 110, 21 CFR 113, and 21 CFR 114) were first published in 1969. Pillsbury's training program to the FDA in 1969, titled "Food Safety through the Hazard Analysis and Critical Control Point System", was the first time that HACCP was used.

HACCP was initially set on three principles, now shown as principles one, two, and four in the section below. Pillsbury quickly adopted two more principles, numbers three and five, to its own company in 1975. It was further supported by the National Academy of Sciences (NAS) that governmental inspections by the FDA go from reviewing plant records to compliance with its HACCP system. A second proposal by the NAS led to the development of the National Advisory Committee on Microbiological Criteria for Foods (NACMCF) in 1987. NACMCF was initially responsible for defining HACCP's systems and guidelines for its application and were coordinated with the Codex Committee for Food Hygiene, that led to reports starting in 1992 and further harmonization in 1997. By 1997, the seven HACCP principles listed below became the standard. A year earlier, the American Society for Quality offered their first certifications for HACCP Auditors. (First known as Certified Quality Auditor-HACCP, they were changed to Certified HACCP Auditor (CHA) in 2004.).

HACCP expanded in all realms of the food industry, going into meat, poultry, seafood, dairy, and has spread now from the farm to the fork.

Principles

1. Conduct a hazard analysis

   Plans determine the food safety hazards and identify the preventive measures the plan can apply to control these hazards. A food safety hazard is any biological, chemical, or physical property that may cause a food to be unsafe for human consumption.

2. Identify critical control points

   A critical control point (CCP) is a point, step, or procedure in a food manufacturing process at which control can be applied and, as a result, a food safety hazard can be prevented, eliminated, or reduced to an acceptable level.

3. Establish critical limits for each critical control point

A critical limit is the maximum or minimum value to which a physical, biological, or chemical hazard must be controlled at a critical control point to prevent, eliminate, or reduce that hazard to an acceptable level.

4. Establish critical control point monitoring requirements

   Monitoring activities are necessary to ensure that the process is under control at each critical control point. In the United States, the FSIS requires that each monitoring procedure and its frequency be listed in the HACCP plan.

5. Establish corrective actions

   These are actions to be taken when monitoring indicates a deviation from an established critical limit. The final rule requires a plant's HACCP plan to identify the corrective actions to be taken if a critical limit is not met. Corrective actions are intended to ensure that no product is injurious to health or otherwise adulterated as a result if the deviation enters commerce.

6. Establish procedures for ensuring the HACCP system is working as intended

   Validation ensures that the plants do what they were designed to do; that is, they are successful in ensuring the production of a safe product. Plants will be required to validate their own HACCP plans. FSIS will not approve HACCP plans in advance, but will review them for conformance with the final rule.

   Verification ensures the HACCP plan is adequate, that is, working as intended. Verification procedures may include such activities as review of HACCP plans, CCP records, critical limits and microbial sampling and analysis. FSIS is requiring that the HACCP plan include verification tasks to be performed by plant personnel. Verification tasks would also be performed by FSIS inspectors. Both FSIS and industry will undertake microbial testing as one of several verification activities.

   Verification also includes 'validation' – the process of finding evidence for the accuracy of the HACCP system (e.g. scientific evidence for critical limitations).

7. Establish record keeping procedures

   The HACCP regulation requires that all plants maintain certain documents, including its hazard analysis and written HACCP plan, and records documenting the monitoring of critical control points, critical limits, verification activities, and the handling of processing deviations. Implementation involves monitoring, verifying, and validating of the daily work that is compliant with regulatory requirements in all stages all the time. The differences among those three types of work are given by Saskatchewan Agriculture and Food.

Standards

The seven HACCP principles are included in the international standard ISO 22000 FSMS 2005. This standard is a complete food safety and quality management system incorporating the elements of prerequisite programmes(GMP & SSOP), HACCP and the quality management system, which together form an organization's Total Quality Management system.

Other Standards, such as Safe Quality Food Institute's SQF Code, also relies upon the HACCP methodology as the basis for developing and maintaining Food Safety (Level 2) and Food Quality (Level 3) Plans and programs in concert with the fundamental prerequisites of Good Manufacturing Practices.

Training

Training for developing and implementing HACCP Food Safety management system are offered by several quality assurance companies. However, ASQ does provide a Trained HACCP Auditor (CHA) exam to individuals seeking professional training. In the UK the Chartered Institute of Environmental Health (CIEH) and Royal Society for Public Health offer HACCP for Food Manufacturing qualifications, accredited by the QCA (Qualifications and Curriculum Authority).

Application

- Fish and fishery products
- Fresh-cut produce
- Juice and nectar products
- Food outlets
- Meat and poultry products
- School food and services

Water Quality Management

The use of HACCP for water quality management was first proposed nearly 20 years ago. Thereafter, a number of water quality initiatives applied HACCP principles and steps to the control of infectious disease from water, and provided the basis for the Water Safety Plan (WSP) approach in the third edition of the World Health Organization (WHO) report, which has been described as "a way of adapting the HACCP approach to drinking water systems".

Water Quality Management Programme Guidelines

Programme Modernization: According to Ongley, 1998, a series of steps could be taken to execute a more useful transition – from technical programmmes to policy to management decisions. Various aspects of the modernization process have been discussed by Ongley in ESCAP (1997).

- Policy reform - A consultative process must define all the policy tenets and should review the execution of the said policy tenets.
- Legal reform – Legal reform with respect to water quality management is one of the most crucial elements. This could be addressed by the creation of national data standards as well as the creation of a national process to analyze and review collected data.
- Institutional reform – This is a complex issue and has got no simple answers. Still, there are some key principles that can be helpful for institutional reform in the light of water

quality management. One of them is water quality monitoring as a service function. Apart from that, both technical efficiency and capacity issues emerge as major factors in a reformed water quality programmes.

- Technical reform – This is one area that garners the most attention as well as investment. However, such a kind of reform most targets facility modernization, including other co-factors like data programmes/networks, technical innovation, data management/data products and remediation.

HACCP for Building Water Systems

Hazards associated with water systems in buildings include physical, chemical and microbial hazards. In 2013, NSF International, a public health and safety NGO, established education, training and certification programs in HACCP for Building Water Systems. The programs, developed with the guidance of subject matter experts Aaron Rosenblatt and William McCoy center on the use of the use of HACCP principles adapted to the specific requirements of domestic (hot and cold) and utility (HVAC, etc.) water systems in buildings, in order to prevent plumbing-associated hazards from harming people. Hazards addressed include scalding, lead, and disinfection byproducts as well as a range of clinically important pathogens, such as Legionella, pseudomonas, non-tubercular Mycobacteria (NTM), acinetobacter, Elizabethkingia and Naegleria. Early adopters of HACCP for Building Water Systems include leading healthcare institutions, notably Mayo Clinic (Rochester, MN).

ISO 22000

ISO 22000 is like a standard designed to help augment HACCP on issues related to food safety. Although several companies, especially big ones, have either implemented or are on the point of implementing ISO 22000, there are many others which are hesitant to do so. The main reason behind that is the lack of information and the fear that the new standard is too demanding in terms of bureaucratic work, from abstract of case study.

ISO 22000

ISO 22000 is a standard developed by the International Organization for Standardization dealing with food safety. It is a general derivative of ISO 9000.

Food Safety

Food safety is linked to the presence of food-borne hazards in food at the point of consumption. Since food safety hazards can occur at any stage in the food chain it is essential that adequate control be in place. Therefore, a combined effort of all parties through the food chain is required.

ISO 22000 Standard

The ISO 22000 international standard specifies the requirements for a food safety management system that involves the following elements:

- interactive communication
- system management
- prerequisite programs
- HACCP principles

Critical reviews of the above elements have been conducted by many scientists [99]. Communication along the food chain is essential to ensure that all relevant food safety hazards are identified and adequately controlled at each step within the food chain. This implies communication between organizations both upstream and downstream in the food chain. Communication with customers and suppliers about identified hazards and control measures will assist in clarifying customer and supplier requirements.

Recognition of the organization's role and position within the food chain is essential to ensure effective interactive communication throughout the chain in order to deliver safe food products to the final consumer.

The most effective food safety systems are established, operated and updated within the framework of a structured management system and incorporated into the overall management activities of the organization. This provides maximum benefit for the organization and interested parties. ISO 22000 has been aligned with ISO 9001 in order to enhance the compatibility of the two standards.

ISO 22000 can be applied independently of other management system standards or integrated with existing management system requirements.

ISO 22000 integrates the principles of the Hazard Analysis and Critical Control Point (HACCP) system and application steps developed by the Codex Alimentarius Commission. By means of auditable requirements, it combines the HACCP plan with prerequisite programmes. Hazard analysis is the key to an effective food safety management system, since conducting a hazard analysis assists in organizing the knowledge required to establish an effective combination of control measures. ISO 22000 requires that all hazards that may be reasonably expected to occur in the food chain, including hazards that may be associated with the type of process and facilities used, are identified and assessed. Thus it provides the means to determine and document why certain identified hazards need to be controlled by a particular organization and why others need not.

During hazard analysis, the organization determines the strategy to be used to ensure hazard control by combining the prerequisite programmes and the HACCP plan.

ISO is developing additional standards that are related to ISO 22000. These standards will be known as the ISO 22000 family of standards. At the present time, the following standards will make up the ISO 22000 family of standards:

ISO 22000 - Food safety management systems - Requirements for any organization in the food chain.

ISO 22001 - Guidelines on the application of ISO 9001:2000 for the food and drink industry (replaces: ISO 15161:2001).

ISO/TS 22002- Prerequisite programmes on food safety—Part 1: Food manufacturing

ISO TS 22003 - Food safety management systems for bodies providing audit and certification of food safety management systems.

ISO TS 22004 - Food safety management systems - Guidance on the application of ISO 22000:2005.

ISO 22005 - Traceability in the feed and food chain - General principles and basic requirements for system design and implementation.

ISO 22006 - Quality management systems - Guidance on the application of ISO 9002:2000 for crop production.

ISO 22000 is also used in the Food Safety Systems Certification (FSSC) Scheme FS22000. FS22000 is a Global Food Safety Initiative (GFSI) approved scheme.

ISO 9001 vs ISO 22000

In comparison with ISO 9001, the standard is a more procedural orientated guidance than a principle based one. Apart from that, ISO 22000 is an industrial-specific risk management system for any type of food processing and marketing, which can be closely incorporated with the quality management system of ISO 9001. The detailed similarities and differences of the two standards can be found elsewhere.

Potential Justification

In 2004, European Office of Crafts, Trades and Small and Medium-sized Enterprises for Standardisation addressed that the standard is only suitable for large sized companies and small food businesses will not be able to seek such a high standard due to the lack of resources to pursue the certification. The agency suggests to create an alternative for small food business to achieve the same objective.EFSA is now making their efforts on the food legislations that are adaptable for the SMEs in food supply chains. A few critics also proposed that organizations which seek the standard certification should also do the same to the ISO 14001 along with the ISO 9001, as they consider that large amounts of risks are mainly from the primary production in the supply chains rather than the later stages of food processing

Food Grading

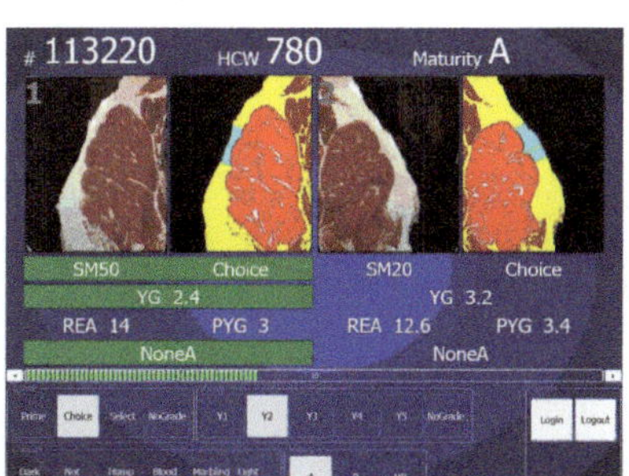

A screenshot from the electronic grading system showing USDA Choice, Yield Grade 2 beef. The left is the natural color view of the cut; the right is the instrument enhanced view that details the amount of marbling, size, and fat thickness.

Optical sorting achieves non-destructive, 100 percent inspection in-line at full production volumes

Food grading involves the inspection, assessment and sorting of various foods regarding quality, freshness, legal conformity and market value. Food grading often occurs by hand, in which foods are assessed and sorted. Machinery is also used to grade foods, and may involve sorting products by size, shape and quality. For example, machinery can be used to remove spoiled food from fresh product.

By Food Type

Beef

Inspected beef carcasses tagged by the USDA

Beef grading in the United States is performed by the United States Department of Agriculture's (USDA) Agricultural and Marketing Service. There are eight beef quality grades, with U.S. Prime being the highest grade and U.S. Canner being the lowest grade. Beef grading is a complex and detailed process, requiring graders to think and calculate quickly with great accuracy. Using technology to complement and supplement the onsite human graders generates an efficient and more precise process.

Beer

In beer grading, the letter "X" is used on some beers, and was traditionally a mark of beer strength, with the more Xs the greater the strength. Some sources suggest that the origin of the mark was in the breweries of medievalmonasteries Another plausible explanation is contained in a treatise entitled "The Art of Brewing" published in London in 1829. It says; *"The duties on ale and beer, which were first imposed in 1643... at a certain period, in distinguishing between small beer and strong, all ale or beer, sold at or above ten shillings per barrel, was reckoned to be* strong and *was, therefore, subjected to a higher duty. The cask which contained this strong beer was then first marked with an X signifying ten; and hence the present quack-like denominations of XX (double X) and XXX (treble X) on the casks and accounts of the strong-ale brewers".*

In mid-19th century England, the use of "X" and other letters had evolved into a standardised grading system for the strength of beer. Today, it is used as a trade mark by a number of brewers in the United Kingdom, the Commonwealth and the United States.

European Bitterness Units scale, often abbreviated as EBU, is a scale for measuring the perceived bitterness of beer, with lower values being generally "less bitter" and higher values "more bitter". The scale and method are defined by the European Brewery Convention, and the numerical value should be the same as of the International Bitterness Units scale (IBU), defined in co-operation with the American Society of Brewing Chemists. However, the exact process of determining EBU and IBU values differs slightly, which may in theory result with slightly smaller values for EBU than IBU.

The International Bittering Units scale, or simply IBU scale, provides a measure of the bitterness of beer, which is provided by the hops used during brewing. Bittering units are measured through the use of a spectrophotometer and solvent extraction.

Butter

Coconut Milk

Several grades of coconut milk exist: from thick at 20-22% fat to thin at 5-7% fat level.

Coffee

Dark-roasted coffee beans

Coffee growers, traders, and roasters grade beans based on size, color, and a variety of other characteristics. Coffees of exceptional quality are traded as "specialty coffees" and fetch a higher price in the international market.

After the roast, Coffee grading involves assessment of roasted coffee seed colorization and then labeling as light, medium light, medium, medium dark, dark, or very dark. A more accurate method of discerning the degree of roast involves measuring the reflected light from roasted seeds illuminated with a light source in the near infrared spectrum. This elaborate light meter uses a process known as spectroscopy to return a number that consistently indicates the roasted coffee's relative degree of roast or flavor development.

Eggs

In the United States, egg grading is performed by the USDA, and is based upon the interior quality of the egg and the appearance and condition of the egg shell. Eggs of any quality grade may differ in weight (size). Egg grading is performed by candling, which involves observing the interior of eggs by placing them in front of a bright light.

Guar Gum

Guar gum grading involves analysis for coloration, viscosity, moisture, granulation, protein content and insolubles ash.

Honey

Honey grading in the United States is performed voluntarily based upon USDA standards (USDA does offer inspection and grading "as on-line (in-plant) or lot inspection...upon application, on a fee-for-service basis.") . Honey is graded based upon a number of factors, including water content, flavor and aroma, absence of defects and clarity. Honey is also classified by color though it is not a factor in the grading scale. U.S. honey grade scales are Grade A, Grade B, Grade C and Grade substandard.

Lobster

A lobster

In the U.S., lobster grading involves denoting lobsters as new-shell, hard-shell or old-shell, and because lobsters which have recently shed their shells are the most delicate, there is an inverse relationship between the price of American lobster and its flavour. New-shell lobsters have paper-thin shells and a worse meat-to-shell ratio, but the meat is very sweet. However, the lobsters are so delicate that even transport to Boston almost kills them, making the market for new-shell lobsters strictly local to the fishing towns where they are offloaded. Hard-shell lobsters with firm shells, but with less sweet meat, can survive shipping to Boston, New York and even Los Angeles, so they command a higher price than new-shell lobsters. Meanwhile, old-shell lobsters, which have not shed since the previous season and have a coarser flavour, can be air-shipped anywhere in the world and arrive alive, making them the most expensive. One seafood guide notes that an eight-dollar lobster dinner at a restaurant overlooking fishing piers in Maine is consistently delicious, while "the eighty-dollar lobster in a three-star Paris restaurant is apt to be as much about presentation as flavor".

Maple Syrup

Maple syrup from Quebec, graded as No. 1, Light (enlarge image to read the label atop the can)

In Canada, the grading of maple syrup involves classification by the Canadian Food Inspection Agency (CFIA) of maple syrups into one of three grades, each with several colour classes: Canada No. 1, including Extra Light, Light, and Medium; No. 2 Amber; and finally No. 3 Dark or any other ungraded category. Producers in Quebec and Ontario may follow either federal or provincial grading guidelines, which differ slightly.

The United States uses different grading standards. Maple syrup is divided into two major grades: Grade A and Grade B. Grade A is further divided into three subgrades: Light Amber (sometimes known as Fancy), Medium Amber, and Dark Amber. The Vermont Agency of Agriculture Food and Markets uses a similar grading system of color, and is roughly equivalent, especially for lighter syrups, but using letters: "AA", "A", etc. The Vermont grading system differs from the US system in maintaining a slightly higher standard of product density (measured on the Baumé scale). New Hampshire maintains a similar standard, but not a separate state grading scale. The Vermont-graded product has 0.9 percent more sugar and less water in its composition than US-graded. One grade of syrup not for table use, called commercial or Grade C, is also produced under the Vermont system. Vermont inspectors enforce strict syrup grading regulations, and can fine producers up to US$1000 for labelling syrup incorrectly.

Milk

In the United States, there are two grades of milk, with Grade A primarily used for direct sales and consumption in stores, and Grade B used for indirect consumption, such as in cheese making or other processing.

The differences between the two grades are defined in the Wisconsin administrative code for Agriculture, Trade, and Consumer Protection, chapter 60. Grade B generally refers to milk that is cooled in milk cans, which are immersed in a bath of cold flowing water that typically is drawn up from an underground water well rather than using mechanical refrigeration.

Oranges

The USDA has established the following grades for Florida oranges, which primarily apply to oranges sold as fresh fruit: US Fancy, US No. 1 Bright, US No. 1, US No. 1 Golden, US No. 1 Bronze, US No. 1 Russet, US No. 2 Bright, US No. 2, US No. 2 Russet, and US No. 3. The general characteristics graded are color (both hue and uniformity), firmness, maturity, varietal characteristics, texture, and shape. Fancy, the highest grade, requires the highest grade of color and an absence of blemishes, while the terms Bright, Golden, Bronze, and Russet concern solely discoloration.

Floridanavel oranges

Peas

Peas

Pea grading involves sorting peas by size, in which smallest peas are graded as the highest quality for their tenderness.Brines may be used, in which peas are floated in them, from which their density can be determined.

Potatoes

In the U.S., potato grading for Idaho potatoes is performed in which No. 1 potatoes are the highest quality and No. 2 are rated as lower in quality due to their appearance (e.g. blemishes or bruises, pointy ends). Density assessment can be performed by floating them in brines. High density potatoes are desirable in the production of dehydrated mashed potatoes, potato crisps and french fries.

Rice

The main criteria used by many countries and millers in rice grading are degree of milling, appearance (color), damaged (broken) and percentage of chalky kernels. In the United States rice is marketed according to three main properties size, color and condition (kernels damage), these properties are directly related to quality, milling percentage and other processing conditions. All properties are considered important in grading. For instance, chalky kernels are not desirable because they give lower milling yields after processing and easily break during handling.

Spices

Cinnamon

In Sri Lanka, cinnamon grading is performed by dividing cinnamon quills into four groups, which are then further divided into specific grades.

Vanilla

People grading vanilla beans in Sambava, Madagascar

Several vanilla fruit grading systems are in use. Each country which produces vanilla has its own grading system, and individual vendors, in turn, sometimes use their own criteria for describing the quality of the fruits they offer for sale.

Tea

In the western black tea industry, tea leaf grading is the process of evaluating products based on the quality and condition of the tea leaves themselves. The highest grades are referred to as "orange pekoe", and the lowest as "fannings" or "dust". This grading system is based upon the size of processed and dried black tea leaves. Despite a purported Chinese origin, these grading terms are typically used for teas from Sri Lanka, India and countries other than China; they are not generally known within Chinese-speaking countries.

Black Tea

Black tea grading is usually based upon one of four scales of quality. Whole-leaf teas are the highest quality, followed by broken leaves, fannings, and dusts. Whole-leaf teas are produced with little or no alteration to the tea leaf. This results in a finished product with a coarser texture than that of bagged teas. Whole-leaf teas are widely considered the most valuable, especially if they contain leaf tips. Broken leaves are commonly sold as medium-grade loose teas. Smaller broken varieties may be included in tea bags.

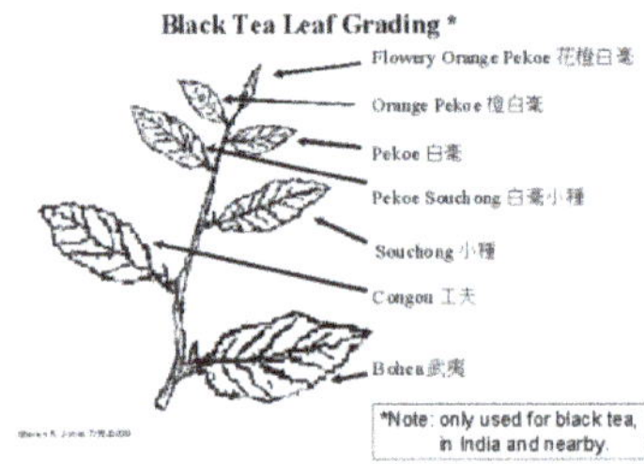

A black tea leaf grading diagram

Fresh tea leaves of different sizes

Rooibos

Rooibos grades are largely related to the proportion of "needle" or leaf to stem content in the mix. A higher leaf content will result in a darker liquor, richer flavour and less "dusty" aftertaste. The high-grade rooibos is exported and does not reach local markets, with major consumers being in the EU, particularly Germany, where it is used in creating flavoured blends for loose-leaf tea markets. In development within South Africa are a small number of specialty tea companies producing similar blends.

Organic green rooibos tea leaves

Food Additives and Preservatives

Carrageenan

There are two basic grades of carrageenan, refined carrageenan (RC) and semi-refined carrageenan (SRC). In the United States, RC and SRC are both labeled as carrageenan. In the European Union, RC is designated by the E number E-407, and SRC is E-407a. RC has a 2% maximum for acid-insoluble material and is produced through an alcohol precipitation process or potassium chloride gel press process. SRC contains a much higher level of cellulosic content and is produced in a less complex process. Indonesia, the Philippines, and Chile are three main sources of raw material and extracted carrageenan.

Lye

Lye is used to cure foods such as lutefisk, olives (making them less bitter), canned mandarin oranges, hominy, lye rolls, century eggs, and pretzels. It is also used as a tenderizer in the crust of baked Cantonese moon cakes, and in lye-water "zongzi" (glutinous rice dumplings wrapped in bamboo leaves), chewy southern Chinese noodles popular in Hong Kong and southern China, and Japanese ramen noodles. In the United States, food-grade lye must meet the requirements outlined in the Food Chemicals Codex (FCC), as prescribed by the U.S. Food and Drug Administration (FDA). Lower or more caustic grades of lye are commonly used as drain de-cloggers and oven cleaners. As a result, such grades should not be used for food preparation, as they would be too poisonous for human consumption.

Sodium Bisulfate

Sodium bisulfate is used as a food additive to leaven cake mixes (make them rise) as well as being used in meat and poultry processing and most recently in browning prevention of fresh-cut produce. The food-grade product meets the requirements set out in the Food Chemicals Codex. It is denoted by E number E514ii in the EU and is approved for use in Australia and New Zealand where it is listed as additive 514. Food-grade sodium bisulfate is used in a variety of food products, including beverages, dressings, sauces, and fillings. It has many synonyms (bisulfate of soda, sodium bisulfate, sodium acid sulfate, monosodium hydrogen sulfate, monosodium salt, sodium hydrogen sulfate, sodium hydrosulfate, sodium pyrosulfate, sulfuric acid, sulfuric acid sodium salt (1:1)).

By Country

- In India, AGMARK is a certification mark employed on agricultural products, assuring that they conform to a set of standards approved by the Directorate of Marketing and Inspection, an agency of the Government of India.

Beef carcass Classification

Countries regulate the marketing and sale of Beef by observing criteria of CattleCarcasses at the Abattoir and classifying the carcasses. This classification, sometimes optional, can suggest a market demand for a particular animal's attributes and therefore the price owed to the producer.

United States Grading System

The United States Department of Agriculture's grading system, which has been designed to reward marbling, has eight different grades: Prime, Choice, Select, Standard, Commercial, Utility, Cutter and Canner. Prime has the highest marbling content when compared to other grades, and is capable of fetching a premium at restaurants and supermarkets. Choice is the grade most commonly sold in retail outlets, and Select is sold as a cheaper, but still nutritious, option in many stores. Prime, Choice, Select and Standard are commonly used in the younger cattle (under 42 months of age), and Commercial, Utility, Canner and Cutter are used in older cattle carcasses which are not marketed as wholesale beef "block" meat, but as material used in ground products and cheaper steaks for family restaurants.

In the United States, the USDA Agricultural Marketing Service (AMS) operates a voluntary beef grading program. The meat processor pays for a trained AMS meat grader to grade whole carcasses at the abattoir. Users are required to comply with Food Safety and Inspection Service (FSIS) grade labeling procedures. The official USDA grade designation can appear in one or any combination of the following ways: container markings, individual bags, legible roller brand appearing on the meat itself, or by a USDA shield stamp that incorporates the quality and/or yield grade.

The United States uses eight beef quality grades. The grades are based on two main criteria: the degree of marbling (intramuscular fat) in the beef, and the maturity (estimated age of the animal at slaughter). Some meat scientists object to the current scheme of USDA grading since it is not based on direct measurement of tenderness, although marbling and maturity are indicators of tenderness. Most other countries' beef grading systems mirror the U.S. model, except for those in the European Union (EU). The EU employs a grading scheme that emphasizes carcass shape and amount of fat covering instead of marbling and aging. The differences in grading yield incompatible value judgments of beef value in the United States and the EU. Most beef offered for sale in supermarkets in the United States is graded U.S. Choice or Select. U.S. Prime beef is sold to hotels and upscale restaurants, and usually marketed as such.

- U.S. Prime – Highest in quality and intramuscular fat, limited supply. Currently, about 2.9% of carcasses grade as Prime.
- U.S. Choice – High quality, widely available in foodservice industry and retail markets. Choice carcasses are 53.7% of the fed cattle total. The difference between Choice and Prime is largely due to the fat content in the beef. Prime typically has a higher fat content (more and well distributed intramuscular "marbling") than Choice.
- U.S. Select (formerly *Good*) – lowest grade commonly sold at retail, acceptable quality, but is less juicy and tender due to leanness.
- U.S. Standard – Lower quality, yet economical, lacking marbling.
- U.S. Commercial – Low quality, lacking tenderness, produced from older animals.
- U.S. Utility
- U.S. Cutter
- U.S. Canner

Utility, Cutter, and Canner grade are rarely used in foodservice operations and primarily used by processors and canners.

Beef grading service began in 1917 as a way to determine both the quality and the quantity of beef that would come from each carcass. Stamping the grades began in May 1927. Each carcass can be stamped with a yield or quality stamp, or a combination of both. The standards have been revised many times since the original standards were formulated. A few notable changes include combining Prime and choice grades into Prime, and changing the Good grade to choice, this change occurred in 1950. In 1980 conditions were set forth to establish guidance on grading protocol. This included a 10-minute bloom time before the grader evaluates the carcass. Most beef plants will allow a longer time for bloom depending on the speed of the grading chain.

In 1997, the official standards were revised to restrict the Select grade to A maturity carcasses, and to raise the minimum marbling score to qualify for Choice to modest for B maturity cattle. These changes were implemented to improve the uniformity and consistency of the grading system. Yield grades are intended to estimate the pounds of boneless closely trimmed retail cuts from the carcass. Closely trimmed refers to approximately ¼ inch of external fat. Yield grade is determined by considering 4 carcass characteristics: external fat, Kidney, pelvic and heart fat (KPH), Ribeye area (REA), and Hot carcass weight (HCW). The amount of external fat is measured at the ribbed surface between the 12th and 13th ribs. The ribbing of carcasses is described in the U.S. standards for beef grading. External fat is measured at a distance of ¾ the length of the ribeye from the chine bone end. This initial number can be adjusted up or down depending on any abnormal fat deposits. As the amount of external fat increases, the percent of retail cuts decreases.

Kidney fat is assessed subjectively and is expressed as a percentage of the carcass weight. As the percentage of KPH increases, the percent of retail cuts decreases. The ribeye area is measured at the ribbed surface, it can be estimated subjectively or measured with a device approved by the AMS. As ribeye area increase, percent retail cuts increases. Hot carcass weight is used to determine yield grade. As carcass weight increases, percent retail cuts decrease. The following equation is used to determine yield grade:

There are five grades, 1-5. Yield grade one carcasses are of the highest cutability, while yield grade 5 yields the lowest cutability.

Beef sold in U.S. restaurants and supermarkets is usually described by its USDA grade; however, in the early twenty-first century many restaurants and retailers began selling beef on the strength of brand names and the reputation of a specific breed of cattle, such as black Angus.

European Union EUROP Classification

The EUROP grid method of carcass classification was implemented in 1981. European Economic Community Regulations (EEC) No. 1208/81 and No. 2930/81 were enacted to facilitate the application of a community scale for the classification of carcases of adult bovine animals. This was to ensure the uniform classification of the carcases of adult bovine animals in the EEC and make the definitions of conformation classes and fat classes more precise. The need arose for a common grading scale when member states of the EEC began operating in the common beef market in 1968 (EEC) No. 805/68 and price reporting to the EC became mandatory.

In the UK, the Meat and Livestock Commission (MLC Services Ltd) is responsible for the classification of over 80% of the cattle slaughtered in Britain. The EUROP grid consists of a 5-point scale in which each conformation and fat class is subdivided into low medium and high classes resulting in 15 classes. In the UK, the fat classes range from 1-5 with classes 4 and 5 having a high and low sub-class which results in a seven-point scale for fatness (figure 1). It is argued by the MLC that this subdivision allows a more precise description of the carcase.

The price a farmer receives for a beast sent for slaughter is calculated by multiplying the carcase weight by the classification price for a particular category of animal (heifer, steer, bull, cow etc.). This classification is subjectively assigned by the meat grader according to the EUROP system where E is excellent, U is Very Good, R is Good, O is Fair and P is Poor. Likewise for the fat class, where 1 is Low, 2 is Slight, 3 is Average, 4 is High, and 5 is Very High. A typical classification would be R4L where the R refers to a "Good" carcass with an "Average" to "High" covering of fat according to the MLC.

The grader is usually an independent classifier who also monitors carcass dressing specification. Most classifiers are employed by MLC services and they are audited quarterly by the Rural Payments Agency (RPA) which is a government organisation. More recently Video Image analysis has been used to classify beef carcasses according to the EUROP scale. There are several machines that can do this, several of which were trialled in Ireland. The Republic of Ireland has used video image analysis for assignment of the EUROP classification grid since 2004.

Two main problems that are often cited in reference to the EUROP grid are its subjective application and its lack of consideration for meat eating quality.

References

- Wyman, Carolyn (2001). Jell-o: A Biography: the History And Mystery of America's Most Famous Dessert. Diane Publishing Company. ISBN
- Ward, A.G.; Courts, A. (1977). The Science and Technology of Gelatin. New York: Academic Press. ISBN 0-12-735050-0.
- Francis, Frederick J., ed. (2000). "Gelatin". Encyclopedia of Food Science and Technology (2nd ed.). John Wiley & Sons. pp. 1183–1188. ISBN 9780471192558.
- McGee, Harold (2004). On Food and Cooking: The Science and Lore of the Kitchen. New York: Scribner, pp. 291–296. ISBN 0-684-80001-2.
- Gareis, Herbert; Schrieber, Reinhard (2007). Gelatine Handbook: Theory and Industrial Practice. Weinheim: Wiley-VCH. ISBN 3-527-31548-9.
- Perdue, R (2009), "Vacuum Packaging", in Yam, K L, Encyclopedia of Packaging Technology, Wiley (published 2010), ISBN 978-0-470-08704-6
- Soroka, W (2008). Illustrated Glossary of Packaging Terms. Institute of Packaging Professionals. p. 3. ISBN 1-930268-27-0.
- Saravacos, George D.; Maroulis, Zacharias B. (2011).Food Process Engineering Operations. CRC Press. pp. 198-199. ISBN 1439877858.
- Johnson, Paul (2007). "Lobster". Fish Forever: The Definitive Guide to Understanding, Selecting, and Preparing Healthy, Delicious, and Environmentally Sustainable Seafood. John Wiley & Sons. pp. 163–175. ISBN 978-0-7645-8779-5.
- Havkin-Frenkel, Daphna; Belanger, Faith C. (2011). Handbook of Vanilla Science and Technology. Chichester, UK: Wiley-Blackwell. pp. 142–145. ISBN 978-1-4051-9325-2.
- HACCP Accreditation Service". aibonline.org. Archived from the original on 6 June 2010. Retrieved 6 June 2012.
- Serna-Saldivar, S.O. 2012. Cereal grains Laboratory Reference and Procedures Manual. Food Preservation Technology Series. LLC NW. CRC Press. Taylor and Francis Group.

6

Food Storage: An Integrated Study

The storing of food and then delivering it to the customers is termed as food storage. Food is stored by humans as well as by animals and has several purposes. Some of the examples of food storage are food banks, food storage containers and refrigerators. The following section unfolds its aspects in a critical yet systematic manner.

Food Storage

Yup'ik elevated food cache (qulvarvik), Hooper Bay, Alaska, 1929. Photograph by Edward S. Curtis

Food logistics is both a traditional domestic skill and is important industrially. The transport and storing of food as well as timely delivery to consumers is important in order to secure the procurement of food. Food is stored by almost every human society and by many animals. Storing of food has several main purposes:

- Storage of harvested and processed plant and animal food products for distribution to consumers
- Enabling a better balanced diet throughout the year
- Reducing kitchen waste by preserving unused or uneaten food for later use
- Preserving pantry food, such as spices or dry ingredients like rice and flour, for eventual use in cooking
- Preparedness for catastrophes, emergencies and periods of food scarcity or famine

- Religious reasons (Example: LDS Church leaders instruct church members to store food)
- Protection from animals or theft

U.S. Federal Emergency Management Agency (FEMA) food storage containers stacked on shipping pallets in Texas, 2008.

History of Food Storage

Domestic Food Storage

Tupperware kitchen storage containers designed for a variety of uses.

The safe storage of food for home use should strictly adhere to guidelines set out by reliable sources, such as the United States Department of Agriculture. These guidelines have been thoroughly researched by scientists to determine the best methods for reducing the real threat of food poisoning from unsafe food storage. It is also important to maintain proper kitchen hygiene, to reduce risks of bacteria or virus growth and food poisoning. The common food poisoning illnesses include Listeriosis, Mycotoxicosis, Salmonellosis, E. coli, Staphylococcal food poisoning and Botulism. There are many other organisms that can also cause food poisoning.

Plastic storage containers can be used to store food.

There are also safety guidelines available for the correct methods of home canning of food. For example, there are specific boiling times that apply depending upon whether pressure canning or waterbath canning is being used in the process. These safety guidelines are intended to reduce the growth of mold and bacteria and the threat of potentially-fatal food poisoning.

Food Storage Safety

Freezers and Thawing Food

Freezer temperature should be maintained below 0 °F. Food should never be thawed at room temperature, this increases the risk of bacterial and fungal growth and accordingly the risk of food poisoning. Once thawed, food should be used and never refrozen. Frozen food should be thawed using the following methods:

- Microwave oven
- During cooking
- In cold water (place food in watertight, plastic bag; change water every 30 minutes)
- In the refrigerator

Throw out foods that have been warmer than 40 °F for more than 2 hours. If there is any doubt at all about the length of time the food has been defrosted at room temperature, it should be thrown out. Freezing does not destroy microbes present in food. Freezing at 0 °F does inactivate microbes (bacteria, yeasts and molds). However, once food has been thawed, these microbes can again become active. Microbes in thawed food can multiply to levels that can lead to foodborne illness. Thawed food should be handled according to the same guidelines as perishable fresh food.

Food frozen at 0 °F and below is preserved indefinitely. However, the quality of the food will deteriorate if it is frozen over a lengthy period. The United States Department of Agriculture, Food Safety and Inspection Service publishes a chart showing the suggested freezer storage time for common foods.

Refrigeration

It is important to note that safe food storage using refrigeration requires adhering to temperature guidelines:

For safety, it is important to verify the temperature of the refrigerator. Refrigerators should be set to maintain a temperature of 40 °F or below. Some refrigerators have built-in thermometers to measure their internal temperature. For those refrigerators without this feature, keep an appliance thermometer in the refrigerator to monitor the temperature. This can be critical in the event of a power outage. When the power goes back on, if the refrigerator is still 40 °F, the food is safe. Foods held at temperatures above 40 °F for more than 2 hours should not be consumed. Appliance thermometers are specifically designed to provide accuracy at cold temperatures. Be sure refrigerator/freezer doors are closed tightly at all times. Don't open refrigerator/freezer doors more often than necessary and close them as soon as possible.

Storage Times for Refrigerated Food

The United States Department of Agriculture, Food Safety and Inspection Service publishes recommended storage times for refrigerated food.

Storing Oils and Fats

Oils and fats can begin to go rancid quickly when not stored safely. Rancid cooking oils and fats do not often smell rancid until well after they have spoiled. Oxygen, light and heat all contribute to cooking oils becoming rancid. The higher the level of polyunsaturated fat that an oil contains, the faster it spoils. The percentage of polyunsaturated fat in some common cooking oils is: safflower (74%); sunflower (66%); corn (60%); soybean (37%); peanut (32%); canola (29%); olive (8%).

To help preserve oils from rancidification, they should be refrigerated once opened. Opened, refrigerated cooking oils should be used within a few weeks, when some types begin to go rancid. Unopened oils can have a storage life of up to one year, but some types have a shorter shelf-life even when unopened (such as sesame and flaxseed).

Dry Storage of Foods

Vegetables

A large root cellar at the Oxon Hill Manor farm in Maryland

The guidelines vary for safe storage of vegetables under dry conditions (without refrigerating or freezing). This is because different vegetables have different characteristics, for example, tomatoes contain a lot of water, while root vegetables such as carrots and potatoes contain less. These factors, and many others, affect the amount of time that a vegetable can be kept in dry storage, as well as the temperature needed to preserve its usefulness. The following guideline shows the required dry storage conditions:

- Cool and dry: onion
- Cool and moist: root vegetable, potato, cabbage
- Warm and dry: winter squash, pumpkin, sweet potatoes, dried hot peppers

Many cultures have developed innovative ways of preserving vegetables so that they can be stored for several months between harvest seasons. Techniques include pickling, home canning, food dehydration, or storage in a root cellar.

Grain

Grain, which includes dry kitchen ingredients such as flour, rice, millet, couscous, cornmeal, and so on, can be stored in rigid sealed containers to prevent moisture contamination or insect or rodent infestation. For kitchen use, glass containers are the most traditional method. During the 20th century plastic containers were introduced for kitchen use. They are now sold in a vast variety of sizes and designs.

Metal cans are used (in the United States the smallest practical grain storage uses closed-top #10 metal cans). Storage in grain sacks is ineffective; mold and pests destroy a 25 kg cloth sack of grain in a year, even if stored off the ground in a dry area. On the ground or damp concrete, grain can spoil in as little as three days, and the grain might have to be dried before it can be milled. Food stored under unsuitable conditions should not be purchased or used because of risk of spoilage. To test whether grain is still good, it can be sprouted. If it sprouts, it is still good, but if not, it should not be eaten. It may take up to a week for grains to sprout. When in doubt about the safety of the food, throw it out as quickly as possible.

Spices and Herbs

Spices and herbs are today often sold prepackaged in a way that is convenient for pantry storage. The packaging has dual purposes of both storing and dispensing the spices or herbs. They are sold in small glass or plastic containers or resealable plastic packaging. When spices or herbs are homegrown or bought in bulk, they can be stored at home in glass or plastic containers. They can be stored for extended periods, in some cases for years. However, after 6 months to a year, spices and herbs will gradually lose their flavour as oils they contain will slowly evaporate during storage.

Spices and herbs can be preserved in vinegar for short periods of up to a month, creating a flavoured vinegar.

Alternative methods for preserving herbs include freezing in water or unsalted butter. Herbs can be chopped and added to water in an ice cube tray. After freezing, the ice cubes are emptied into a plastic freezer bag for storing in the freezer. Herbs also can be stirred into a bowl with unsalted butter, then spread on wax paper and rolled into a cylinder shape. The wax paper roll containing the butter and herbs is then stored in a freezer, and can be cut off in the desired amount for cooking. Using either of these techniques, the herbs should be used within a year.

Meat

Unpreserved meat has only a relatively short life in storage. Perishable meats should be refrigerated, frozen, dried promptly or cured. Storage of fresh meats is a complex discipline that affects the costs, storage life and eating quality of the meat, and the appropriate techniques vary with the kind of meat and the particular requirements. For example, dry ageing techniques are sometimes used to tenderize gourmet meats by hanging them in carefully controlled environments for up to 21 days, while game animals of various kinds may be hung after shooting. Details depend on personal tastes and local traditions. Modern techniques of preparing meat for storage vary with the type of meat and special requirements of tenderness, flavour, hygiene, and economy.

Semi-dried meats like salamis and country style hams are processed first with salt, smoke, sugar, acid, or other "cures" then hung in cool dry storage for extended periods, sometimes exceeding a year. Some of the materials added during the curing of meats serve to reduce the risks of food poisoning from anaerobic bacteria such as species of Clostridium that release botulinum toxin that can cause botulism. Typical ingredients of curing agents that inhibit anaerobic bacteria include nitrates. Such salts are dangerously poisonous in their own right and must be added in carefully controlled quantities and according to proper techniques. Their proper use has however saved many lives and much food spoilage.

Like the semi-dried meats, most salted, smoked, and simply-dried meats of different kinds that once were staples in particular regions, now are largely luxury snacks or garnishes; examples include jerky, biltong, and varieties of pemmican, but ham and bacon for instance, still are staples in many communities.

Food Rotation

Food rotation is important to preserve freshness. When food is rotated, the food that has been in storage the longest is used first. As food is used, new food is added to the pantry to replace it; the essential rationale is to use the oldest food as soon as possible so that nothing is in storage too long and becomes unsafe to eat. Labelling food with paper labels on the storage container, marking the date that the container is placed in storage, can make this practice simpler. The best way to rotate food storage is to prepare meals with stored food on a daily basis.

For Emergency Preparation

Guides for surviving emergency conditions in many parts of the world recommend maintaining a store of essential foods; typically water, cereals, oil, dried milk, and protein rich foods such as beans, lentils, tinned meat and fish. A food storage calculator can be used to help determine how much of these staple foods a person would need to store in order to sustain life for one full year. In addition to storing the basic food items many people choose to supplement their food storage with frozen or preserved garden-grown fruits and vegetables and freeze-dried or canned produce. An unvarying diet of staple foods prepared in the same manner can cause appetite exhaustion, leading to less caloric intake. Another benefit to having a basic supply of food storage in the home is for the potential cost savings. Costs of dry bulk foods (before preparation) are often considerably less than convenience and fresh foods purchased at local markets or supermarkets. There is a significant market in convenience foods for campers, such as dehydrated food products.

Commercial Food Logistics

Grain and beans are stored in tall grain elevators, almost always at a rail head near the point of production. The grain is shipped to a final user in hopper cars. In the former Soviet Union, where harvest was poorly controlled, grain was often irradiated at the point of production to suppress mold and insects. In the U.S., threshing and drying is performed in the field, and transport is nearly sterile and in large containers that effectively suppresses pest access, which eliminates the need for irradiation. At any given time, the U.S. usually has about two weeks worth of stored grains for the population.

Silos connected to a grain elevator on a farm in Israel.

Fresh fruits and vegetables are sometimes packed in plastic packages and cups for fresh premium markets, or placed in large plastic tubs for sauce and soup processors. Fruits and vegetables are usually refrigerated at the earliest possible moment, and even so have a shelf life of two weeks or less.

In the United States, livestock is usually transported live, slaughtered at a major distribution point, hung and transported for two days to a week in refrigerated rail cars, and then butchered and sold locally. Before refrigerated rail cars, meat had to be transported live, and this placed its cost so high that only farmers and the wealthy could afford it every day. In Europe much meat is transported live and slaughtered close to the point of sale. In much of Africa and Asia most meat is for local populations is raised, slaughtered and eaten locally, which is believed to be less stressful for the animals involved and minimizes meat storage needs. In Australia and New Zealand, where a large proportion of meat production is for export, meat enters the cold chain early, being stored in large freezer plants before being shipped overseas in freezer ships.

Food Bank

Volunteers pass out food items from a food bank run by Feeding America

A food bank or foodbank is a non-profit, charitable organization that distributes food to those who have difficulty purchasing enough food to avoid hunger.

In North America and Australia, food banks usually operate on the "warehouse" model. They act as food storage and distribution depots for smaller front line agencies; and usually do not themselves give out food directly to the hungry. After the food is collected, sorted, and reviewed for quality, these food banks distribute it to non-profit community or government agencies, including *food pantries*, *food closets*, soup kitchens, homeless shelters, orphanages, and schools.

Outside North America and Australia, the "front line" model is often found. Such food banks give out most or all of their food directly to the end users. For both models, the largest sources of food include for-profit growers, manufacturers, distributors and retailers who in the normal course of business have excess food that they cannot sell. Some foodbanks receive a substantial proportion of their food from individual donors, including their volunteer workers. There is considerable overlap with food salvage, food rescue and gleaning.

The world's first food bank was established in the US in 1967, and since then many thousands have been set up all over the world. In Europe, which until recently had little need for food banks due to extensive welfare systems, their numbers grew rapidly after the lasting global inflation in the price of food which began in late 2006, and especially after the financial crisis of 2007–08 began to further worsen economic conditions for those on low income.

The growth of food banks has been broadly welcomed, most especially by those on the right of the political spectrum, but also by many on the left, who see them as evidence of active community that is independent of the state. However, academics and commentators have expressed concern that the rise of foodbanks may erode political support for welfare provision. Researchers have reported that food banks can be inefficient compared with state run services, and that some people feel ashamed at having to turn to them.

Standard Model

The warehouse of the Capital Area Food Bank

With thousands of food banks operating around the world, there are many different models.

A major distinction between food banks is whether or not they operate on the "front line" model, giving out food directly to the hungry, or whether they operate with the "warehouse" model, supplying food to intermediaries like food pantries, soup kitchens and other front-line organisations. In the US, Australia and to an extent in Canada, the standard model is for food banks to act as warehouses rather than as suppliers to the end user, though there are exceptions. In other countries, food banks usually do hand out food parcels direct to hungry people, providing the service that in the US is offered by *food pantries*.

Another distinction is between the charity model and the labour union model. At least in Canada and the US, food banks run by charities often place relatively more weight on the salvaging of food that would otherwise go to waste, and on encouraging voluntarism, whereas those run by unions can place greater emphasis on feeding the hungry by any means available, on providing work for the unemployed, and on education, especially on explaining to users their civil rights.

In the US, cities will often have a single food bank which acts as a centralized warehouse and will serve several hundred front line agencies. Like a blood bank, that warehouse serves as a single collection and distribution point for food donations. A food bank operates a lot like a for-profit food distributor, but in this case it distributes food to charities, not to food retailers. There is often no charge to the charities, but some food banks do charge a small "shared maintenance" fee to help defray the cost of storage and distribution.

For many US food banks, most of their donated food comes from food left over from the normal processes of for-profit companies. It can come from any part of the food chain, e.g. from growers who have produced too much or whose food is not sufficiently visually appealing; from manufacturers who overproduced; or from retailers who over-ordered. Often the product is approaching or past its "sell by" date. In such cases, the food bank liaises with the food industry and with regulators to make sure the food is safe and legal to distribute and eat.

Volunteers weigh food drive donations.

Other sources of food include the general public, sometimes in the form of "food drives", and government programs that buy and distribute excess farm products mostly to help support higher commodity prices. Food banks can also buy food either at market prices or from wholesalers and retailers at discounted prices, often at cost. Sometimes farmers will allow food banks to send gleaners to salvage leftover crops for free once their primary harvest is complete. A few food banks have even taken over their own farms, though such initiatives have not always been successful.

Many food banks don't accept fresh produce, preferring canned or packaged food due to health and safety concerns, though some have tried to change this as part of a growing worldwide awareness of the importance of nutrition. As an example, in 2012, *London Food Bank* (Canada) started accepting perishable food, reporting that as well as the obvious health benefits, there were noticeable emotional benefits to recipients when they were given fresh food.

Summer can be a challenging time for food banks, especially in regions where school children are usually given regular free meals during term time. Spikes in demand can coincide with periods where donations fall due to folk being on holiday.

North America

History

In the U.S. and often also in Canada, food banks don't typically give food direct to the hungry. Instead they act as warehouses, supplying front line agencies like this Californian soup kitchen. (Pic taken in 2009, and shows members of the United States Navy serving visitors.)

The world's first food bank was the St. Mary's Food Bank Alliance in Arizona, founded by John van

Hengel in 1967. According to sociology professor Janet Poppendieck, hunger within the US was widely considered to be a solved problem until the mid-1960s. By the mid sixties, several states had ended the free distribution of federal food surpluses, instead providing an early form of food stamps which had the benefit of allowing recipients to choose food of their liking, rather than having to accept whatever happened to be in surplus at the time. However, there was a minimum charge and some people could not afford the stamps, leading to severe hunger. One response from American society to the rediscovery of hunger was to step up the support provided by soup kitchens and similar civil society food relief agencies - some of these dated back to the Great Depression and earlier. In 1965, while volunteering for a community dining room, van Hengel learned that grocery stores often had to throw away food that had damaged packaging or was near expiration. He started collecting that food for the dining room but soon had too much for that one program. He thought of creating a central location from which any agency can receive donations. Described as a classic case of "if you build it they will come", the first food bank was created with the help of St. Mary's Basilica. Food banks spread across the United States, and to Canada. By 1976, the precursor to Feeding America had been established. As of the early 21st century, their network of 200+ foodbanks provides support for 90,000 projects. Other large networks exist such as AmpleHarvest.org, created by CNN Hero Gary Oppenheimer which lists some 7,100 food pantries (1 out of every 5 in America) that can utilize overproduction of fresh produce.

It was not however until the 1980s that U.S. food banks began to enjoy rapid growth. A second response to the "rediscovery" of hunger in the mid sixties had been extensive lobbying of politicians to improve welfare. Until the 1980s, this approach had greater impact. In the 1970s, U.S. federal expenditure on hunger relief grew by about 500%, with food stamps distributed free of charge to those in greatest need. According to Poppendieck, welfare was widely considered preferable to grass roots efforts, as the latter could be unreliable, did not give recipients consumer-style choice in the same way as did food stamps, and risked recipients feeling humiliated by having to turn to charity. In the early 1980s, president Reagan's administration scaled back welfare provision, leading to a rapid rise in activity from grass roots hunger relief agencies. According to a comprehensive government survey completed in 2002, over 90% of food banks were established in the US after 1981. Poppendieck says that for the first few years after the change, there was vigorous opposition from left, who argued that state welfare was much more suitable for meeting recipients needs. But in the decades that followed, food banks have become an accepted part of America's response to hunger. Demand for the services of US food bank increased further in the late 1990s, after the "end of welfare as we know it" with President Clinton's Personal Responsibility and Work Opportunity Act. In Canada, foodbanks underwent a period of rapid growth after the cutbacks in welfare that took place in the mid-1990s. As early as the 1980s, food banks had also begun to spread from the United States to the rest of the world. The first European food bank was founded in France during 1984. In the 1990s and early 2000s, food banks were established in South America, Africa and Asia, in several cases with van Hengel acting as a consultant. In 2007, *The Global FoodBanking Network* was formed.

Food Aid for Pets

Some U.S. cities have organizations that provide dog and cat food for pets whose owners qualify for food assistance. For example, *Daffy's Pet Soup Kitchen* in Lawrenceville, Georgia is considered the largest pet food aid agency in Georgia, distributing over 800,000 pounds of dog and cat food in 2012. *Daffy's Pet Soup Kitchen* was started in 1997 by Tom Wargo, a repairman who was working in an elderly woman's home when he noticed her sharing her Meals On Wheels lunch with her pet cat because she couldn't afford cat food. Daffy's was one of seven non-profits recognized by Barefoot Wine in 2013 through a $10,000 donation and by being featured on labels of the vintner's Impression Red Blend wines. *Pet Buddies Food Pantry* in Woodstock, Georgia is another example of an establishment that provides food aid for pets.

Post 2007 Financial Crisis

Following the financial crisis of 2007–08, and the lasting inflation in the price of food that began in late 2006, there has been a further increase in the number of individuals requesting help from American and Canadian food banks. By 2012, according to *Food Banks Canada*, over 850,000 Canadians needed help from a food bank each month. For the United States, *Gleaners Indiana Food bank* reported in 2012 that there were now 50 million Americans struggling with food insecurity (about 1 in 6 of the population), with the number of individuals seeking help from Food banks having increased by 46% since 2005. According to a 2012 UCLA Center for Health Policy Research study, there has been a 40% increase in demand for Californian food banks since 2008, with even married couples who both work sometimes requiring the aid of food banks. Dave Krepcho, director of the *Second Harvest Food Bank* in Orlando has said that even college educated professional couples have begun to turn to food pantries.

By mid-2012, US food banks had expressed concerns on the expected difficulty in feeding the hungry over the coming months. Rapidly rising demand has been coinciding with higher food prices and with a decrease in donations, partly as the food industry is becoming more efficient and so has less mislabelled and other slightly defective food to give away. Also there has been less surplus federal food on offer. Additionally, there have been recent decreases in Federal funding, and Congress have been debating possible further cuts, including potentially billions of dollars from the Supplemental Nutrition Assistance Program (Food stamp programme). In September, Feeding America launched *Hunger Action Month*, with events planned all over the nation. Food banks and other agencies involved hope to raise awareness of the fact that about one in six Americans are struggling with hunger, and to get more Americans involved in helping out.

Europe

The first European food bank was founded in France during 1984. The first Italian food bank was established in 1989. Similar to the UK's experience, foodbanks have become much more common across continental Europe since the crisis that began in 2008, and especially since austerity began to take effect from late 2010.

In Spain, food banks can operate on the warehouse model, supplying a network of surrounding soup kitchens and other food relief agencies. The Spanish federation of food banks (FESBAL) helped to feed about 800,000 people during 2008-2011, according to the Carrefour Foundation. By October 2014, Spain had 55 food banks in total, with the numbers who depend on them having increased to 1.5 million.

In Belgium, food banks helped about 121,000 people during 2012. That was an increase of about 4,500 compared with 2011, the biggest increase since the start of the 2008 crisis. Belgian food banks account for about 65% of all food aid given out within the country.

The number of food banks has increased rapidly even in Germany, a country that has weathered the crisis relatively well, and has not needed to implement severe austerity. In 2012, professor Sabine Pfeiffer of Munich University of Applied Sciences said there has been an "explosion" of food bank usage.

The Most Deprived Persons Programme

While many European food banks are run by civil society with no government assistance, there is a project funded by the EU, the *Most deprived persons programme* (MDP), which specialises in supplying food to marginalised people who are not covered by the benefit system and who may be reluctant to approach the more formal food banks. Food is largely given out by Catholic churches. However, the EU is due to end funding for the MDP in 2013. Similar to the US, the EU no longer expects to need to buy much food to help farmers, as with high global food prices, market surpluses are now much less frequent, So there is less food available to hand out to food banks. In October 2012, the European Commission proposed a new Fund to replace the *Most deprived persons programme.*

United Kingdom

Though Foodbanks were rarely seen in the UK in the second half of the 20th century, their use has started to grow, especially in the 2000s, and have since dramatically expanded. The increase in the dependency on food banks has been blamed on the growing unemployment in the UK as well as the 2008 recession. The OECD found that people answering yes to the question 'Have there been times in the past 12 months when you did not have enough money to buy food that you or your family needed?' decreased from 9.8% in 2007 to 8.1% in 2012, leading some to say that the rise was due to both more awareness of food banks, and the government allowing Jobcentres to refer people to food banks when they were hungry (the previous Labour government had not allowed this).

As of January 2014 there were close to 1,000 UK food banks. The largest group co-ordinating UK foodbanks was The Trussell Trust, a Christian charity based in Salisbury. About 43% of the UK's foodbanks were run by Trussell, about 20% by smaller church networks such as Besom and Basics, about 31% were independent, and about 4% were run by secular food bank networks such as *Fare Share* and *Food Cycle.*

Before the financial crisis, food banks were "almost unheard of" in the UK. In 2004, Trussell only ran two food banks, but by 2007–2008, there were 22 food banks in the Trussell Trust Foodbank Network and by early 2011, The Trussell Trust supported 100. As of May 2012, they had 201. By August, 252. The rate of increase had been rising rapidly. In 2011, about one new food bank was being opened per week. In early 2012, about two were being opened each week. By July, The Trussell Trust had reported that the rate of new openings had increased to three per week. In August, the rate of new openings spiked at four per week, with three new food banks being opened in that month for Nottingham alone. By October 2012, the rate of increase had fallen back to about two or three per week.

Most UK food banks are hosted by churches in partnership with the wider community. They operate on the "frontline" model, giving out food directly to the hungry. Over 90% of the food given out is donated by the public, including schools, churches, businesses and individuals. The Trussell

Trust had aimed to provide short-term support for people whose needs have not yet been addressed by official state welfare provision; those who had been "falling into the cracks in the system". The Trussell franchise has procedures which aim to prevent long-term dependency on their services, and to ensure that those in need are referred to qualified outside agencies. The charity suggests that the *credit crunch* caused an upsurge in the number of people needing emergency food. Since 2010, demand for foodbanks continued to increase, and at a more rapid rate, partly as austerity began to take effect, and partly as those on low incomes began to draw down savings and run out of friends of whom they were willing to ask for help. Unlike *soup kitchens*, most UK food banks are unable to help people who come in off the street without a referral - instead they operate with a referral system. Vouchers are handed out to those in need by various sorts of frontline care professionals, such as social workers, health visitors, Citizens Advice Bureau, Jobcentres and housing officials. The voucher can typically be exchanged at the food bank for a package of food sufficient to last three days. The year to April 2013 saw close to 350,000 referrals to Trussel foodbanks, more than double the amount from the previous year.

A number of food banks have been set up outside of the Trussell system, in part as they don't like having to turn away people without referrals, although Trussell Trust foodbanks do help clients in need without vouchers to get one as quickly as possible. There is also FareShare, a London-based charity which operates some 19 depots on the US-style warehouse model. Rather than giving out food directly to individuals, FareShare distributes food to over 700 smaller agencies, mainly smaller independent operations like soup kitchens and breakfast clubs. Great emphasis is placed on reducing food waste as well as relieving food poverty. Fareshare operates on a business basis, employing a number of managers to oversee operations alongside their army of volunteers. Employee costs constituted over 50% of their expenditure in both 2011 and 2012. Additionally, charities receiving food do incur a charge for deliveries.

Another charity which operates on the US-style warehouse model and with a similar emphasis placed on reducing food waste as well as relieving food poverty is the Oxford Food Bank, which has a single base delivering to around 30 charities in the Oxford area. Although a much smaller enterprise than FareShare, it has a significantly lower pro-rata cost base as it employs no staff, with the whole operation up to director level run entirely by volunteers. This allows it to provide food at no cost to the recipient charities with all operating costs covered by grants and donations, supplying an estimated £25 of food at retail value for each £1 received in donations. In December 2012, it also started distributing food to needy families direct with the aid of local community centres and social services.

People who turn to food banks are typically grateful both for the food and for the warmth and kindness they receive from the volunteers. However, sometimes food banks have run out of supplies by the time they arrive. Some find it humiliating to have to ask for food, and that the packages they receive don't always seem nutritious. Some food banks have tried to respond with innovative programmes; *London Street Foodbank* for example has begun asking donors to send in supermarket vouchers so that those they serve will be able to choose food that best meets their nutritional needs.

Comparison to Other Countries

Food bank use in Germany and France is much higher than in Britain. In 2014, 1.5 million people a week used food banks in Germany and there are twice as many food banks in France as there are in Britain.

UK Food Bank Users

According to a May 2013 report by Oxfam and Church Action on Poverty, about half a million Britons had used food banks. The Trussel Trust reports that their food banks alone helped feed 346,992 people in 2012-13.

Close to half of those needing to use food banks have had issues with their benefits. Sanctioning benefits was the single most frequent reason for food bank referrals and there has been criticism over sanctions being imposed for allegedly spurious reasons.

A joint report from the Trussell Trust, the Church of England, and the charities Oxfam and Child Poverty Action Group found that food bank users were more likely to live in rented accommodation, be single adults or lone parents, be unemployed, and have experienced a "sanction," where their unemployment benefits were cut for at least one month

Delay in paying housing benefit disability benefit and other benefits and general bureaucratic issues with benefits can force people to use food banks. Many further people who need food banks have low-income jobs, but struggle to buy food after making debt repayments and all expenses. Low paid workers, part-time workers and those with zero-hour contracts are particularly vulnerable to financial crisis and sometimes need food banks. As had been predicted, demand for food banks further increased after cuts to welfare came into effect in April 2013, which included the abolishment of Crisis loans. In April 2014 Trussell reported that they had handed out 913,000 food parcels in the last year, up from 347,000 the year before. Several councils have begun looking at funding food banks to increase their capability, as cuts to their budgets mean they will be less able to help vulnerable people directly.

The UK Government

Archbishop of Canterbury, Justin Welby urged the government to provide £150 million which he feels is needed to help prevent hunger in the UK. According to an all-party parliamentary report released in December 2014, key reasons for the increased demand for UK foodbanks are delays in paying benefits, welfare sanctions, and the recent reversal of the post WWII trend for poor people's incomes to rise above or in line with increased costs for housing, utility bills and food. However, the government has responded by saying that in fact the extra usage is likely due to more awareness of food banks as the proportion of benefits paid on time has risen from 88-89% under Labour, to 96-97% in 2014.

In 2013 the British government blocked a £22 million European Union fund to help finance food banks in the UK. This disappointed Labour MEP, Richard Howitt who assisted in negotiating the fund. Howlett stated:

It is very sad that our government is opposing this much-needed help for foodbanks on the basis that it is a national responsibility, when in reality it has no intention of providing the help itself. The only conclusion is that Conservative anti-European ideology is being put before the needs of the most destitute and deprived in our society.

Britain's prime Minister David Cameron said in the House of Commons in 2012 that he welcomed the efforts of food banks. Caroline Spelman, his Secretary of State for Environment, Food and

Rural Affairs, has described food banks as an “excellent example” of active citizenship. Labour MP Kate Green has a different view, feeling that the rise of food banks reflects people being let down by the state welfare system, saying “I feel a real burning anger about them ... People are very distressed at having to ask for food; it’s humiliating and distressing.” Cookery writer and poverty campaigner Jack Monroe wrote that those referred to food banks or given vouchers were “the lucky ones with a good doctor or health visitor who knows us well enough to recognise that something has gone seriously wrong” and expressed concern for those who lack this support.

Germany

There are over 900 food banks in Germany, up from just 1 in 1993. In 2014, 1.5 million people a week used food banks in Germany

Asia

Several Asian countries have begun to use foodbanks; these include South Korea, Japan and Taiwan.

Singapore

In Singapore the Food Bank concept is being pioneered by the Food Bank Singapore.

India

Delhi food bank is an organisation that feeds, empowers and transforms lives in the New Delhi / NCR Region. They hold that their shared capabilities can make the basic aspiration of universal access to food a reality. They attempt to pursue this vision through high quality and standards for processes leveraged by technology to get the right aid to the right people at the right time.

Hong Kong

The first food bank in Hong Kong is Feeding Hong Kong. It was founded in 2009. Food Angel is also a food bank in Hong Kong.

Africa

The Egyptian Food Bank was established in Cairo in 2006, and less than ten years later, food banks run on similar principles spread to other Arab countries in North Africa and the Middle East.

In Sub-Saharan Africa, there are charity-run food banks that operate on a semi-commercial system that differs from both the more common “warehouse” and “frontline” models. In some rural LDCs such as Malawi, food is often relatively cheap and plentiful for the first few months after the harvest, but then becomes more and more expensive. Food banks in those areas can buy large amounts of food shortly after the harvest, and then as food prices start to rise, they sell it back to local people throughout the year at well below market prices. Such food banks will sometimes also act as centres to provide small holders and subsistence farmers with various forms of support.

Formed in 2009, *FoodBank South Africa* (FoodBank SA) is South Africa's national foodbanking network and a member of *The Global FoodBanking Network*. FoodBank SA's vision is "A South Africa without hunger and malnutrition".

Worldwide

Since the 1980s foodbanking has spread around the world. There are over 25 countries with active food bank groups under the umbrella of The Global FoodBanking Network. Countries in the international network include Australia, Israel, Turkey, Russia, India, Taiwan, Colombia, Brazil, Argentina, Chile, Guatemala, El Salvador, Nicaragua, Hong Kong, Singapore, South Korea and the UK. There are also several countries with foodbanks but which have not yet joined the network, either as they don't yet meet the required criteria or as they have not applied.

An alternative facility offering food to the hungry can be found worldwide wherever there are sizable Sikh communities. Long before foodbanks were invented, Langar has been making free vegetarian food available to Sikhs and non-Sikhs alike.

Reactions

Olivier De Schutter, a senior United Nations official, has cautioned Europe against allowing food banks to become a permanent partial replacement for welfare provision, as is the case in the U.S. and Canada.

The rise of food banks has been broadly welcomed. Not only do they provide a solution to the problem of hunger that doesn't require resources from the state, but they can be viewed as evidence of increasing community spirit and of active, caring citizenship. In the UK for example, Patrick Butler, society editor for The Guardian, has written that the rise of foodbanks has been most enthusiastically welcomed by the right, but also by many on the left of the political spectrum, who were often "nervously excited" about them. However, there has been considerable concern expressed by some researchers and politicians. Drawing on the United States's experience after the rapid rise of food banks in the 1980s, American sociology professor Janet Poppendieck warned that the rise of food banks can contribute to a long term erosion of human rights and support for entitlements. Once food banks become well established, it can be politically impossible to return responsibility for meeting the needs of hungry people to the state. Poppendieck says that the logistics of running food banks can be so demanding that they prevent kind hearted people from having time to participate in public policy advocacy; yet she also says if they can be encouraged to lobby politicians for long term changes that would help those on low income, they often have considerable credibility with legislators. As of 2012, senior US food banks workers have expressed a preference to remain politically neutral, which political activists have suggested may relate to their sources of funding.

Rachel Loopstra from University of Toronto has said foodbanks are often inefficient, unreliable and unable to supply nutritional food. She said a survey in Toronto found that only 1 in 5 families suffering from food insecurity would turn to food banks, in part as there is a stigma associated with having to do so. Elizabeth Dowler, Professor of Food & Social Policy at Warwick University, said that most British people prefer the state to take responsibility for helping the hungry. Hannah Lambie-Mumford, from Sheffield University, echoed the view that some users of food banks find having to ask for food humiliating, and also that food banks volunteers should be encouraged to advocate for long term solutions to the underlying causes of poverty and hunger.

Olivier De Schutter, a senior United Nations official charged with ensuring governments honour their obligation to safeguard their citizen's Right to food, has expressed alarm at the rise of food banks. He has reminded the governments of the advanced economies in Europe, Britain and Canada that they have a "duty to protect" their citizens from hunger, and suggested that leaving such an obligation to food banks may be an abuse of human rights.

Delhi Food Banking Network

Delhi Food Banking Network is a food bank system based in Delhi, India. Its goal is feeding and providing nutrition that helps break the cycle of poverty with linkages for healthcare, education, livelihoods for an integrated development objective catering to the vulnerable populations of Delhi and the National Capital Region.

The vulnerable populations being serviced by the Delhi FoodBank include feeding programs that serve school programs, charitable hospitals, orphanages, the destitute, beggars, homeless, faith based organisations and many others.

The Foodbank operates by matching food from donors to those that need this the most.

FoodBanking Technology

The Delhi Food Bank is leveraging mobile technologies through short code services to getfood and givefood.

Technology helps the management solutions for FoodBanking — specifically for process and workflow management — with the ability to adapt for sourcing, procurement, warehousing, delivery, monitoring, evaluation and training.

Through the sourcing process, donors are able to view the products that have been donated and view the inventory of those or despatch information to feeding programs. Historical transaction details are also available for a rapid reporting process. Recipients are able to view the details of donated products to their feeding programs and history of transactions. The warehouse managers are able to efficiently receive, manage the inventory and despatch through the automated processes.

The reporting process at each level of technology workflow enables users to track and generate customised reports based on their requirements. This also helps the monitoring, evaluation and learning by which feeding programs upload information of beneficiaries they have serviced.

India FoodBanking Network

In India the FoodBanking concept is being pioneered by the India FoodBanking Network to augment ongoing Government and Non Profit feeding programs, with the establishment of the Delhi FoodBank. IFBN acquires donated food in the form of grains, pulses, oil, spices etc. sourced on the basis of community food habits and makes these available through a network of community owned agencies. These institutional feeding programs include school feeding programs, shelter homes, Old age homes substance abuse clinics, after-school programs and cater to various needs.

India FoodBanking Network, of the Food Security Foundation India, is a national level project started under the leadership of Sam Pitroda, Advisor to the Prime Minister for Public Information and Infrastructure and Innovation, with the support of organisations like Cargill India, the DLF Foundation, the Reliance Foundation, the Global Alliance for Improved Nutrition and the Global FoodBanking Network, and is implemented and made operational by the Aidmatrix Foundation. IFBN works with a vision to support thousands of existing feeding programs and create a minimum one FoodBank in each district of India by 2020.

The Delhi Food Banking Network is a banking system that feeds, empowers and transforms lives in the New Delhi / NCR Region. We believe that our shared capabilities can make the most basic aspiration of universal access to food a reality. This is through high quality and standards for processes leveraged by technology to get the right aid to the right people at the right time.

Food Storage Container

Warehouse storing many kinds of food

Food storage containers are widespread in use throughout the world and have probably been in use since the first human civilizations.

Consumer pantry, also diverse

Early Civilizations

In early civilizations cereal grains such as maize, wheat, barley etc. were stored in large airy buildings, often raised up from the ground to reduce infestation by pests and vermin. Ancient Egyptian and early Hebrew writings include reference to such buildings and their successors can still be seen in use in less developed countries and regions.

Smaller quantities of food were stored in baskets made from woven grasses or leaves and such designs have remained in use to the present day.

Modern Containers

In the modern developed world, a very wide range of food packaging and containers is now available made from many materials.

Plastic containers

Many products use low-density polyethylene formed into plastic bags or plastic boxes. There are many makers of plastic boxes some such as Tupperware and Lock&Lock are known throughout much of the developed world.

Metal containers

Longer term storage or storage of items needing a higher degree of protection from the elements may use sheet metal. A common form of such storage is the biscuit tin.

Refrigerator

Perhaps the most ubiquitous domestic item of food storage is the fridge or fridge-freezer in which a wide variety of foodstuffs are contained and preserved through the use of low temperatures.

Disposable

Disposable foam food containers may be used for food or beverages.

Industrial Food Storage Containers

Wherever food is harvested, manufactured or distributed there is a need for containers to enable the food to travel securely and in good condition to the shop, warehouse or distribution depot. For many foods, especially those in their own individual containers such as canned vegetables, the common container is the corrugated fiberboard box, Corrugated boxes are available in many sizes.

Foam Food Container

A foam food container is a disposable container for various foods, including beverages. Food types may be processed (instant noodles), raw (meat from supermarkets), or cooked (delicatessen).

The containers are commonly used to serve takeout food from restaurants, and are also available by request for diners who wish to take the remainder of their meal home.

Siu mei with rice in a foam takeout container

Khanom krok, Thai doughnuts with egg and coconut fillings in a foam container.

Construction

Foam take-out containers are made from EPS foam, or another type of polystyrene foam, and produced by injecting the foam into a mold. They are usually white in color, although they may be printed or impressed with a company logo or other message.

Styrofoam is a trademark of The Dow Chemical Company for closed-cell extruded polystyrene foam currently made for thermal insulation and craft applications. The term is often incorrectly used as a generic term for expanded (not extruded) polystyrene foam, such as disposable coffee cups, coolers, or cushioning material in packaging, which are typically white and are made of expanded polystyrene beads.

This is a different material from the extruded polystyrene used for Styrofoam insulation. The polystyrene foam used for craft applications, which can be identified by its roughness and by the fact that it "crunches" when cut, is moderately soluble in many organic solvents, cyanoacrylate, and the propellants and solvents of spray paint, and is not specifically identified as expanded or extruded. Another tradename for expanded polystyrene is thermacol, originated by BASF.

Varieties

Foam takeout containers come in different varieties. These may include:

- A rectangular shaped clamshell style container with an attached lid, which comes in various sizes. The larger ones usually feature several compartments, allowing different foods to be kept separated from each other. This variety usually features several small projections on the lid of the container, which fit through slots on the bottom half to "lock" it, therefore keeping the cover closed. These containers are leak-resistant only if kept upright, and often have a square imprint area for labeling.

- A cylindrical style container with a separate, translucent or opaque plastic lid which can seal tightly to resist leaks. The container may or may not taper somewhat towards the bottom. Both overall size, and the ratio of height to diameter can vary greatly. Such containers usually hold soups and stews; however, smaller varieties are often used to hold sauces and condiments.

Chinese Cuisine

This is the most commonly used takeout box for Chinese cuisines in East and Southeast Asia. It is standard for Cantonese cuisine in Hong Kong and many parts of China. It is sometimes used overseas in various restaurants, particularly in the United States and Canada.

Issues

Environmental Issues

Because foam takeout containers are entirely made out of polystyrene foam, these containers affect the environment as they do not biodegrade easily. However, Methanogenic consortia degrade styrene, using it as a carbon source. *Pseudomonas putida* can also convert styrene oil into various biodegradable polyhydroxyalkanoates. Some cities have gone as far as banning the use of foam take-out containers, notably San Francisco, Seattle and Portland, Oregon. In 2013, the mayor of New York City proposed banning foam food containers, for both health and environmental reasons. This plan began to be implemented in early 2015, however a September court ruling has since put the ban on hold.

Health Issues

It is debated whether styrene may migrate into food which is stored in foam food containers for even a small amount of time. Some researchers argue that polystyrene containers do pose a health risk, while industry defenders argue that trace amounts of styrene are already naturally present in food. Styrene foam containers can melt if the food or liquid is of a sufficient temperature. Some containers have been tested and labelled for safe use in microwave ovens; although the absence of such labeling does not mean a container is unsafe for this use, caution should still be taken.

Styrene is considered by both the EPA and IARC to be a possible carcinogen. It poses a health risk to workers during production of styrene and polystyrene items, with industries using a compliance program to deal with liabilities.Prolonged exposure to high amounts of styrene may affect the central nervous system, causing diseases like Parkinson's disease and multiple sclerosis.

Juicebox (Container)

A juice box, also called a carton or popper, is a small container used to conveniently carry and consume drinks (most often juice). They are frequently made of paperboard with an aluminum foil lining, but variations exist. Juice boxes are most popular with children, although other uses include emergency drinking water and wine.

A juice box

Ruben Rausing first created a product in 1963 that consisted of a box that would be used for containing liquids, more specifically, milk. His creation was named the Tetra Brik, and gained popularity because the product was efficient and a major space saver compared to the canisters that were previously used. The juice box was officially incorporated in the U.S. market in 1980. After its introduction, the product gained almost instant popularity and the market began to grow at a fast rate. According to an article on the website E notes, in 1986, only six years after the product's introduction, juice boxes accounted for 20% of the United States juice market, as more and more companies were introducing their own lines of juice boxes.

Packaging and Marketing

Beetroot juice box packing, Germany

A juice box is made of liquid packaging board which is usually six layers of paper, polyethylene, and aluminum foil. Paper is used to shape the product and give the box an extra source of strength, Polyethylene forms a liquid-tight seal and keeps the product dry, polyethylene is the layer used to print the information and graphics that are provided on the packaging, aluminum is used for the purpose of keeping light and oxygen out (as well preventing the juice from becoming spoiled without having to use extra preservatives). One of the reasons as to why juice boxes are so popular is their convenience. Juice boxes are portable and can be easily drunk while on-the-go. The shape of the product makes them easy for kids and adults to hold and use.

Juice boxes were originally designed in a very specific fashion, a style that has proved to be successful as it has remained, for the most part, unchanged. Manufacturers chose a box shape because they foresaw this shape as being the most convenient and easily handled. Juice boxes typically come with a covered hole and an attached bendable straw, which was introduced by Lynn Tilley, one of the designers who worked on porting it to America, after there were complaints about it being hard to drink from. These bendable straws make it easier for children to drink and generally results in less of a mess, and the bending nature of the straw helps it not fall into the box. However, there are some juice boxes available for purchase that are equipped with a pull tab; this variation of the juice box is ideal for larger portions that are not consumed in one sitting, because the tab is resealable.

At the juice box's inception in the early 1980s, juice boxes were primarily made up of diluted juice drinks containing many more artificial ingredients other than real fruit juice. However, as the product's popularity grew throughout the years, so did a preference with health and consuming natural products. So juice boxes evolved and began containing real fruit juice, as well as healthy add-ins, such as vitamins, to also appeal to a more health-conscious public.

Square Milk Jug

The square milk jug is a variant of the plastic gallon (3.785 litres) container of milk commonly sold in the United States . The design was introduced in the summer of 2008 and is marketed as environmentally friendly because of the shape's advantages for shipping and storage (better cube efficiency).

Square milk jugs are stackable and 9% more can be shipped in the same space. They also eliminate the need for milk crates, which are required for traditional bell-shaped jugs.

Pouring Technique

The technique used to pour milk from a square milk jug is different from that of a traditional milk jug. A Sam's Club representative suggests "tilt it slowly and pour slowly" and a dairy owner in Yerington, Nevada, where some square milk jugs are filled, described the pouring technique as "rock-and-pour instead of a lift-and-tip".

Complaints

Some consumers have complained that it is impossible to pour from this container without spilling, resulting in waste of time and milk, in addition to water and paper towels being used for cleanup. Social media has numerous complaints. Gains for the producer have resulted in a loss for the consumer.

As of fall, 2016, after about 8 years, this milk jug has been replaced with a design that works much more easily.

Design

Environmental benefits and cost savings are the primary stated benefits of square milk jugs, although there are noticeable trade-offs.

Because more milk fits on each truck, shipping costs can be reduced by half; this reduces the number of trucks on the road each year by 11000 (US estimate). Stores require five shipments per week in traditional containers now only require two, resulting in lower fuel consumption. Instead of being packed in reusable crates, the square jugs are shrink wrapped for shipment; however, the hundreds of square feet of single-use polyethylene film "stretch wrap" used on each pallet raises significant new environmental concerns. If one pallet uses approximately 175 linear feet of film, a 40 x 48 inch pallet 6 feet tall will hold 80 cubic feet or 360 gallons of milk (given 4.5 gallons of milk per cubic foot), and require 1,056 square feet of film, equaling just under 3 square feet of film per gallon. Only around 5% of stretch wrap is actually recycled; the remainder contributes to significant environmental fallout.

Manufacturers claim that this rectangular design has cut water use by 67% to 70%, according to one dairy, since there are no crates to be cleaned; the absence of dirty crates also reduces the risk of jugs carrying contamination. There is also the reduced cost with no crates to be stolen: Pennsylvania dairies alone spent US$6 million replacing milk crates in 2005, mostly from theft.

Impact

Critics of square milk jugs have noted increased spillage and difficulty pouring. This is a problem specific to the Gallon size that has a much larger mass when first opened.

There are also cost advantages resulting from lower shipping costs for the retailer and these can be passed on to the consumer. The cost to a consumer of a gallon of milk in a square milk jug is 10¢ to 20¢ cheaper than a gallon of milk in a traditional milk jug.

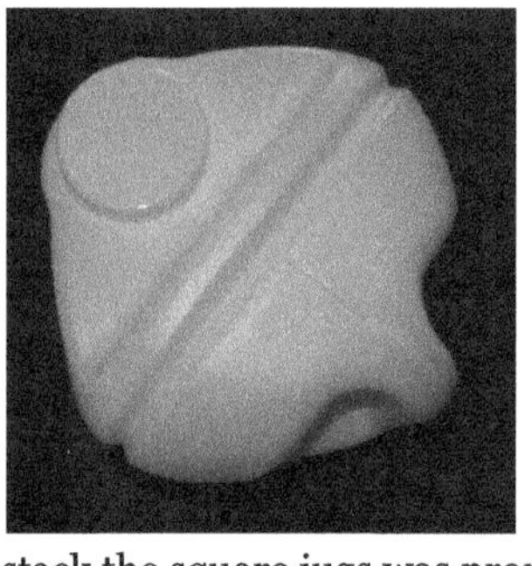

The top of a square milk jug. The ability to stack the square jugs was promoted as a cost saving improvement to the design.

Because square milk jugs can be stacked, they require less labor at the dairy, and can be packaged more quickly, which allows the milk to be shipped sooner, providing fresher milk to the retailer. The wide mouth is also easier to fill which reduces the likelihood of spillage and the sanitary issues caused by milk getting on the container.

The shape of the jug also allows more milk to be stored in a milk cooler. One Sam's Club store reported fitting 224 square gallon jugs in an area that only held 80 traditional one gallon jugs.

Reception

Square milk jugs have been adopted by some Wal-Mart, Costco and Sam's Club stores.

Consumers have criticized the square milk jug for being difficult to pour, especially for children.

When tilted shallowly, the larger opening (larger than a traditional milk jug), combined with its small lip, generates a wide stream of milk. This causes spills and leaks onto the sides of the container.

Some Sam's Club stores offered lessons on pouring, complete with complimentary cookies. Concern was also raised on the issue of the narrower base of the square jug, which allows it to be stored in the door of many refrigerators. Milk stored in this area may be more prone to spoiling due to frequent drastic changes in temperature.

Milk Bottle

Examples of milk bottles from the late 19th century made by the Warren Glass Works Company.

Milk bottles are bottles used for milk. They may be reusable glass bottles used mainly for doorstep delivery of fresh milk by milkmen. Customers are expected to rinse the empty bottles and leave on the doorstep for collection. The standard size of a bottle varies with location; common sizes are 1 pint or 1 quart, although cream may be delivered in smaller bottles.

More recently, plastic bottles have been commonly used for milk. These are often made of high-density polyethylene (HDPE), are used only once, and are easily recycleable.

History

A modern British milk bottle owned by Dairy Crest.

Before milk bottles, milkmen filled the customers' jugs. For many collectors, milk bottles carry a nostalgic quality of a bygone age. The most prized milk bottles are embossed or pyroglazed (painted) with names of dairies on them, which were used for home delivery of milk so that the milk bottles could find their way back to their respective dairies.

Seven modern Dairy Crest milk bottles.

It is not clear when the first milk bottles came into use. However, the New York Dairy Company is credited with having the first factory that produced milk bottles, and one of the first patents for a milk container was held by the Lester Milk Jar. There are many other similar milk containers from around this period, including the Mackworh Pure Jersey Cream crockery type jar, the Manorfield Stock Farm, the Manor, the Pa glass wide mouth jar, and the Tuthill's Dairy Unionville, NY.

Lewis P. Whiteman holds the first patent for a glass milk bottle with a small glass lid and a tin clip. The next earliest patent is for a milk bottle with a dome-type tin cap and was granted September 23, 1884 to Whitemen's brother, Abram V. Whiteman. This bottle has been found with cream line marks and is very valuable. The Whiteman brothers produced milk bottles based on these specifications at the Warren Glass Works Company in Cumberland, Maryland and sold them through their New York sales office.

The Original Thatcher is one of the most desirable milk bottles for collectors. The patent for the glass dome lid is dated April 27, 1886. There are several variations of this early milk bottle and many reproductions. During this time period, many types of bottles were being used to hold and distribute milk. These include a pop bottle type with a wire clamp, used by the Chicago Sterilized Milk Company, Sweet Clover, and others. Fruit jars were also used, but only the Cohansey Glass Manufacturing plant made them with dairy names embossed on them.

The Commonsense Milk Bottle with the first cap seat was developed as an economical means for sealing a reusable milk bottle by the Thatcher Manufacturing Company around 1900. Most bottles produced after this time have a cap seat.

By the 1920s, glass milk bottles had become the norm in the UK after slowly being introduced from the US before World War I.

Milk bottles before the 1930s were round. In 1935, slender-neck bottles were introduced in the UK. In the 1940s, a square squat bottle became the more popular style. Milk bottles since the 1930s have used pyroglaze or ACL (Applied Color Label) to identify the bottles. Before the 1930s, names were embossed on milk bottles using a slug plate. The name was impressed on the slug plate, then the plate was inserted into the mold used to make the bottle – the result was the embossed name on the bottle. In 1980 a new bottle, nicknamed "dumpy," was introduced in the UK where it remains the standard now.

From the 1960s onward in the United States, with improvements in shipping and storage materials, glass bottles have almost completely been replaced with either LDPE coated paper cartons or recyclable HDPE plastic containers (such as square milk jugs), depending on the brand. These paper and plastic containers are lighter, cheaper and safer to both manufacture and ship to consumers.

In 1975, 94% of milk in the UK was in glass bottles, but as of 2012 this number was down to 4%.

However, there are growing concerns among some Americans as to the quality and safety of industrialized milk, and the local, non-homogenized milk industry has seen a popular resurgence in certain markets in the US in the last decade or so. Because of this, the use of glass bottles in local or regional, non-industrial milk distribution has become an increasingly common sight.

Chronology

- 1878 – George Lester patents a glass milk container in New York.
- 1880 – British milk bottles were first produced by the Express Dairy Company. They were delivered by horse-drawn carts and delivered four times a day. The first bottles used a porcelain stopper top held on by wire. Lewis Whiteman patents the glass milk bottle with a glass lid.
- 1894 – Anthony Hailwood developed a pasteurisation process for milk which allowed it to be sterilized and be safely stored for longer periods. Milk could now be delivered once a day.
- 1920 – Advertisements began to appear on milk bottles. A sand-blasting technique was used to etch them on the glass.
- 1930s – Increased prevalence of battery electric vehicles as milk floats
- mid 1950s – Cardboard tops were deemed unhygienic and banned in some locations. Delivery by horse-drawn carts was still common.
- early 1990s – The advertising largely disappeared with the introduction of infrared bottle scanners designed to check cleanliness.

Present Day

In some locations around the world, different colored tops on milk bottles indicate the fat content. Unpasteurized is often green-topped. However other colour designations may be used by some dairies. Bottles may also be marked, stamped or embossed with the name of the dairy.

In the United Kingdom, the aluminium foil tops on glass milk bottles are normally coloured:

Modern dairies may also use refillable plastic bottles, as well as plastic bottle tops. The colour coding for plastic milk bottle tops can be different from that of glass bottles. In the United Kingdom, the plastic tops on plastic milk bottles are normally coloured:

Colour	Meaning
Gold	Channel Island milk
Blue	Homogenized whole milk
Green	Semi-skimmed milk
Red	Skimmed milk
Orange	1% Fat milk

In the United Kingdom, milk sold to the door is mainly measured in Imperial pints (but labelled 568ml), because the glass bottles are 'returnable', which means they were excluded from metrication. However, Sterilized milk sold to the door is in 500ml glass bottles (and is 'non-returnable' and has colour coded lids that match the colour codes normally seen on plastic bottles). Often, in supermarkets they are sold in pints but labelled with their metric equivalent (568ml). Quantities larger than a pint are generally sold in metric units or multiples of a pint.

With lesser milk consumption, milk in Hong Kong is sold in both glass and plastic bottles as well as in cartons. The glass milk bottles are sold in supermarkets, convenience stores and in small restaurants. The glass milk bottle carries a deposit of 1 Hong Kong dollar. Most commonly available in the market in glass milk bottles are Kowloon Dairy and Trappist Dairy with different shaped glass milk bottles with both in school bottle size.

Orange juice and other fruit juices are also sold in doorstep deliveries in the same style of bottle used for milk. Typically these have an aluminium foil top colour coded to indicate the flavour.

In some countries (e.g. Estonia and some provinces of Canada), it is common to buy milk in a milk bag.

While the proportion of sales in milk bags in the United Kingdom is quite low, semi-skimmed milk is sold in bags by Dairy Crest/Milk and More, and in Sainsbury's supermarket it is available in Whole milk, Semi-skimmed milk and Skimmed milk options. However supermarket availability is limited to larger branches.

Refrigerator

Food in a refrigerator with its door open

A side-by-side refrigerator-freezer with an icemaker

A refrigerator (colloquially fridge) is a popular household appliance that consists of a thermally insulated compartment and a heat pump (mechanical, electronic or chemical) that transfers heat from the inside of the fridge to its external environment so that the inside of the fridge is cooled to a temperature below the ambient temperature of the room. Refrigeration is an essential food storage technique in developed countries. The lower temperature lowers the reproduction rate of bacteria, so the refrigerator reduces the rate of spoilage. A refrigerator maintains a temperature a few degrees above the freezing point of water. Optimum temperature range for perishable food storage is 3 to 5 °C (37 to 41 °F). A similar device that maintains a temperature below the freezing point of water is called a freezer. The refrigerator replaced the icebox, which had been a common household appliance for almost a century and a half. For this reason, a refrigerator is sometimes referred to as an icebox in American usage.

The first cooling systems for food involved using ice. Artificial refrigeration began in the mid-1750s, and developed in the early 1800s. In 1834, the first working vapor-compression refrigeration system was built. The first commercial ice-making machine was invented in 1854. In 1913, refrigerators for home use were invented. In 1923 Frigidaire introduced the first self-contained unit. The introduction of Freon in the 1920s expanded the refrigerator market during the 1930s. Home freezers as separate compartments (larger than necessary just for ice cubes) were introduced in 1940. Frozen foods, previously a luxury item, became commonplace.

Freezer units are used in households and in industry and commerce. Commercial refrigerator and freezer units were in use for almost 40 years prior to the common home models. Most households use the freezer-on-top-and-refrigerator-on-bottom style, which has been the basic style since the 1940s. A vapor compression cycle is used in most household refrigerators, refrigerator–freezers and freezers. Newer refrigerators may include automatic defrosting, chilled water and ice from a dispenser in the door.

Domestic refrigerators and freezers for food storage are made in a range of sizes. Among the smallest is a 4 L Peltier refrigerator advertised as being able to hold 6 cans of beer. A large domestic refrigerator stands as tall as a person and may be about 1 m wide with a capacity of 600 L. Refrigerators and freezers may be free-standing, or built into a kitchen. The refrigerator allows the modern family to keep food fresh for longer than before. Freezers allow people to buy food in bulk and eat it at leisure, and bulk purchases save money.

History

Refrigeration Technology

Before the invention of the refrigerator, icehouses were used to provide cool storage for most of the year. Placed near freshwater lakes or packed with snow and ice during the winter, they were once very common. Natural means are still used to cool foods today. On mountainsides, runoff from melting snow is a convenient way to cool drinks, and during the winter one can keep milk fresh much longer just by keeping it outdoors. The word "refrigeratory" was used at least as early as the 17th century

The history of artificial refrigeration began when Scottish professor William Cullen designed a small refrigerating machine in 1755. Cullen used a pump to create a partial vacuum over a container of diethyl ether, which then boiled, absorbing heat from the surrounding air. The experiment even created a small amount of ice, but had no practical application at that time.

Fig: 1.

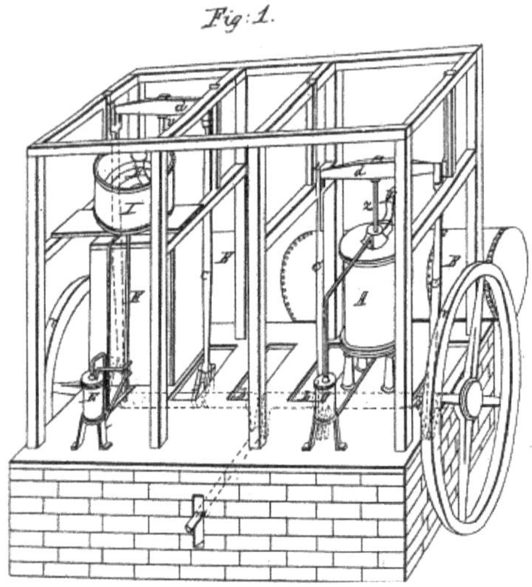

Schematic of Dr. John Gorrie's 1841 mechanical ice machine.

In 1805, American inventor Oliver Evans described a closed vapor-compression refrigeration cycle for the production of ice by ether under vacuum. In 1820, the British scientist Michael Faraday liquefied ammonia and other gases by using high pressures and low temperatures, and in 1834, an American expatriate to Great Britain, Jacob Perkins, built the first working vapor-compression refrigeration system. It was a closed-cycle device that could operate continuously. A similar attempt was made in 1842, by American physician, John Gorrie, who built a working prototype, but it was a commercial failure. American engineer Alexander Twining took out a British patent in 1850 for a vapor compression system that used ether.

The first practical vapor compression refrigeration system was built by James Harrison, a British journalist who had emigrated to Australia. His 1856 patent was for a vapor compression system using ether, alcohol or ammonia. He built a mechanical ice-making machine in 1851 on the banks of the Barwon River at Rocky Point in Geelong, Victoria, and his first commercial ice-making machine followed in 1854. Harrison also introduced commercial vapor-compression refrigeration to breweries and meat packing houses, and by 1861, a dozen of his systems were in operation.

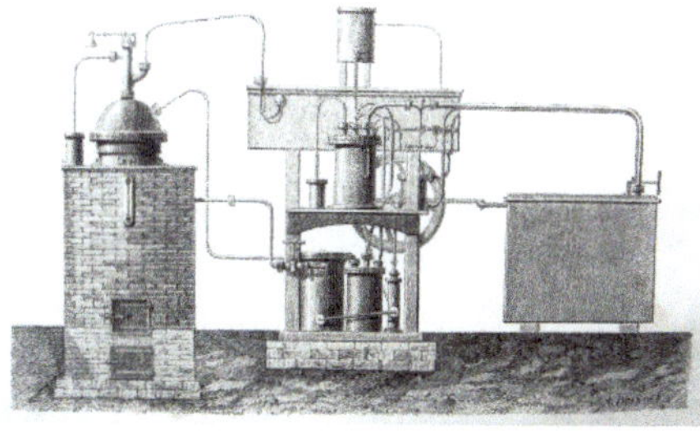

Ferdinand Carré's ice-making device

The first gas absorption refrigeration system using gaseous ammonia dissolved in water (referred to as "aqua ammonia") was developed by Ferdinand Carré of France in 1859 and patented in 1860. Carl von Linde, an engineering professor at the Technological University Munich in Germany, patented an improved method of liquefying gases in 1876. His new process made possible the use of gases such as ammonia, sulfur dioxide (SO_2) and methyl chloride (CH_3Cl) as refrigerants and they were widely used for that purpose until the late 1920s.

Domestic Refrigerator

McCray pre-electric home refrigerator ad (1905) This company, founded in 1887, is still in business.

In 1913, refrigerators for home and domestic use were invented by Fred W. Wolf of Fort Wayne, Indiana with models consisting of a unit that was mounted on top of an ice box. In 1914, engineer Nathaniel B. Wales of Detroit, Michigan, introduced an idea for a practical electric refrigeration unit, which later became the basis for the Kelvinator. A self-contained refrigerator, with a compressor on the bottom of the cabinet was invented by Alfred Mellowes in 1916. Mellowes produced this refrigerator commercially but was bought out by William C. Durant in 1918, who started the Frigidaire Company to mass-produce refrigerators. In 1918, Kelvinator Company introduced the first refrigerator with any type of automatic control. The absorption refrigerator was invented by Baltzar von Platen and Carl Munters from Sweden in 1922, while they were still students at the Royal Institute of Technology in Stockholm. It became a worldwide success and was commercialized by Electrolux. Other pioneers included Charles Tellier, David Boyle, and Raoul Pictet. Carl von Linde was the first to patent and make a practical and compact refrigerator.

These home units usually required the installation of the mechanical parts, motor and compressor, in the basement or an adjacent room while the cold box was located in the kitchen. There was a 1922 model that consisted of a wooden cold box, water-cooled compressor, an ice cube tray and a 9-cubic-foot (0.25 m^3) compartment, and cost $714. (A 1922 Model-T Ford cost about $450.) By 1923, Kelvinator held 80 percent of the market for electric refrigerators. Also in 1923 Frigidaire introduced the first self-contained unit. About this same time porcelain-covered metal cabinets began to appear. Ice cube trays were introduced more and more during the 1920s; up to this time freezing was not an auxiliary function of the modern refrigerator.

General Electric "Monitor-Top" refrigerator, introduced in 1927.

The first refrigerator to see widespread use was the General Electric "Monitor-Top" refrigerator introduced in 1927, so-called because of its resemblance to the gun turret on the ironclad warship USS *Monitor* of the 1860s. The compressor assembly, which emitted a great deal of heat, was placed above the cabinet, and enclosed by a decorative ring. Over a million units were produced. As the refrigerating medium, these refrigerators used either sulfur dioxide, which is corrosive to the eyes and may cause loss of vision, painful skin burns and lesions, or methyl formate, which is highly flammable, harmful to the eyes, and toxic if inhaled or ingested. Many of these units are still functional today, after requiring little more service than a replacement start relay or thermostat if at all. These cooling systems cannot legally be recharged with the hazardous original refrigerants if they leak or break down.

The introduction of Freon in the 1920s expanded the refrigerator market during the 1930s and provided a safer, low-toxicity alternative to previously used refrigerants. Separate freezers became common during the 1940s; the popular term at the time for the unit was a *deep freeze*. These devices, or *appliances*, did not go into mass production for use in the home until after World War II. The 1950s and 1960s saw technical advances like automatic defrosting and automatic ice making. More efficient refrigerators were developed in the 1970s and 1980s, even though environmental issues led to the banning of very effective (Freon) refrigerants. Early refrigerator models (from 1916) had a cold compartment for ice cube trays. From the late 1920s fresh vegetables were successfully processed through freezing by the Postum Company (the forerunner of General Foods), which had acquired the technology when it bought the rights to Clarence Birdseye's successful fresh freezing methods.

The first successful application of frozen foods occurred when General Foods heiress Marjorie Merriweather Post (then wife of Joseph E. Davies, United States Ambassador to the Soviet Union) deployed commercial-grade freezers in Spaso House, the US Embassy in Moscow, in advance of the Davies' arrival. Post, fearful of the USSR's food processing safety standards, fully stocked the freezers with products from General Foods' Birdseye unit. The frozen food stores allowed the Davies to entertain lavishly and serve fresh frozen foods that would otherwise be out of season. Upon returning from Moscow, Post (who resumed her maiden name after divorcing Davies) directed General Foods to market frozen product to upscale restaurants.

Home freezers as separate compartments (larger than necessary just for ice cubes), or as separate units, were introduced in the United States in 1940. Frozen foods, previously a luxury item, became commonplace.

Freezer

Freezer units are used in households and in industry and commerce. Food stored at or below −18 °C (0 °F) is safe indefinitely. Most household freezers maintain temperatures from −23 to −18 °C (−9 to 0 °F), although some freezer-only units can achieve −34 °C (−29 °F) and lower. Refrigerators generally do not achieve lower than −23 °C (−9 °F), since the same coolant loop serves both compartments: Lowering the freezer compartment temperature excessively causes difficulties in maintaining above-freezing temperature in the refrigerator compartment. Domestic freezers can be included as a separate compartment in a refrigerator, or can be a separate appliance. Domestic freezers are generally upright units resembling refrigerators or chests (upright units laid on their backs). Many modern upright freezers come with an ice dispenser built into their door. Some upscale models include thermostat displays and controls, and sometimes flatscreen televisions as well.

Commercial and Domestic Refrigerators

Commercial refrigerator and freezer units, which go by many other names, were in use for almost 40 years prior to the common home models. They used gas systems such as ammonia (R-717) or sulfur dioxide (R-764), which occasionally leaked, making them unsafe for home use. Practical household refrigerators were introduced in 1915 and gained wider acceptance in the United States in the 1930s as prices fell and non-toxic, non-flammable synthetic refrigerants such as Freon-12 (R-12) were introduced. However, R-12 damaged the ozone layer, causing governments to issue a ban on its use in new refrigerators and air-conditioning systems in 1994. The less harmful replacement for R-12, R-134a (tetrafluoroethane), has been in common use since 1990, but R-12 is still found in many old systems today.

A common commercial refrigerator is the glass fronted beverage cooler. These type of appliances are typically designed for specific re-load conditions meaning that they generally have a larger cooling system. This ensures that they are able to cope with a large throughput of drinks and frequent door opening. As a result, it is common for these types of commercial refrigerators to have energy consumption of >4 kWh/day.

Styles of Refrigerators

In the early 1950s most refrigerators were white, but from the mid-1950s through present day designers and manufacturers put color onto refrigerators. In the late-1950s/early-1960s, pastel colors like turquoise and pink became popular, brushed chrome-plating (similar to stainless finish) was available on some models from different brands. In the late 1960s and throughout the 1970s, earth tone colors were popular, including Harvest Gold, Avocado Green and almond. In the 1980s, black became fashionable. In the late 1990s stainless steel came into vogue, and in 2009, one manufacturer introduced multi-color designs.

Frigidaire Imperial "Frost Proof" model FPI-16BC-63, top refrigerator/bottom freezer with brushed chrome door finish made by General Motors Canada in 1963

Production by Country

Country	Production	Year
China	29,871,000	(2005)
United States	11,639,000	(2003)
Italy	7,201,000	(2004)
South Korea	7,122,000	(2004)
Turkey	4,867,000	(2003)
India	3,715,000	(2003)
Brazil	3,544,000	(2003)
Japan	2,821,000	(2005)
Mexico	2,291,000	(2004)
Thailand	2,246,000	(1996)
Germany	2,061,000	(2004)
Hungary	1,625,000	(2004)
Poland	1,618,000	(2005)
Spain	1,269,000	(1995)
Romania	1,169,000	(2005)
Belarus	995,000	(2005)
Slovenia	863,000	(1995)
Egypt	808,000	(2003)
United Kingdom	745,000	(2003)

Country	Production	Year
South Africa	711,000	(2003)
Sweden	639,000	(2004)
Ukraine	562,000	(1995)
France	544,000	(2003)
Australia	423,000	(1995)
Portugal	399,000	(2004)
Bulgaria	353,000	(2005)
Slovakia	330,000	(1995)
Indonesia	291,000	(1995)
Malaysia	187,000	(2003)
Algeria	150,000	(2003)
Lithuania	107,000	(2004)
Finland	104,000	(1995)
Argentina	49,000	(1995)
Republic of Moldova	24,300	(1995)
Uzbekistan	18,600	(1995)
Azerbaijan	13,400	(2005)
Kazakhstan	10,900	(1995)
Tajikistan	50	(1995)

General Technical Explanation

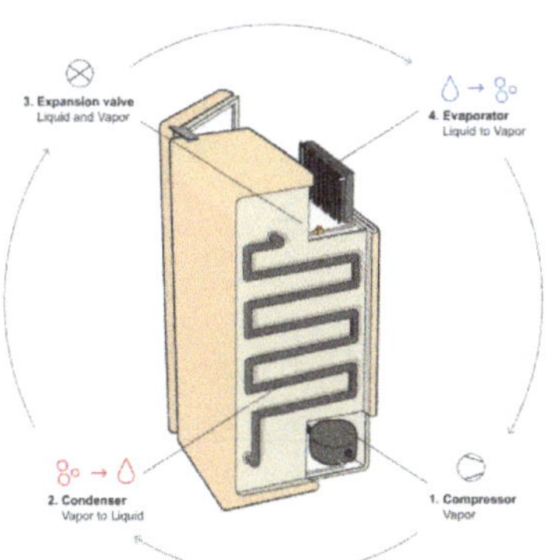

Basic functioning of a refrigerator

A vapor compression cycle is used in most household refrigerators, refrigerator–freezers and freezers. In this cycle, a circulating refrigerant such as R134a enters a compressor as low-pressure vapor at or slightly below the temperature of the refrigerator interior. The vapor is compressed and exits the compressor as high-pressure superheated vapor. The superheated vapor travels under pressure through coils or tubes that make up the *condenser*; the coils or tubes are passively cooled by exposure to air in the room. The condenser cools the vapor, which liquefies. As the refrigerant leaves the condenser, it is still under pressure but is now only slightly above room temperature. This liquid refrigerant is forced through a metering or throttling device, also known as an expansion valve (essentially a pin-hole sized constriction in the tubing) to an area of much lower pressure. The sudden decrease in pressure results in explosive-like flash evaporation of a portion (typically about half) of the liquid. The latent heat absorbed by this flash evaporation is drawn mostly from adjacent still-liquid refrigerant, a phenomenon known as *auto-refrigeration*. This cold and partially vaporized refrigerant continues through the coils or tubes of the evaporator unit. A fan blows air from the refrigerator or freezer compartment ("box air") across these coils or tubes and the refrigerant completely vaporizes, drawing further latent heat from the box air. This cooled air is returned to the refrigerator or freezer compartment, and so keeps the box air cold. Note that the cool air in the refrigerator or freezer is still warmer than the refrigerant in the evaporator. Refrigerant leaves the evaporator, now fully vaporized and slightly heated, and returns to the compressor inlet to continue the cycle.

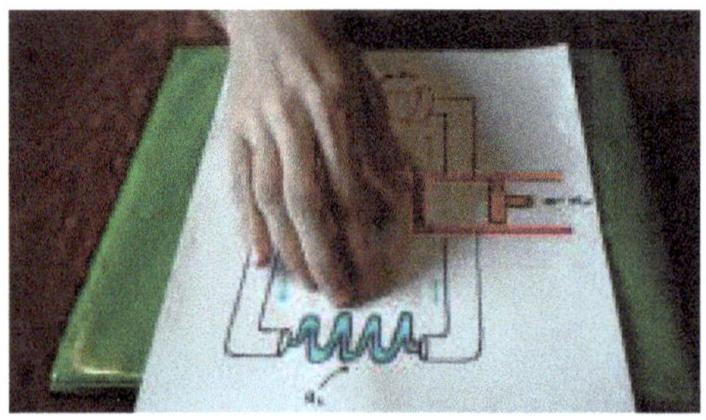

Process and Components of a Conventional Refrigerator

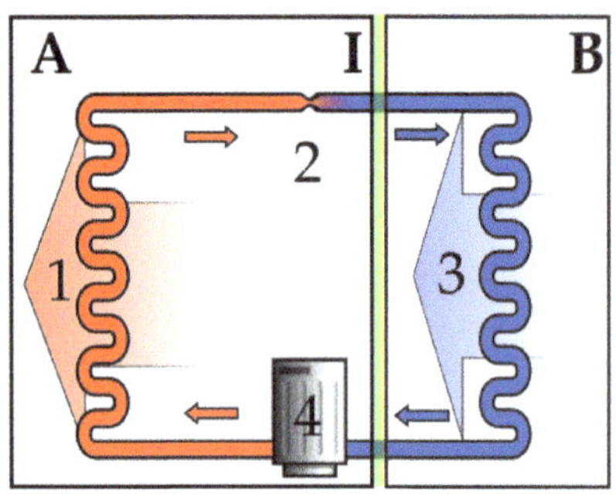

Vapor Compression Cycle – A: hot compartment (kitchen), B: cold compartment (refrigerator box), I: insulation, 1: Condenser, 2: Expansion valve, 3: Evaporator unit, 4: Compressor

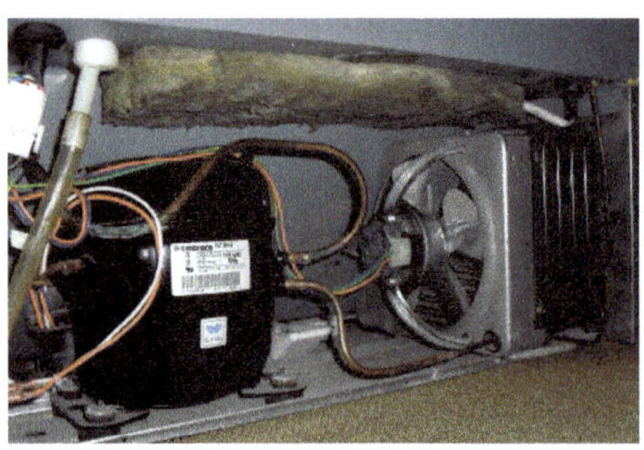

An Embraco compressor and fan-assisted condenser coil

Domestic refrigerators are extremely reliable because the moving parts and fluids are sealed from the atmosphere for life, with no possibility of leakage or contamination. In comparison, mechanically-driven refrigeration compressors, such as those in automobile air conditioning, inevitably leak fluid and lubricant past the shaft seals. This leads to a requirement for periodic recharging and, if ignored, possible compressor failure.

An absorption refrigerator works differently from a compressor refrigerator, using a source of heat, such as combustion of liquefied petroleum gas, solar thermal energy or an electric heating element. These heat sources are much quieter than the compressor motor in a typical refrigerator. A fan or pump might be the only mechanical moving parts; reliance on convection is considered impractical.

The Peltier effect uses electricity to pump heat directly; refrigerators employing this system are sometimes used for camping, or in situations where noise is not acceptable. They can be totally silent (if a fan for air circulation is not fitted) but are less energy-efficient than other methods.

Other uses of an absorption refrigerator (or "chiller") include large systems used in office buildings or complexes such as hospitals and universities. These large systems are used to chill a brine solution that is circulated through the building.

Many modern refrigerator/freezers have the freezer on top and the refrigerator on the bottom. Most refrigerator-freezers—except for manual defrost models or cheaper units–use what appears to be two thermostats. Only the refrigerator compartment is properly temperature controlled. When the refrigerator gets too warm, the thermostat starts the cooling process and a fan circulates the air around the freezer. During this time, the refrigerator also gets colder. The freezer control knob only controls the amount of air that flows into the refrigerator via a damper system. Changing the refrigerator temperature will inadvertently change the freezer temperature in the opposite direction. Changing the freezer temperature will have no effect on the refrigerator temperature. The freezer control may also be adjusted to compensate for any refrigerator adjustment.

This means the refrigerator may become too warm. However, because only enough air is diverted to the refrigerator compartment, the freezer usually re-acquires the set temperature quickly, unless the door is opened. When a door is opened, either in the refrigerator or the freezer, the fan in some units stops immediately to prevent excessive frost build up on the freezer's evaporator coil, because this coil is cooling two areas. When the freezer reaches temperature, the unit cycles off, no matter what the refrigerator temperature is. Modern computerized refrigerators do not use the damper system. The computer manages fan speed for both compartments, although air is still blown from the freezer.

A few manufacturers offer dual compressor models. These models have separate freezer and refrigerator compartments that operate independently of each other, sometimes mounted within a single cabinet. Each has its own separate compressor, condenser and evaporator coils, insulation, thermostat, and door. Typically, the compressors and condenser coils are mounted at the top of the cabinet, with a single fan to cool them both.

This design, where no air passes between the two compartments, provides for more appropriate humidity levels and much tighter temperature control in each compartment. It also requires much less energy to operate, since each compressor & coolant system can be optimized for a specific temperature range. Further, opening the door of one compartment does not affect the temperature of the air or humidity level in the other compartment. Thus, it avoids many of the disadvantages of the much more common single compressor designs described above, although at a higher initial cost and increased system noise. Manufacturers of such designs argue that the increased cost is compensated over time due to reduced energy use and less food waste due to reduced spoilage.

Alternatives to the vapor-compression cycle not in current use include:

- Acoustic cooling
- Air cycle
- Magnetic cooling
- Malone engine
- Pulse tube
- Stirling cycle
- Thermoelectric cooling and thermionic cooling
- Vortex tube
- Water cycle systems.

Features

The inside of a home refrigerator containing a large variety of everyday food items.

Newer refrigerators may include:

- Automatic defrosting
- A power failure warning that alerts the user by flashing a temperature display. It may display the maximum temperature reached during the power failure, and whether frozen food has defrosted or may contain harmful bacteria.
- Chilled water and ice from a dispenser in the door. Water and ice dispensing became available in the 1970s. In some refrigerators, the process of making ice is built-in so the user doesn't have to manually use ice trays. Some refrigerators have water chillers and water filtration systems.
- Cabinet rollers that lets the refrigerator roll out for easier cleaning
- Adjustable shelves and trays
- A status indicator that notifies when it is time to change the water filter

- An in-door ice caddy, which relocates the ice-maker storage to the freezer door and saves approximately 60 litres (2 cu ft) of usable freezer space. It is also removable, and helps to prevent ice-maker clogging.
- A cooling zone in the refrigerator door shelves. Air from the freezer section is diverted to the refrigerator door, to cool milk or juice stored in the door shelf.
- A drop down door built into the refrigerator main door, giving easy access to frequently used items such as milk, thus saving energy by not having to open the main door.
- A Fast Freeze function to rapidly cool foods by running the compressor for a predetermined amount of time and thus temporarily lowering the freezer temperature below normal operating levels. It is recommended to use this feature several hours before adding more than 1 kg of unfrozen food to the freezer. For freezers without this feature, lowering the temperature setting to the coldest will have the same effect.

Early freezer units accumulated ice crystals around the freezing units. This was a result of humidity introduced into the units when the doors to the freezer were opened condensing on the cold parts, then freezing. This frost buildup required periodic thawing ("defrosting") of the units to maintain their efficiency. Manual Defrost (referred to as Cyclic) units are still available. Advances in automatic defrosting eliminating the thawing task were introduced in the 1950s, but are not universal, due to energy performance and cost. These units used a counter that only defrosted the freezer compartment (Freezer Chest) when a specific number of door openings had been made. The units were just a small timer combined with an electrical heater wire that heated the freezer's walls for a short amount of time to remove all traces of frost/frosting. Also, early units featured freezer compartments located within the larger refrigerator, and accessed by opening the refrigerator door, and then the smaller internal freezer door; units featuring an entirely separate freezer compartment were introduced in the early 1960s, becoming the industry standard by the middle of that decade. These older freezer compartments were the main cooling body of the refrigerator, and only maintained a temperature of around −6 °C (21 °F), which is suitable for keeping food for a week.

In the early 1950s, the butter conditioner's patent was filed and published by the inventor Nave Alfred E. This feature was supposed to "provide a new and improved food storage receptacle for storing butter or the like which may quickly and easily be removed from the refrigerator cabinet for the purpose of cleaning." Because of the high interest to the invention, companies in UK, New Zealand, and Australia started to include the feature into the mass fridge production and soon it became a symbol of the local culture. However, not long after that it was removed from production as according to the companies this was the only way for them to meet new ecology regulations and they found it inefficient to have a heat generating device inside a fridge.

Later advances included automatic ice units and self compartmentalized freezing units.

An increasingly important environmental concern is the disposal of old refrigerators— initially because freon coolant damages the ozone layer—but as older generation refrigerators wear out, the destruction of CFC-bearing insulation also causes concern. Modern refrigerators usually use a refrigerant called HFC-134a (1,1,1,2-Tetrafluoroethane), which does not deplete the ozone layer, instead of Freon. A R-134a is now becoming very uncommon in Europe. Newer refrigerants are being used instead. The main refrigerant now used is R-600a, or isobutane. This refrigerant is naturally occurring, and has a smaller effect on the atmosphere if released. There have been reports of refrigerators exploding if the refrigerant leaks gas in the presence of a spark.

Disposal of discarded refrigerators is regulated, often mandating the removal of doors; children playing hide-and-seek have been asphyxiated while hiding inside discarded refrigerators, particularly older models with latching doors. Since 2 August 1956, under U.S. federal law, refrigerator doors are no longer permitted to latch so they cannot be opened from the inside. Modern units use a magnetic door gasket that holds the door sealed but allows it to be pushed open from the inside. This gasket was invented, developed and manufactured by Max Baermann (1903-1984) of Bergisch Gladbach/Germany.

Types of Domestic Refrigerators

Domestic refrigerators and freezers for food storage are made in a range of sizes. Among the smallest is a 4 L Peltier refrigerator advertised as being able to hold 6 cans of beer. A large domestic refrigerator stands as tall as a person and may be about 1 m wide with a capacity of 600 L. Some models for small households fit under kitchen work surfaces, usually about 86 cm high. Refrigerators may be combined with freezers, either stacked with refrigerator or freezer above, below, or side by side. A refrigerator without a frozen food storage compartment may have a small section just to make ice cubes. Freezers may have drawers to store food in, or they may have no divisions (chest freezers).

Refrigerators and freezers may be free-standing, or built into a kitchen.

- Compressor refrigerators are by far the most common type; they make a noticeable noise.
- Absorption refrigerators or thermo-electric Peltier units are used where quiet running is required; Peltier coolers are used in the smallest refrigerators as they have no bulky mechanism.
- Compressor and Peltier refrigerators are powered by electricity. Absorption units can be designed to get power from any heat source. A noticeable difference between the two types is the absence of refrigerant with Peltier coolers (these use a different method of cooling). But Peltier coolers use more electricity because they are thermodynamically inefficient.
- Oil, gas (natural gas or propane) and dual power gas/electricity units are also available (typically found in RV's).
- Solar refrigerators and Thermal mass refrigerators are designed to reduce electrical consumption. Solar refrigerators have the added advantage that they do not use refrigerants that are harmful to the environment or flammable. Typical solar designs are absorption refrigerators that use ammonia as the working gas, and employ large mirrors to concentrate sufficient sunlight to reach the temperature required to free gaseous ammonia from the solvent. Most thermal mass refrigerators are designed to use electricity intermittently. As these units are heavily insulated, cooling load is limited primarily to heat introduced by new items to be refrigerated, and ambient air transfer when the unit is open. Very little power is therefore required if opened infrequently. Refrigeration units for commercial and industrial applications can be made in various size, shape or style to fit customer needs.

Other specialised cooling mechanisms may be used for cooling, but have not been applied to domestic refrigerators.

- Magnetic refrigerators are refrigerators that work on the magnetocaloric effect. The cooling effect is triggered by placing a metal alloy in a magnetic field.
- Acoustic refrigerators are refrigerators that use resonant linear reciprocating motors/alternators to generate a sound that is converted to heat and cold using compressed helium gas. The heat is discarded and the cold is routed to the refrigerator.

Energy Efficiency

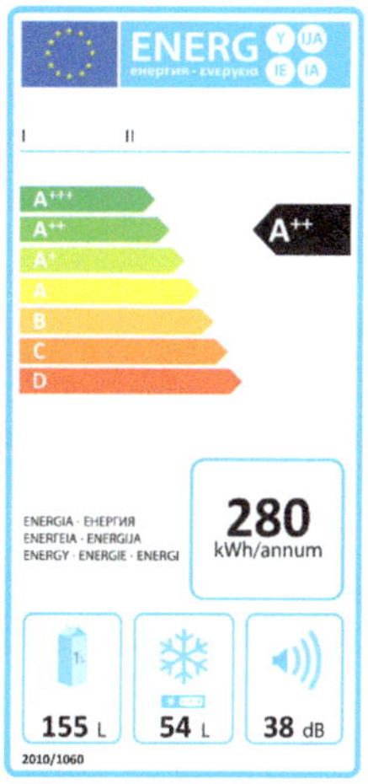

European energy label for a fridge.

In a house without air-conditioning (space heating and/or cooling) refrigerators consumed more energy than any other home device. In the early 1990s a competition was held among the major manufacturers to encourage energy efficiency. Current US models that are Energy Star qualified use 50% less energy than the average models made in 1974. The most energy-efficient unit made in the US consumes about half a kilowatt-hour per day (equivalent to 20 W continuously). But even ordinary units are quite efficient; some smaller units use less than 0.2 kWh per day (equivalent to 8 W continuously). Larger units, especially those with large freezers and icemakers, may use as much as 4 kW·h per day (equivalent to 170 W continuously). The European Union uses a letter-based mandatory energy efficiency rating label instead of the Energy Star; thus EU refrigerators at the point of sale are labelled according to how energy-efficient they are.

Modern refrigerators

For US refrigerators, the Consortium on Energy Efficiency (CEE) further differentiates between Energy Star qualified refrigerators. Tier 1 refrigerators are those that are 20% to 24.9% more efficient than the Federal minimum standards set by the National Appliance Energy Conservation Act (NAECA). Tier 2 are those that are 25% to 29.9% more efficient. Tier 3 is the highest qualification, for those refrigerators that are at least 30% more efficient than Federal standards. About 82% of the Energy Star qualified refrigerators are Tier 1, with 13% qualifying as Tier 2, and just 5% at Tier 3.

Besides the standard style of compressor refrigeration used in normal household refrigerators and freezers, there are technologies such as absorption refrigeration and magnetic refrigeration. Although these designs generally use a much larger amount of energy compared to compressor refrigeration, other qualities such as silent operation or the ability to use gas can favor these refrigeration units in small enclosures, a mobile environment or in environments where unit failure would lead to devastating consequences.

Many refrigerators made in the 1930s and 1940s were far more efficient than most that were made later. This is partly attributable to the addition of new features, such as auto-defrost, that reduced efficiency. Additionally,after World War 2, refrigerator style became more important than efficiency. This was especially true in the 1970s, when side-by-side models with ice dispensers and water chillers became popular. However, the reduction in efficiency also arose partly from reduction in the amount of insulation to cut costs. Because of the introduction of new energy efficiency standards, refrigerators made today are much more efficient than those made in the 1930s; they consume the same amount of energy while being three times as large.

The efficiency of older refrigerators can be improved by defrosting (if the unit is manual defrost) and cleaning them regularly, replacing old and worn door seals with new ones, adjusting the thermostat to accommodate the actual contents (a refrigerator needn't be colder than 4 °C (39 °F) to store drinks and non-perishable items) and also replacing insulation, where applicable. Some sites recommend you clean condenser coils every month or so on units with coils on the rear. It has been proven that this does very little for improving efficiency, however, the unit should be able to "breathe" with adequate spaces around the front, back, sides and above the unit. If the refrigerator uses a fan to keep the condenser cool, then this must be cleaned, at the very least, yearly.

Frost-free refrigerators or freezers use electric fans to cool the appropriate compartment. This could be called a "fan forced" refrigerator, whereas manual defrost units rely on colder air lying at the bottom, versus the warm air at the top to achieve adequate cooling. The air is drawn in through an inlet duct and passed through the evaporator where it is cooled, the air is then circulated throughout the cabinet via a series of ducts and vents. Because the air passing the evaporator is supposedly warm and moist, frost begins to form on the evaporator (especially on a freezer's evaporator). In cheaper and/or older models, a defrost cycle is controlled via a mechanical timer. This timer is set to shut off the compressor and fan and energize a heating element located near or around the evaporator for about 15 to 30 minutes at every 6 to 12 hours. This melts any frost or ice build up and allows the refrigerator to work normally once more. It is believed that frost free units have a lower tolerance for frost, due to their air-conditioner like evaporator coils. Therefore, if a door is left open accidentally (especially the freezer), the defrost system may not remove all frost, in this case, the freezer (or refrigerator) must be defrosted.

If the defrosting system melts all the ice before the timed defrosting period ends, then a small device (called a defrost limiter) acts like a thermostat and shuts off the heating element to prevent too large a temperature fluctuation, it also prevents hot blasts of air when the system starts again, should it finish defrosting early. On some early frost-free models, the defrost limiter also sends a signal to the defrost timer to start the compressor and fan as soon as it shuts off the heating element before the timed defrost cycle ends. When the defrost cycle is completed, the compressor and fan are allowed to cycle back on.

Frost-free refrigerators, including some early frost free refrigerator/freezers that used a cold plate in their refrigerator section instead of airflow from the freezer section, generally don't shut off their refrigerator fans during defrosting. This allows consumers to leave food in the main refrigerator compartment uncovered, and also helps keep vegetables moist. This method also helps reduce energy consumption, because the refrigerator is above freeze point and can pass the warmer-than-freezing air through the evaporator or cold plate to aid the defrosting cycle.

Regarding total life-cycle costs, many governments offer incentives to encourage recycling of old refrigerators. One example is the Phoenix refrigerator program launched in Australia. This government incentive picked up old refrigerators, paying their owners for "donating" the refrigerator. The refrigerator was then refurbished, with new door seals, a thorough cleaning and the removal of items, such as the cover that is strapped to the back of many older units. The resulting refrigerators, now over 10% more efficient, were then distributed to low income families.

Effect on Lifestyle

The refrigerator allows the modern family to keep food fresh for longer than before. The most notable improvement is for meat and other highly perishable wares, which needed to be refined to gain anything resembling shelf life. (On the other hand, refrigerators and freezers can also be stocked with processed, quick-cook foods that are less healthy.) Refrigeration in transit makes it possible to enjoy foodstuffs from distant places.

Dairy products, meats, fish, poultry and vegetables can be kept refrigerated in the same space within the kitchen (although raw meat should be kept separate from other foodstuffs for reasons of hygiene).

Freezers allow people to buy food in bulk and eat it at leisure, and bulk purchases save money. Ice cream, a popular commodity of the 20th century, could previously only be obtained by traveling to where the product was made and eating it on the spot. Now it is a common food item. Ice on demand not only adds to the enjoyment of cold drinks, but is useful for first-aid, and for cold packs that can be kept frozen for picnics or in case of emergency.

Temperature Zones and Ratings

Commercial for electric refrigerators in Pittsburgh, Pennsylvania, 1926

Some refrigerators are now divided into four zones to store different types of food:

- −18 °C (0 °F) (freezer)
- 0 °C (32 °F) (meat zone)
- 5 °C (41 °F) (cooling zone)
- 10 °C (50 °F) (crisper)

The capacity of a refrigerator is measured in either liters or cubic feet. Typically the volume of a combined refrigerator-freezer is split with 1/3rds to 1/4th of the volume allocated to the freezer although these values are highly variable.

Temperature settings for refrigerator and freezer compartments are often given arbitrary numbers by manufacturers (for example, 1 through 9, warmest to coldest), but generally 3 to 5 °C (37 to 41 °F) is ideal for the refrigerator compartment and −18 °C (0 °F) for the freezer. Some refrigerators must be within certain external temperature parameters to run properly. This can be an issue when placing units in an unfinished area, such as a garage. European freezers, and refrigerators with a freezer compartment, have a four star rating system to grade freezers.

- [⍰] : min temperature = −6 °C (21 °F). Maximum storage time for (pre-frozen) food is 1 week
- [⍰⍰] : min temperature = −12 °C (10 °F). Maximum storage time for (pre-frozen) food is 1 month
- [⍰⍰⍰] : min temperature = −18 °C (0 °F). Maximum storage time for (pre-frozen) food is between 3 and 12 months depending on type (meat, vegetables, fish, etc.)
- [*⍰⍰⍰] : min temperature = −18 °C (0 °F). Maximum storage time for pre-frozen or frozen-from-fresh food is between 3 and 12 months

Although both the three and four star ratings specify the same storage times and same minimum temperature of −18 °C (0 °F), only a four star freezer is intended for freezing fresh food, and may include a "fast freeze" function (runs the compressor continually, down to as low as −26 °C (−15 °F)) to facilitate this. Three (or fewer) stars are used for frozen food compartments that are only suitable for storing frozen food; introducing fresh food into such a compartment is likely to result in unacceptable temperature rises. This difference in categorisation is shown in the design of the 4-star logo, where the "standard" three stars are displayed in a box using "positive" colours, denoting the same normal operation as a 3-star freezer, and the fourth star showing the additional fresh food/fast freeze function is prefixed to the box in "negative" colours or with other distinct formatting.

Most European refrigerators include a moist cold refrigerator section (which does require (automatic) defrosting at irregular intervals) and a (rarely frost free) freezer section.

Sweet Potato Storage

The sweet potato (*Ipomoea batatas*) is a very important crop for subsistence farmers in Africa and other developing countries. Due to its relatively short growing period, tolerance to drought and high yield from poor soils, it is used as a famine reserve for many of these households. However it is a highly perishable food source that is susceptible to destruction by microorganisms, metabolic spoilage, physical destruction and pests. As such they are not generally stored for extended periods of time following harvest. This is a major barrier for the optimal use of the crop and causes a lot of waste.

The majority of subsistence farmers who eat sweet potato do not utilize a storage technique of any kind, instead using a method of continuous cultivation and leaving the tubers in the ground until they are ready to be used. While less common, simple methods of storage do exist. The common ones include pit, clamp, and indoor storage and all extend the storage time of sweet potatoes by some degree. Maintaining proper storage temperature is probably the most significant barrier to good storage in subsistence farmers.

Pretreatment of sweet potato can help to minimize risk of losses. Curing can toughen the skin and heal minor physical damages while drying can reduce spoilage and inactivate metabolic degradation.

Optimum storage of sweet potatoes occurs at 12–16 °C (54–61 °F), 85–90% relative humidity (RH) and requires proper ventilation to remove excess carbon dioxide (CO_2) and bring in oxygen (O_2) for respiration. At these conditions sweet potatoes have been shown to last 5 months to a maximum of a year compared to 2–3 months normally.

Susceptibility to Spoilage

Sweet potatoes are a highly perishable crop and difficult to store for extended periods of time. This is mainly due to their high moisture content, metabolic activity following harvesting and thin, permeable skin. Following harvest, sweet potatoes are susceptible to spoilage by physical, physiological, pathological, pests and environmental means.

Physical

Physical losses refer to sweet potatoes damaged by mechanical force during harvest, handling, and transportation of the tubers. This includes injury by way of cuts, bruising, and skinning. Delicate skin and poor harvesting and handling techniques are the main cause of these issues. Damage to tubers by physical means can be minimized careful handling and storing in boxes and cartons instead of cloth sacks.

Physiological

Physiological processes within the plant can contribute highly to spoilage. Natural transpiration causes water loss and results in an unappealing pithy texture. Respiration also continues to occur after harvesting and a buildup of CO_2 around the crop can cause early spoilage. Sprouting will also occur during long term storage, which is useful for planting but will utilize nutrients in the tuber. Sprouting can be inhibited by preharvest spraying with maleic hydrazide, treatment of tubers with methyl ester of napthalene acetic acid (MENA) in acetone on paper spread between tubers and 0.5–4.0% thiourea solution.

Pathological and pests

Sweet potatoes are susceptible to a number of fungi, bacteria and viruses. Infection can cause the production of toxins, pathogenic bacteria, bitter flavours, tissue breakdown, and surface blemishes. Two common sources of infection are from *Rhizopus oryzae* (soft rot) and *Botryodiplodia* (Java black rot). These are responsible for 78% of sweet potato infections in Bangladesh. The pest of greatest concern for farmers is the sweet potato weevil (*Cylas spp.*). This can enter the tuber prior to harvesting and proliferate in storage and has been known to cause up to 50% loss in yields. Simple techniques can be used to minimize the threat of the sweet potato weevil. It has been shown that storage of the potatoes in temperatures under 20 °C (68 °F) can kill up to 89.5% of weevils in an infected crop. Dipping of tubers in hot water also been shown as an effective intervention to control infestations . Damage to tuber by physical and physiological means can increase chance of pathological infection and pests.

Environmental

The length of time that sweet potatoes may be stored is highly dependent on temperature and relative humidity. Storage in temperatures less than 10 °C (50 °F) will cause chill injury in sweet potatoes. This is characterized by discoloration of the insides of the tuber, which can range from brown to black instead of the usual orange. Formation of hard areas also occurs that remain even after cooking. Other signs of chill injury include production of off flavours, internal cell structure breakdown and increased rotting due to reduced skin formation, which allows fungal infection. On the other end of the scale, temperatures above 16 °C (61 °F) will cause a rise in the rates of respiration and sprouting in tubers as well as increased activity by microorganisms. Relative humidity under 80% will cause excessive water loss while RH above 90% will cause condensation to form, which accelerates rotting.

Ideal Storage Conditions

To minimize potential spoilage, precise storage conditions should be met. Sweet potatoes should be stored at between 12 and 16 °C (54 and 61 °F) with a RH of 85-90%. Proper ventilation of the storage area is a key factor in maximizing potential storage time. Respiration causes the production of CO_2, which will act to increase the rate of spoilage. Sweet potatoes can convert approximately 57dm3 of oxygen/ton/day to carbon dioxide and need ventilation to compensate for this. The O_2 in the storage environment should not fall below 7% and CO_2 must not exceed 10%. Due to lack of proper equipment, maintaining adequate temperature conditions are the largest hurdle that farmers in a developing nation will have to face.

Not all sweet potatoes are suitable for storage. Tubers that have been mechanically damaged, infected with pests or pathogens, or are rotten or sprouting should not be kept. These factors will increase rate of spoilage and can increase losses in the rest of the stored crop. Under these optimum conditions sweet potatoes have been shown to keep for 5 months up to a max of about a year.

Storage Methods

In developed countries proper refrigeration technology is used to keep sweet potatoes for long periods of time. Many poor farmers do not have access to refrigeration or even electricity so must rely on other storage methods. There are several methods of storage that only require simple and cheap building materials. The most commonly used are pit storage, clamp storage and indoor storage. These techniques are used only sporadically however as most farmers practice sequential harvesting in which tubers are left in-ground until needed. Such as the Stanley Potato House.

In-ground

In subsistence farming, sweet potatoes are commonly left in the ground and eaten or sold directly following harvest, this is called piece-meal or sequential harvesting. Sweet potatoes are delicate and easily damaged. In-ground storage is used to protect the tubers while reducing the work required to set up storage facilities. In areas with reliable rainfall patterns farmers are able to maintain a supply of fresh sweet potatoes for most of the year by continuous cultivation and harvesting. In drier places however, such as northern Uganda, this is not a practical solution. This is not a long-term option and the plants will degrade after a short period of time. There are numerous disadvantages to this method as it takes up field space that could otherwise be used for growing more crops.

Pit Storage

This is a simple, cheap method that involves digging a hole in the ground for storing the potatoes. Pit storage differs from in-ground as tubers are collected and kept together and considerations are made to control the storage environment.

Construction of the pit can vary according to what materials are available but commonly used are; grass, soil, wood, lime, sawdust and ash. Grass is typically used to line the bottom and sides of the pit in order to insulate against temperature change and absorb moisture. Before usage, grass should be flamed in order to destroy any pests that may be hiding within. Soil is used to seal the roof of the pit and as filler. Wood and plant material can be used to strengthen walls as well as create a roof covering the pit. Roofs built over the pit structure are beneficial in that they can help keep rain out and provide shade to lower temperature. Lime may have some effectiveness in absorbing CO_2 and removing it form the environment. Sawdust is used as a cushioning material and to help control condensation on the tubers. Wood ash can applied to potatoes prior to storage and has shown some effectiveness in protecting against insect attacks and mould.

In order to prevent rotting the pit should ideally be dug into a slope to promote drainage. It should also be situated at least 0.35 metres (1.1 ft) above the water table during dry season and must remain above during wet/rainy season and be approximately 0.5 metres (1.6 ft) below the surface.

There can be ventilation problems with this storage method. Commonly the pit will be sealed shut in order to maintain a good RH. The draw back to this is that CO_2 will collect and cause spoilage. To help this, a sizable headspace should be maintained in the pit to help promote airflow.

Pits can be reused but they should be cleaned, the soil turned over and disinfected with fire or sulphur to rid the area of any microorganisms.

Losses are still fairly high for this technique, though it has been shown in some instances to keep roots for 2–4 months.

Clamp and Mound Storage

Another simple and low cost method, clamp storage consists of covered piles of sweet potatoes. After selecting for the tubers in the best condition, they are stacked in a heap on a layer of grass and covered in layers of grass and soil. As with pit storage ash, lime and sawdust can be used for added effect. The piles may be made at ground level or in shallow or deep trenches. Drainage should be considered and ruts may be made in the ground to lead off water. The clamps may be covered by a roof or kept in a building for added protection. To minimize losses due to respiration a ventilation shaft can be added. Results are fairly poor with this technique and estimated storage time is 2–3 months.

Indoor Storage

Sweet potatoes may be harvested and stored in building. This could include in the living area or in a granary built specifically to store produce. In home storage is typically done in straw woven baskets, cloth bags or wooden boxes. Baskets and boxes have been shown to be more effective at minimizing mechanical damage. If possible tubers should be kept off of the ground to keep them away from rodents and other pests. This is an effective technique for maintaining proper ventilation though depending on the type of building maintaining proper storage temperature and RH may be difficult.

Granaries or other storage buildings typically consist of a round hut with walls made of straw, mud, clay and wood and a conical straw roof. These are commonly supported above ground by a system of legs to keep the crop dry and away from animals, rodents and pests.

Pretreatment of Sweet Potatoes

Typically sweet potatoes are stored and eaten fresh. However there are some simple methods used to increase their storability that can be used in conjunction with other storage methods. Drying and curing of sweet potatoes are two common methods to prepare them prior to storage.

Drying of tubers is done on those that are too damaged to be stored fresh but still have edible material on them. This involves slicing tubers to a thickness of approximately 2–4 mm and then laying them out in the sun for four days or until they are rid of most of their moisture. During drying the potatoes can be covered in prickly bushes or thorns to ward off animals. Dried slices can be kept in-doors or in raised silos until eaten. Drying removes moisture, reduces bacterial growth, and inactivates metabolic processes and enzymatic decomposition.

Curing is a technique done to toughen up the outer layer of skin (periderm). It is a measure that prevents against excessive moisture loss, entry of microorganisms into the plant and even facilitates healing of mechanically damaged tubers. Curing is done by exposing the whole potatoes to a moderately high temperature for several days immediately following harvest.; 30–32 °C (86–90 °F) and 80–95% RH for 4–10 days has been found to be adequate curing conditions A common method to cure without use of incubation equipment is to wrap tubers in black polythene sheets and leave out in the sun for 5 days. The black sheet keeps moisture in and collects heat to reach necessary conditions for curing. Immediately following curing the temperature must be dropped quickly in order to avoid sprouting damage.

References

- Graham Riches (1986). "passim. Models of Food Banks". Food banks and the welfare crisis. Lorimer. ISBN 0888103638.

- Elizabeth Henderson and Robyn Van En (1986). "Chapt 19". Sharing the Harvest: A Citizen's Guide to Community Supported Agriculture. Chealsea Green Publishing. ISBN 193339210X.
- Janet Poppendieck (1999). "Introduction, Chpt 1". Sweet Charity?: Emergency Food and the End of Entitlement. Penguine. ISBN 0140245561.
- Leslie Crutchfield and Heather McLeod Grant (2007). "Chpt 3". Forces for Good: The Six Practices of High-Impact Nonprofits. Jossey-Bass. ISBN 0470580348.
- Andrew Walter (2012). William A Dando, ed. Food and Famine in the 21st Century. ABC-CLIO. pp. 171–181. ISBN 978-1-59884-730-7.
- Yam, K. L., "Encyclopedia of Packaging Technology", John Wiley & Sons, 2009, ISBN 978-0-470-08704-6.
- Arora, Ramesh Chandra. "Mechanical vapour compression refrigeration". Refrigeration and Air Conditioning. New Delhi, India: PHI Learning. p. 3. ISBN 81-203-3915-0.
- Dennis R. Heldman (29 August 2003). Encyclopedia of Agricultural, Food, and Biological Engineering (Print). CRC Press. p. 350. ISBN 978-0-8247-0938-9.
- Marotz, Lynn R. (2008). Health, Safety, and Nutrition for the Young Child. Wadsworth Publishing. pp. 491–2. ISBN 978-1-4283-2070-3.
- Marotz, Lynn R. (2008). Health, Safety, and Nutrition for the Young Child. Wadsworth Publishing. p. 482. ISBN 978-1-4283-2070-3.
- Mitchell, Deborah (2004). Safe foods: the A-to-Z guide to the most wholesome foods for you and your family. Penguin. pp. Ch. 15. ISBN 978-1-101-21015-4.
- Editors of Creative Publishing. Dressing & Cooking Wild Game. Publisher: Creative Publishing international 1999 ISBN 978-0865731080.

Permissions

All chapters in this book are published with permission under the Creative Commons Attribution Share Alike License or equivalent. Every chapter published in this book has been scrutinized by our experts. Their significance has been extensively debated. The topics covered herein carry significant information for a comprehensive understanding. They may even be implemented as practical applications or may be referred to as a beginning point for further studies.

We would like to thank the editorial team for lending their expertise to make the book truly unique. They have played a crucial role in the development of this book. Without their invaluable contributions this book wouldn't have been possible. They have made vital efforts to compile up to date information on the varied aspects of this subject to make this book a valuable addition to the collection of many professionals and students.

This book was conceptualized with the vision of imparting up-to-date and integrated information in this field. To ensure the same, a matchless editorial board was set up. Every individual on the board went through rigorous rounds of assessment to prove their worth. After which they invested a large part of their time researching and compiling the most relevant data for our readers.

The editorial board has been involved in producing this book since its inception. They have spent rigorous hours researching and exploring the diverse topics which have resulted in the successful publishing of this book. They have passed on their knowledge of decades through this book. To expedite this challenging task, the publisher supported the team at every step. A small team of assistant editors was also appointed to further simplify the editing procedure and attain best results for the readers.

Apart from the editorial board, the designing team has also invested a significant amount of their time in understanding the subject and creating the most relevant covers. They scrutinized every image to scout for the most suitable representation of the subject and create an appropriate cover for the book.

The publishing team has been an ardent support to the editorial, designing and production team. Their endless efforts to recruit the best for this project, has resulted in the accomplishment of this book. They are a veteran in the field of academics and their pool of knowledge is as vast as their experience in printing. Their expertise and guidance has proved useful at every step. Their uncompromising quality standards have made this book an exceptional effort. Their encouragement from time to time has been an inspiration for everyone.

The publisher and the editorial board hope that this book will prove to be a valuable piece of knowledge for students, practitioners and scholars across the globe.

Index

L

M

N

O

P

R

S

T

V

W